Marcin Hernes · Jarosław Wątróbski
Editors

Emerging Challenges in Intelligent Management Information Systems

Proceedings of 26th European Conference on Artificial Intelligence ECAI 2023 - IMIS 2023 Workshop

Springer

Editors
Marcin Hernes
Center for Intelligent Management Systems,
Department of Process Management
Wroclaw University of Economics
and Business
Wrocław, Poland

Jarosław Wątróbski
Institute of Management
University of Szczecin
Szczecin, Poland

ISSN 2367-3370 ISSN 2367-3389 (electronic)
Lecture Notes in Networks and Systems
ISBN 978-3-031-66760-2 ISBN 978-3-031-66761-9 (eBook)
https://doi.org/10.1007/978-3-031-66761-9

This Springer imprint is published by the registered company Springer Nature Switzerland AG
The registered company address is: Gewerbestrasse 11, 6330 Cham, Switzerland

Preface

This volume contains the proceedings of the ECAI 2023 Workshop on Intelligent Management Information Systems (IMIS 2023). IMIS 2023 was part of the 26th European Conference on Artificial Intelligence ECAI 2023, held in Krakow from September 30, 2023, to October 4, 2023.

Following the topics of IMIS 2023, this book discusses emerging challenges related to implementing artificial intelligence in Management Information Systems. The book is devoted to models, methods, and approaches addressing the development of artificial intelligence solution for improving the functionality of information systems that supports management. The main focus is put on machine learning, including deep learning to support business processes, artificial intelligence for financial systems and cryptocurrencies, intelligent human-computer interfaces, knowledge management in business organizations, hybrid artificial intelligence, and multiple criteria decision analysis methods. The book has an interdisciplinary character; therefore, it is intended for a broad scope of readers, including researchers, students, managers, and employees of business organizations, software developers, IT, and management specialists.

The volume is divided into four major parts covering the main issues related to the topic.

Part I deals with the application of artificial intelligence in information systems.

Joanna Kołodziejczyk ("Uncovering Patterns in Training Skills with ABA: Rule Extraction from the SYSABA Database") proposes a methodological framework to prepare datasets for extracting association rules from Individual Education Programs (IPET) employed in Polish therapeutic centers teaching children with developmental disorders. These IPETs are sourced from the System for Applied Behavior Analysis (SYSABA) and comprise various elements, including targeted skills, teaching techniques, and recording methods. Exposing the fundamental patterns within IPETs significantly contributes to understanding relations in ABA-based treatments.

Adrianna Kozierkiewicz, Marcin Pietranik, and Aleksandra Stawarz ("Ensemble Learning Approach To Fake News Detection Problem") propose a novel approach to fake news detection problem based on ensemble learning methods. The proposed approach allows combining several different strategies to detect fake news detections into a unified, cohesive methodology. The performed experiments that used the widely known "Liar, Liar Pants on Fire" benchmark dataset yielded promising results that surpassed outcomes described in the literature.

Michele Persiani, Madeleine Blusi, Helena Norberg, and Juan Carlos Nieves ("SMAP: Smart Personalized Medication Assistant to Help Nurses in Medication Adherence") report on the current development of an intelligent assistant supporting health professionals in the task of preparing pillboxes. They propose supporting this process through an assistant residing on a holographic device worn during practices. In this way, nurses can rely on an online holographic layer supporting the task, showing the availability of drugs and online information comfortably. The authors introduce the argumentative framework and provide an initial formulation for deciding which drugs are equivalent and where they are in the warehouse.

Agnieszka Biernat-Jarka, Monika Gębska, Wiesław Świderek, and Sławomir Jarka ("The Use of Modern Technologies in Decision-Making by Polish Farmers") present issues related to using intelligent decision-making support systems to improve the productivity and profitability of farming and stimulate sustainable development. The research distinguished three segments of farmers: firstly, enthusiasts of modern technologies, who had no problems using them; the second group was sceptics, who did not use modern technologies; and the third one was the farmers, who used modern technologies, but often had problems with their use.

Artur Skoczylas, Bogdan Franczyk, Wiesława Gryncewicz, and Ryszard Zygała ("Decision Support in Production Environment Using Artificial Intelligence") examine the essence and scale of using Intelligent Decision Support Systems about solutions dedicated to operational decisions in production-focused company. The research also examines other aspects of the solutions and presents some insights into areas of application and the methods used to build such solutions.

Ryszard Zygała, Wieslawa Gryncewicz, Agnieszka Pilch, Mykola Dyvak, and Andriy Melnyk ("Identification Of Barriers To AI Implementation In Medical Information Systems") identified organizational and legal, technical, economic, and human factor-related barriers to implementing artificial intelligence in medical information systems. Awareness and consideration of these barriers to implementing AI in health systems can help eliminate (or minimize) the risk of project failure.

Filip Grzonkowski and Paweł Weichbroth ("Exploring Cause-And-Effect Relationships Between Public Company Press Releases And Their Stock Prices") designed and implemented a method of exploring the cause-and-effect relationships between company announcements and the stock prices on the NASDAQ stock exchange. The mechanism of sentiment analysis was prepared based on the supervised machine learning approach. The implemented method was used to analyze the sentiment of a set of stock exchange announcements, and it was correlated with changes in the share prices of selected companies. The association rules were extracted using the Apriori algorithm based on the collected data.

Łukasz Lemieszewski, Piotr Borkowski, Aleksandra Radomska-Zalas, Larisa Dobryakova, and Evgeny Ochin ("Cybersecurity Of The Unmanned Marine Vehicles In The Conditions Of Partial Or Complete Interruption of Multi-GNSS Signals by Jamming and/or Spoofing") discuss suppressing Unmanned Marine Vehicles (UMVs) navigation and countermeasures. The cybersecurity of UAWs under conditions of partial or complete interruption of several GNSS signals due to jamming and/or spoofing cannot be calculated by analytical methods, but the results of predicting the movement of the UMVs under extreme conditions can be obtained using simulation methods at the design stage of the UMVs.

Part II is devoted to advanced machine learning methods to support business processes.

Iga Rudawska and Grzegorz Wojarnik ("Applying Convolutional Neural Networks to Stock Market Forecasting – A Case Study of Stock Volume Prediction") explore the possibility of using Convolutional Neural Networks (CNN) to process graphics containing stock charts and demonstrate the potential of predicting volume changes. They developed a CNN-based model that supports predicting stock volume changes in the capital markets. The authors also compared the quality measures of the generated predictive model for different types of stock charts.

Tomasz Zema, Filip Wójcik, Adam Sulich, and Marcin Hernes ("Forecasting Fuel Retail Sales Volume Using Machine Learning For Sustainable Decision-Making") analyze forecasting fuel retail sales volume problem. The three machine learning models are proposed and examined to address this research problem. The best results were obtained by the LightGBM model, utilizing both future covariates and fore-casted fuel prices.

Marek Karwanski, Boleslaw Borkowski, Wieslaw Szczesny, and Artur Wilinski ("The Effectiveness of AI Methods Compared to Statistical Analysis and Data Modeling in the Construction of Scoring Models") performed a comparative analysis of the effectiveness of two analytical cultures, statistical analysis and data modeling (DMC) and algorithmic AI analysis (AMC) using machine learning methods. They proposed variable selection methods in the DMC and AMC models by analyzing the importance of variables.

Justyna Szydłowska-Samsel ("Supporting Algorithmic Trading with Machine Learning: Progress in Backend Technology") tested Python optimization on the machine learning models frequently used in algorithmic trading. The results are compared between four different Python versions.

Paweł Golec, Marcin Hernes, Tomasz Sajewski, Ewa Walaszczyk, Artur Rot, Marcin Fojcik, Tomasz Turek, and Damian Dziembek ("Forecasting E-Learning Course Purchases Using Deep Learning Based On Customer Retention") developed a method for forecasting e-learning course purchases using deep learning based on customer retention. The multilayer perceptron and long short-term memory are used. Using a developed approach, there is no need to acquire additional data characterizing the user, such as demographic data. From the business point of view, it is a crucial benefit for new customers who are not registered in the customer relationship management system but use this system anonymously via the website.

Dorota Jelonek, Magdalena Graczyk-Kucharska, Magdalena Wyrwicka, and Robert Olszewski ("Customer Segmentation and Anticipation of Consumer Behaviors based on Machine Learning and CART") propose the utilization of machine learning and CART for modeling customer needs to facilitate market segmentation among the younger generation, focusing on young individuals studying in the IT field. Implementing such modeling in practice can optimize decision-making, minimize financial losses, and improve resource efficiency. The conducted research demonstrated that to comprehend the motivations of the surveyed individuals, it is crucial to consider several completely different independent variables in the process of modeling the spatial distribution of expected remuneration, including economic parameters.

Krzysztof Lutosławski, Marcin Hernes, Artur Rot, and Cezary Olejarczyk ("Prediction Of Customer Purchases Using LSTM Deep Neural Network") develop a deep learning model using the long short-term memory (LSTM) neural network to predict customer purchases in the next week, including transaction count and overall transaction gross value, based on their historical transaction data. In the research, to achieve the best predictions, various sizes of the training dataset, different model architectures, and its hyperparameters, as well as the hyperparameters of the learning algorithm, were tested. The obtained model can be implemented before validating its performance on another instance of production data.

Part III is devoted to problems related to the social aspects of artificial intelligence in management.

Ewa Ziemba, Ewa Maruszewska, Dariusz Grabara, and Katarzyna Renik ("Acceptance of ChatGPT among Accounting and Finance Higher Education Students") examine the acceptance and usage of ChatGPT among students in the field of accounting and finance in higher education. They adopt and verify Venkatesh's theoretical model of acceptance and use of technology (UTAUT) with a sample of 237 accounting and finance students. The results demonstrate that behavioral intentions towards ChatGPT are influenced by effort expectancy, social influence, and performance expectancy. The study emphasizes the necessity for additional investigation into adopting ChatGPT in educational settings.

Małgorzata Łatuszyńska, Anna Borawska, and Mariusz Borawski ("Virtual Reality Games in Promoting Waste-Sorting Behaviour") present a research procedure that could enable a comparison of immersive virtual reality and desktop games concerning promoting waste-sorting behavior in terms of their effectiveness with the use of cognitive neuroscience techniques. They propose a methodology for investigating the most effective ways of promoting waste-sorting behavior.

Mateusz Piwowarski, Malgorzata Nermend, and Kesra Nermend ("Psychophysiological Research In The Design Of Useful Interfaces For E-Learning Materials") analyze the impact of e-learning system components at the level of cognitive and emotional human functions. They present a methodological approach based on measuring and analyzing eye movement activity and facial expression for designing valuable e-learning materials.

Piotr Stolarski and Włodzimierz Lewoniewski ("Analysis of the Successful and Bankrupt Digital Currency Exchanges based on Open Data") proposed two models for the prediction of the successful and bankrupt digital currency exchanges based on financial data and measures collected from Wikipedia and Wikidata. They indicated how intermediary entities providing services related to innovative digital asset markets are characterized by the dominant open online sources. The analysis is necessary to better select intermediary entities for transactions carried out with digital means of exchange of value. It thus should significantly increase the awareness of entities operating in these markets and allow for the reduction of specific risks.

Part IV is devoted to intelligent multiple criteria decision analysis methods.

Andrii Shekhovtsov and Wojciech Sałabun ("The New Algorithm For Effective Reducing The Number Of Pairwise Comparisons In The Decision Support Methods") conducted research on developing an algorithm that efficiently reduces the number of pairwise comparison questions to an expert. The algorithm is based on the triads' consistency and achieves more than ninety percent reduction in questions for big expert judgment matrices. Although the original algorithm is proposed for the Characteristic Objects METhod (COMET), it could be generalized and used with such methods as

RANking COMparison (RANCOM) and Analytic Hierarchy Process (AHP). The efficiency of the algorithm is proved using a simulation approach. The results of those simulations, as well as the pseudocodes, are presented in the chapter.

Aleksandra Bączkiewicz and Jarosław Wątróbski ("Multi-Criteria Assessment of the Sustainable Share of Renewable Energy Sources in European Countries Using the SSP-TOPSIS Method") developed the Strong Sustainability Paradigm Based Technique for Order Preference by Similarity to Ideal Solution (SSP-TOPSIS) method incorporating criteria compensation reduction modeling. The proposed method is based on the well-known Technique for Order Preference by Similarity to the Ideal Solution (TOPSIS) algorithm, and the innovative criterion compensation reduction step makes the technique suitable for assessments that require consideration of the robust sustainability paradigm.

Witold Chmielarz and Marek Zborowski ("Application Of The Multi-Criteria VMCM Method, To Investigate The Dynamics Of Changes In The Level Of Early Childhood Education In EU") compared the customer ratings provided by Generation Z representatives regarding e-banking websites of the most popular banks in Poland and to examine the impact of the multiple criteria decision analysis (MCDA) methods on the objectivity of these ratings. The obtained results indicate the possibility of interchangeable use of the presented assessment methods, which may be helpful for business practitioners when analyzing and designing banking services.

Małgorzata Nermend, Mateusz Piwowarski, and Kesra Nermend ("Application Of The Multi-Criteria VMCM Method To Investigate, The Dynamics Of Changes In The Level Of Early Childhood Education In EU Countries") present an original approach to assessing the dynamics of changes in the level of early childhood education in EU countries by using an author's Vector Measure Construction Method (VMCM) method. The presented method is dedicated to the study of complex socio-economic phenomena described by many factors over time. It is a relatively new approach to studying the level of changes in the level of early childhood education. The research shows that the educational situation in the European countries studied varied.

Paula Bajdor ("A Fuzzy Multi-Criteria Approach For The Quality Assessment Of Online Shopping Platforms") presents applying the fuzzy TOPSIS method as appropriate when the respondents' answers were given using an ordinal scale. The method used made it possible to create a ranking of these criteria, indicating which, according to the respondents, are the most important and which play a minor role in determining the level of quality of the functioning of the online platform. Appropriate management of an online platform must consider the emphasis on increasing its quality, and the ranking of criteria, which is the effect of the method used, is, in this case, a helpful reference point for people responsible for managing online platforms.

Małgorzata Nermend, Kinga Flaga-Gieruszyńska, and Mariusz Borawski ("Use of the Technique for Order Preference by Similarity Ideal Solution method for the assessment of education level in Poland related to EU countries") present an attempt to use the multi-criteria TOPSIS method to assess the level of pre-primary to tertiary education in the countries of the European Union in the dynamic approaches. The developed method allows each country to monitor its development in education in relation to the rest of the member EU countries, which can be helpful in rapid intervention in improving this process.

Robert Król, Aleksandra Bączkiewicz, and Jarosław Wątróbski ("Application of Machine Learning Algorithms for Identification of Key Criteria Groups in Public Tendering Proceedings in Poland") conducted research to examine the feasibility of performing a grouping of criteria used in public tender proceedings in Poland using machine learning. A potential application for such grouping is the creation of a dictionary of criteria. In the research, it was proven that it is possible to join criteria into groups successfully.

We want to express our gratitude to all the authors for their interesting, novel, and inspiring contributions. Peer-reviewers also deserve deep appreciation because their insightful and constructive remarks and suggestions have considerably improved many contributions. Last but not least, we wish to thank prof. Janusz Kacprzyk for his help in implementing and finishing this large publication project on time, maintaining the highest publication standards.

October 2023

Marcin Hernes
Jarosław Wątróbski

Organization

International Programme Committee

Wolfgang Dorner	Technische Hochschule Deggendorf, Germany
Marcin Fojcik	Western Norway University of Applied Sciences, Norway
Mykola Dyvak	Ternopil National Economic University, Ukraine
Bogdan Franczyk	University of Leipzig, Germany
Jan Stępniewski	Université Paris 13, France
Philippe Krajsic	University of Leipzig, Germany
Artur Rot	Wrocław University of Economics, Poland
Ewa Walaszczyk	Wrocław University of Economics, Poland
Wiesława Gryncewicz	Wrocław University of Economics, Poland
Monika Eisenbardt	Wrocław University of Economics, Poland
Dorota Jelonek	Czestochowa University of Technology, Poland
Paweł Weichbroth	Gdańsk University of Technology, Poland
Dragan Pamučar	University of Belgrade, Serbia
Muhammet Deveci	University College London, United Kingdom
Witold Chmielarz	University of Warsaw, Poland
Mykola Dyvak	West Ukrainian National University, Ukraine
Adrianna Kozierkiewicz	Wrocław University of Science and Technology, Poland
Kesra Nermend	University of Szczecin, Poland
Boleslaw Borkowski	Warsaw University of Life Sciences, Poland
Andriy Melnyk	West Ukrainian National University, Ukraine
Paweł Ziemba	University of Szczecin, Poland
Artur Karczmarczyk	West Pomeranian University of Technology, Poland
Zbigniew Pastuszak	Maria Curie-Sklodowska University, Poland
Joanna Kołodziejczyk	National Institute of Telecommunications, Poland
Richard Chbeir	University of Pau and the Adour Region, France
Joanna Paliszkiewicz	Warsaw University of Life Sciences, Poland
Wojciech Sałabun	West Pomeranian University of Technology, Poland
Józef Żurada	University of Louisville, USA

Contents

Advanced Machine Learning Methods for Support Business Processes

Social Aspects of Artificial Intelligence in Management

Application of Artificial Intelligence in Information Systems

Uncovering Patterns in Training Skills with ABA: Rule Extraction from the SYSABA Database

Joanna Kołodziejczyk(✉)

National Institute of Telecommunications, ul. Szachowa 1, 04-894 Warsaw, Poland
j.kolodziejczyk@il-pib.pl

Abstract. This study applies a refined methodological framework to extract association rules from the System for Applied Behavior Analysis (SYSABA) dataset, representing years of therapeutic preferences, course of treatment, and experience in developing and utilizing Individual Education Programs (IPET) in Polish therapeutic centers for children with developmental disorders. These IPETs comprise various elements, including targeted skills, teaching techniques, and recording methods. This study aims to uncover how IPETs are constructed and utilized by therapists in training new skills. Though utilizing a well-established Apriori algorithm, the study's novelty stems from its tailored approach to a unique dataset, revealing uncovers salient patterns within IPETs, such as a pronounced preference for the Discrete Trial Training (DTT) technique, bi-monthly incremental task learning assessments, and customized reinforcement. The comprehensive examination of IPET patterns in Poland offers pivotal insights to enhance therapeutic decision-making strategies. This research contributes significantly to ABA-based treatments by understanding these patterns, ensuring more informed interventions and better patient outcomes.

Keywords: Applied Behavior Analysis · Discrete Trial Training · Individual Education Program management · Apriori Algorithm · Association Rules

1 Introduction

Applied Behavior Analysis (ABA), grounded in empirical observations and skill (behavior) training through reinforcement, is a robust, evidence-informed methodology. Its wide applicability is especially recognized in therapeutic interventions for neurodiverse individuals, notably those with Autism Spectrum Disorder (ASD) [8,11].

SYSABA, an information system designed to aid Applied Behavior Analysis (ABA) teaching, was developed and launched across Polish therapy centers for individuals with developmental disorders in 2018 [7]. This interdisciplinary project integrates insights from ABA therapists and computer science

M. Hernes and J. Wątróbski (Eds.): ECAI 2023, LNNS 1079, pp. 3–14, 2024.
https://doi.org/10.1007/978-3-031-66761-9_1

researchers. As a freeware, SYSABA includes modules for managing therapy centers and performing detailed therapy analyses adhering to ABA principles. Specifically, SYSABA facilitates the standardization of therapeutic processes, creation of extensive patient profiles, provision of parent support for individual therapy, therapist education through the platform, guidance in therapy program design, and a graphical evaluation of measurements recorded during therapeutic sessions.

In ABA therapy, formulating treatment plans tailored to the patient's needs is critical for behavior analysts. Consequently, a crucial functionality of SYSABA is the facility to develop an IPET. Over the past five years, the SYSABA database has stored approximately 2300 training IPETs for therapist education and over 4700 as active therapy components. This repository serves as a shared resource for all therapists, allowing them to utilize the insights of their predecessors and construct therapeutic programs with the exclusive use of predeveloped modules specifically approved by supervisors as templates.

The potential of the SYSABA database as a reservoir of therapeutic processes motivates the current study's objective: to mine this data to uncover embedded knowledge. Considering the preference of human comprehension for rule-generating models, this study employs the Apriori algorithm, a classic data mining tool for extracting frequent item sets and relevant rules. Within the SYSABA context, the *items* can be viewed as IPET elements – such as teaching techniques, measurement frequencies, skill target categories, etc. – while the *transactions* can be interpreted as the individual IPETs, each encompassing a combination of various items.

The use of the Apriori algorithm on the SYSABA dataset reveals prevalent therapeutic strategies in IPETs, contributing valuable insight into predominant therapy techniques. This unique approach, enhancing treatment decision-making, results in a distilled set of best practices based on observed IPET patterns. The inherent interpretability of the Apriori algorithm, offering intelligible rules for therapists, underscores this study's novelty and substantial contribution to the field of ABA.

The remainder of this paper is organized as follows: the diverse techniques utilized in ABA are initially surveyed. Subsequently, a dataset description is presented, outlining all attribute categories and their respective domains. The applied methodology in this study is then presented. Finally, the research findings are exhibited and discussed, revealing the hidden knowledge.

2 Background and Related Work

Various Applied Behavior Analysis (ABA) techniques and intervention modes have been studied, demonstrating a wide spectrum of beneficial approaches and data collection methodologies. For instance, Matson et al. [10] reviewed techniques such as graduated guidance, interruption of incorrect actions, and positive reinforcement as interventions for adaptive living skill deficits in individuals with autism. Similarly, Lerman et al. [9] analyzed data collection methods during discrete-trial teaching, endorsing a two-or-more session approach to assess skill mastery. Smith et al. [12] emphasized the necessity for extensive hours of DTT per week for children with autism, necessitating specialized teacher training. Other works explored various prompting strategies [3], the application of constant time delay and the system of least prompts in teaching ASD students [2], and the effectiveness of the most-to-least prompting procedure on the dressing skill of students with autism [6].

Given this short overview, therapists have plenty of options for treatment planning, resulting in a broad range of potential Individual Education Program (IPET) configurations. While previous studies typically focus on a few cases and investigate therapy effectiveness, this paper aims to understand how these research findings are applied on a larger scale in Polish therapeutic centers to identify common therapeutic strategies and discerning combinations of tools frequently used together.

3 Dataset of Individual Education-and-Therapeutic Programs (IPETs) for Skills Acquisition

The study used a SYSABA-derived dataset updated from May 2019 to May 2023, including 4781 instances. This dataset was created by integrating 11 relations from the database, resulting in 24 attributes for IPET descriptions, as detailed in Table 1. Most attributes are qualitative variables, with unique value counts as #Labels. Non-qualitative attributes will be discretized and potentially filtered as discussed later.

3.1 Data Preprocessing

The data preprocessing phase in this research comprised several pivotal steps: attribute consolidation for dimensionality reduction, discretization of non-categorical variables, and statistical evaluation for efficient data filtering.

Dimensionality Reduction: This step necessitates certain observations and manual modifications.

The `technique` attribute initially comprised 37 distinct values. A closer examination uncovered values redundancies because therapists can add new ABA teaching techniques. Unfortunately, this led to duplication, which was manually resolved, yielding a concise list of 11 distinct techniques.

Further dimensionality reduction was accomplished through attribute linkage. The `is_ioa` and `ioa_freq` attributes were combined into a single `ioa` based on the observation that if an interobserver agreement was planned (`ioa_freq` = 1), then `ioa_freq`, which is the agreement period was set up. Merging resulted in 6 unique attribute values. The attributes `record_when` and `record_when_unit` were joined to a single `record` attribute, encompassing 14 unique values.

The `category` column, a subset of `subcategories`, was discarded.

Table 1. Initial set of IPET features originally selected for analysis (#Labels refers to the number of values in the categorical variable or post-discretization)

No	Name	Description	Type	#Labels	Filtered (y/n)
1	`target`	target skill(LTO*)	cat	146	n
2	`category`	target's category label	cat	5	n
3	`subcategory`	target's subcategory label	cat	15	n
4	`technique`	ABA teaching techniques	cat	37/11	n
5	`method`	teaching/training methodology	cat	5	n
6	`sequence`	a way of educating a task	cat	6	n
7	`prompts`	Are prompts applied?	bool	2	n
8	`reinforcement`	type of reinforcement at intervention	cat	8	n
9	`init_reinforcement`	reinforcement at the beginning of an intervention	cat	4	y
10	`soc_reinforcement`	Is a social reinforcement used?	bool	2	n
11	`trial_counts`	number of trials at intervention	int	3	n
12	`initial`	baseline setup	cat	2	y
13	`generalization`	generalization method	cat	2	y
14	`pass`	threshold to pass a task in %	int	3	n
15	`num_pass`	number of passes in a row or time period	int	3	n
16	`close_criteria`	closing criteria for IPET	cat	2	y
17	`verif_period`	IPET verification period in days	int	7	n
18	`is_ioa`	Is IOA** planned?	bool	2	join w/19, n
19	`ioa_freq`	period of IOA accomplished	cat	5/6	join w/18, n
20	`record_when`	how often data are record	int	14	join w/21, y
21	`record_when_unit`	record's unit	cat	3	join w/20, y
22	`is_template`	Is IPET a template?	bool	2	y
23	`prog_status`	a program status	cat	8	n
24	`ref_count`	number of references to IPET	int	4	n

* Long Term Objective ** Interobserver Agreement

Discretization: To standardize the attributes and meet the prerequisites of the Apriori algorithm, all numerical attributes underwent discretization:

- `trial_count`: representing the count of trials in the DTT method, ranging from 1 to 50 or NaN. The median value of 5 served as the breakpoint for discretization, resulting in 3 categories
- `pass`: indicates the proficiency threshold for skill acquisition. Customarily set at 80% (the attribute's median value), it was divided into three values (Fig. 2). The prior cleaning removed values below 20%, as these were not considered that the student is allowed to pass with such a low level.
- `num_pass`: variable represents the consecutive successful attempts needed for task completion; it was divided into three values
- `verif_period`: denoting the period of student progress assessment, it was discretized into standard intervals such as 1, 2, 3 weeks, and 1, 2, 3, and 6 months
- `ref_cont`: this internal database index reflects program activity by counting linked objects. From 8 to 6242, it was discretized based on quantiles to maintain sample distribution equality

Statistical Evaluation. This analysis was performed for each variable to facilitate knowledge acquisition, assisting in data filtering and dimensionality reduction, thus enhancing rule extraction efficiency. Considerable imbalance was observed in variables `close_criteria`, `initial`, `is_template`, and `generalization`, as depicted in Fig. 1 histograms.

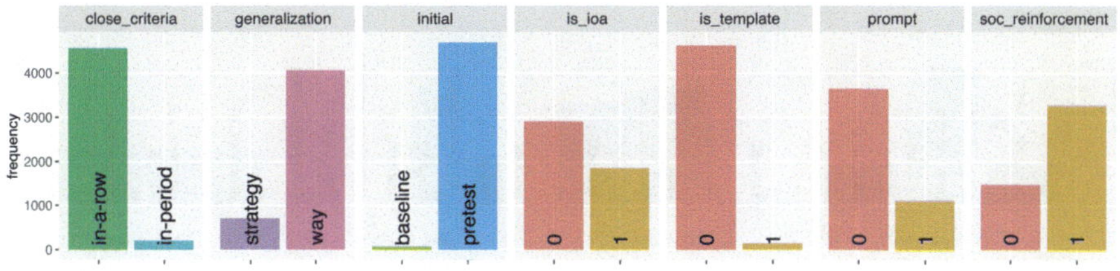

Fig. 1. Distribution of dichotomous variables in the dataset: `close_criteria`, `generalization`, `initial`, `is_ioa`, `is_template`, `prompt`, and `soc_reinforcement`.

The `close_criteria` variable indicates the type of students' pass observations as time (`in-period`) or frequency (`in-a-row`). Only in 4.3% of the instances was the observation period `in-period` suggesting that the most common method of closing an IPET was by tallying the number of successful program completions in consecutive sessions.

For the `initial` variable, only 1.6% of all samples necessitated a baseline rather than a pretest at the therapy's commencement. Setting a baseline involves gathering a sequence (minimum two) of pre-treatment data collected by observing skills before initiating therapy. The pretest, markedly the more prevalent approach, records pre-treatment data during a single session.

Only 3% of all programs, amounting to 151, are designated template programs accessible to all therapists. This small amount indicates a rigorous IPET evaluation of supervisors.

The `generalization` variable is a compulsory component of IPETs, and has two types: strategy and synthetic (`way`). Synthetic generalization, indicating generalization across individuals and locations, was chosen in 85% of all IPETs, equivalent to 4064 instances.

Final Dataset. The final dataset consisted of 4775 instances with 22 attributes. Due to the considerable number of categories, the values for the variables `target`, `subcategory`, `technique`, and `record_when` are not displayed within the confines of this paper. The non-dichotomous variables discussed in Sect. 3.1 are depicted in Fig. 2. In the following discussion, specific values will be revealed if they appear in the derived rules.

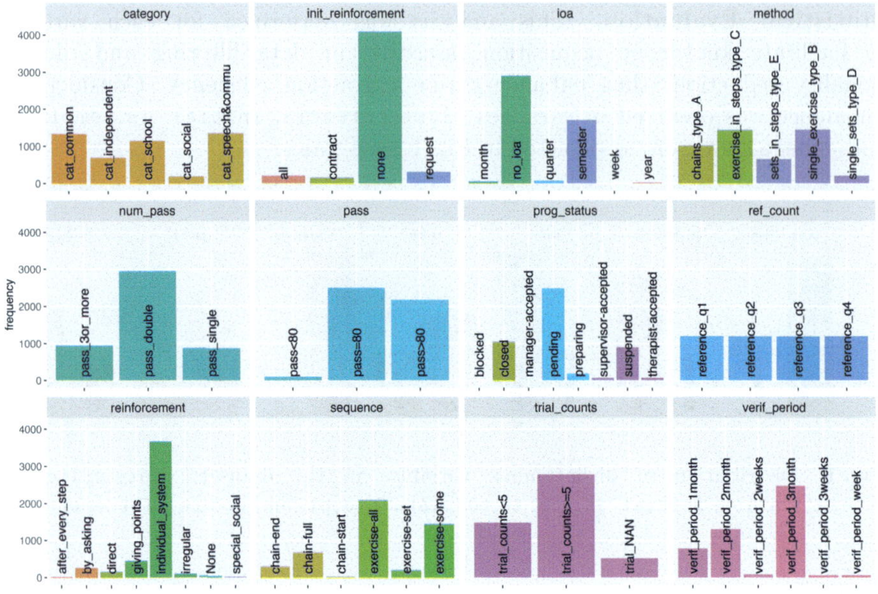

Fig. 2. Distribution of selected non-dichotomous variables, also after discretization.

4 Methodology

Given the rich and diverse dataset from SYSABA, the Apriori algorithm was selected for its proven effectiveness in such contexts.

4.1 Apriori Algorithm

Definition 1. *Let* $V = \{v_1, v_2, v_3, \ldots, v_m\}$ *represent a set of items.*

An association rule is expressed as $X \Rightarrow Y$*, where* X *and* Y *are sets of items, with* $X \subset V$*,* $Y \subset V$ *and* $X \cap Y = \emptyset$*.*

Given $D = dataset$, a set of transactions, each transaction T is a subset of items where $T \subseteq V$. A transaction T is said to support an item v_i if $v_i \in T$.

Three crucial metrics for a rule $X \Rightarrow Y$ are:

1. Support: It represents the frequency of X and Y occurring together as a number or proportion of total transactions.
2. Confidence: It measures the conditional probability of an item given another, defined as the ratio of transactions containing X and Y to those containing X.
3. Lift: It gauges the strength of association between X and Y, considering the frequency of both X and Y in D. Defined as the ratio of observed to the expected frequency of X and Y co-occurring, assuming they are independent. A lift greater than 1 suggests a higher association between X and Y.

The Apriori algorithm extracts frequent item sets, and relevant association rules, operating on the principle that all subsets of a frequent item set must also be frequent [1]. The algorithm seeks to discover rules with support and confidence greater than or equal to pre-set thresholds. Experiments were conducted using the *arules* [5] and *arulesViz* [4] packages in R.

4.2 Filtering Ubiquitous Items

Items ubiquitous across all transactions often offer limited insights into data mining and association rule learning due to their reduced discriminative power and minimal influence on association rules. Furthermore, such items can inflate the count of frequent itemsets and dilute measures of interestingness like support, confidence, and lift.

A filtering process was implemented using support (frequency) to identify ubiquitous items. A cutoff threshold was set at 80%, meaning all items with a frequency above 80% were removed. Table 2 lists all values above this threshold and the overall percentage. The subsequent filtering removed these items.

Table 2. Ubiquitous variables in the SYSABA dataset, filtered before mining

Attribute	Value/Item	% of all IPETs	% of templates
`init_reinforcement`	`none`	86.75	85.74
`record`	`1_week`	88.54	80.13
`close_criteria`	`in-a-row`	95.60	88.08
`initial`	`pretest`	98.32	99.34
`generalization`	`way`	85.11	–
`is_template`	`template_no`	96.84	–
`prompts`	`prompts_no`	–	91.39
`verif_period_cat`	`verif_period_3month`	–	92.05

5 Results and Discussion

The following section outlines two distinct approaches: extracting association rules from the entire IPET set and exclusively from template IPETs (post-cleaning and filtration).

5.1 Rules from All IPETs

The subsequent analysis involved 4775 IPETs/transactions with $V = 248$ values. The transaction array had a density of 0.06249367. Transactions comprised

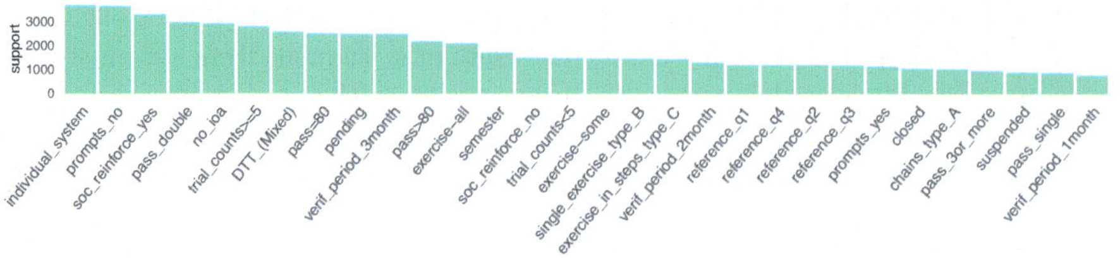

Fig. 3. Top 30 the most frequent variables in all IPETs. A support is the number of occurrence in transactions.

15 values for 3144 instances and 20 for two (representing the most item-dense transaction). The top 30 elements from the set of 248 are illustrated in Fig. 3.

Given the sparse nature of the transaction matrix, low support (e.g., 0.20) is required to extract rules while maintaining a reasonably high confidence (e.g., 0.85). This setup generates 298 rules, all with a lift greater than 1, indicating the association between variables is not accidental.

Table 3 presents the top 20 rules sorted by lift. Observing these rules, one finds patterns such as **R1:** `single_exercise_type_B` ⇒ `exercise-some` implies that if the teaching mode is of type B (where skill tasks are taught independently, each task is individually recorded, graphed, and assessed over time), it is likely associated with the recording of individual exercises/tasks from the IPET plan. **R2:** `exercise_in_steps_type_C` ⇒ `exercise-all`, suggesting that when the teaching method allows decomposing the task into multiple hierarchically ordered learning steps, i.e., tasks can be taught in parallel, and the steps for a given task will be presented on a single graph divided by pretests/baseline, and verification of efficiency over time (plotting of trends) will be possible for each step separately, then a therapist will record all steps of students' performance.

Furthermore, the presence of `verif_period_2month` in the left-hand side (10, 11, and 12 LHS), along with `soc_reinforce_yes` and `no_ioa` variables, indicates that IPETs with social reinforcement and a verification period of two months often have no IOA. The variable `individual_system` describes the reinforcement style prescribed in the IPET which is defined individually for every student and is involved in LHS 7, and appears as a RHS of a rule with 1.15 Lift. Reinforcements are obligatory in IPET.

The subsequent discussion focuses on the most explanatory rules, selected based on their high lifts and associations with the maximum number of variables, featuring a rule length of six variables. The three highest lifts rules are:

R3: {`individual_system, pass=80, prompts_no, soc_reinforce_yes, trial_counts> =5`}⇒ `DTT_(Mixed)` (lift = 1.636, support: 1091)
R4: {`individual_system, pass=80, no_ioa, soc_reinforce_yes, trial_counts>= 5`} ⇒ `DTT_(Mixed)` (lift = 1.622, support: 993)
R5: {`individual_system, prompts_no, no_ioa, soc_reinforce_yes, trial_counts>= 50`} ⇒ `DTT_(Mixed)` (lift = 1.589, support: 975)

Table 3. The table presents a set of 20 rules with the highest lift. The right-hand side (RHS) represents the consequence of the rule, and the left-hand side (LHS) is the antecedent.

RHS	Lift for each rule: 1–14 number is the suitable LHS													
	1	2	3	4	5	6	7	8	9	10	11	12	13	14
exercise-some								3.30						
single_exercise_type_B									3.30					
exercise-all	2.28	2.28	2.28	2.28	2.28	2.28	2.28							
no_ioa										1.46	1.43			
soc_reinforce_yes											1.30	1.32	1.26	1.24
individual_system														1.15
prompts_no								1.12	1.12				1.11	1.13

LHS - antecedent	LHS number
{exercise_in_steps_type_C}	1
{exercise_in_steps_type_C,pass=80}	2
{DTT_(Mixed),exercise_in_steps_type_C}	3
{exercise_in_steps_type_C,trial_counts>= 5}	4
{exercise_in_steps_type_C,no_ioa}	5
{exercise_in_steps_type_C,soc_reinforce_yes}	6
{exercise_in_steps_type_C,individual_system}	7
{single_exercise_type_B}	8
{exercise-some}	9
{soc_reinforce_yes,verif_period_2month}	10
{verif_period_2month}	11
{no_ioa,verif_period_2month}	12
{trial_counts>= 5 }	13
{DTT_(Mixed)}	14

Rule **R3** implies that when therapists plan an individual system of reinforcement during interventions, social reinforcement (in the form of feedback such as, e.g., smiles)) is utilized, and prompts are not expected (autonomous student's work). Additionally, if each skill is attempted at least five times and 80% of trials are successful, the teaching approach is likely the DTT technique with records collected discontinuously (Mixed). Rules **R4** and **R5** explicitly reveal that the Interobserver Agreement (IOA) is not planned for learning evaluation.

The adjustment of support to 0.15 and confidence to 0.8 resulted in 4137 rules generated where the three being the most comprehensive of which incorporated seven elements (lift=3.302, support: 502). Notably, one of these rules was:

R6: { DTT_(Mixed), individual_system, exercise-some, prompts_no, soc_reinforce_yes, trial_counts>= 5 }⇒ single_exercise_type_B

The rule R6 merges elements from both Rule **R1** and **R3**, except for the **pass = 80** variable.

5.2 Rules from IPET Templates

In SYSABA, a supervisor may mark high-quality IPET as a template for therapists to apply in training easily. The present study reduces IPET to templates only and examined 151 IPETs/transactions with $V = 150$ values, achieving a transaction array density of 0.09748344. The top five most frequent values were: `individual_system` (115), `soc_reinforce_yes` (108), `pass_double` (98), `trial_counts>= 5` (93), and `reference_q1` (91). 87 IPETs contained 14 values, while the most populated IPET, containing 17 occurred in count 8.

The support threshold was set at 0.4 and the confidence level at 0.85 to facilitate rules extraction. The longest pattern identified, possessing the highest lift of 1.726 and support of 64 samples, was:

R7: {`individual_system,soc_reinforce_yes,trial_counts>= 5`} ⇒ `DTT_(Mixed)` (lift=1.726, support: 64)

All variables present in rule **R7** are a subset of rule **R3** derived from all IPETs, thus outlining a similar scheme for the DTT technique.

Twenty rules were produced, as visualized in Fig. 4. Here, node shading corresponds to lift values, and their sizes are proportional to support, with variables denoted in blue. Blue arrows are inputs, and red are outputs.

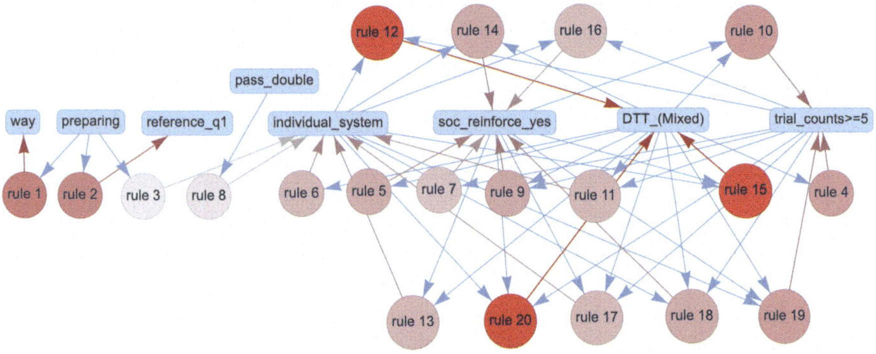

Fig. 4. Graph presenting 20 rules extracted from IPETs limited only to templates

The graph highlights emerging variables such as IPET status in the `preparing` phase and the prevalence of `references_q1`, representing the number of links to the IPET. It also presents the `way` of generalization across diverse contexts and individuals. The common teaching technique pattern involves using DTT in an `individual_system`, supplemented by social reinforcement shaped by therapist-student interactions. In these IPETs, the standard number of teaching attempts is five or more (`trial_counts>= 5`), and success is typically defined as two consecutive completions of a task (`pass_double`). Utilizing ubiquitous filtering revealed that, in contrast to all IPETs, templates promote a 3-month verification period.

5.3 Discussion

The unveiled patterns constitute tacit knowledge embedded through therapists' practical actions and the creation of their programs. The extraction of this knowledge reveals a relatively low variance in the selection of techniques and teaching methods used in skill-acquisition programs. On one hand, the Discrete Trial Teaching (DTT) technique which is prevalent, well-established, and highly regarded in the ABA therapist community [9], is understandably favored. Regarding teaching methods, Types B and C, which decompose the taught skill into smaller components, are the most prevalent. This approach aligns with the logical assumption that knowledge is easier to assimilate when built from essential and minor experiences. The chains' (technique type A) lack of popularity, a particular mode of teaching primary and life skills, is intriguing, suggesting that few programs adopt this teaching style. Chains are underrepresented in the database because most IPETs are connected with preschool and school skills.

Noteworthy patterns in recording results include the use of five or more repetitions, defining skill acquisition based on 80% attempts, and passing the task twice consecutively. Interestingly, it appears that therapists often refrain from providing prompts to students during sessions.

The discovered correlations can be translated into practical applications within SYSABA to facilitate program creation. With appropriately designed conditions, it could be possible to auto-fill certain forms a therapist must complete when creating a program. Given the consistent co-occurrence of specific program components, they could potentially be set as default values.

6 Conclusion and Future Work

This study extracted association rules from the SYSABA database containing IPETs for skills to uncover patterns and relationships within Polish therapeutic centers that employ ABA principles. Among many discovered rules, seven key rules were examined in detail. The general findings indicate: DTT is the predominant technique for skill acquisition, reinforcement is always individually tailored with the frequent addition of social reinforcement, therapists avoid IOA, use a minimum of five trials per session, and prefer an 80% pass threshold. Specifically, in IPET templates, achieving two consecutive passes is popular.

Statistical analysis further reveals patterns. Over 98% of IPETs employ a single pretest, and 95% evaluate IPET skills by tracking consecutive successes. Notably, less than 3% of IPETs reach the template database, indicating a stringent selection process.

The insights from IPET analysis could be valuable for therapeutic centers by elucidating prevailing methodologies and ABA techniques. This study suggests the potential of SYSABA to shape recommendations and aid in default IPET setup, thereby supporting therapeutic decisions.

The Apriori algorithm's non-dependence on controlled variables, focusing instead on frequency discovery, is both an advantage and a limitation. Notably, the presented rules do not include teaching targets or their categories. For future work, the use of decision rule extraction is suggested. This method may allow discovering of additional knowledge since it enables hypothesis-driven conclusions within rules.

References

1. Agrawal, R., Imieliński, T., Swami, A.: Mining association rules between sets of items in large databases. ACM SIGMOD Rec. **22** (1993). https://doi.org/10.1145/170036.170072
2. Ault, M.J., Wolery, M., Gast, D.L., Doyle, P.M., Eizenstat, V.: Comparison of response prompting procedures in teaching numeral identification to autistic subjects. J. Autism Dev. Disord. **18** (1988). https://doi.org/10.1007/BF02211880
3. Collins, B.C., Lo, Y.Y., Park, G., Haughney, K.: Response prompting as an ABA-based instructional approach for teaching students with disabilities. TEACHING Except. Child. **50**(6), 343–355 (2018)
4. Hahsler, M.: arulesViz: visualizing association rules and frequent itemsets (2023). https://CRAN.R-project.org/package=arulesViz. R package version 1.5-2
5. Hahsler, M., Buchta, C., Gruen, B., Hornik, K.: arules: mining association rules and frequent itemsets (2023). https://CRAN.R-project.org/package=arules. R package version 1.7-6
6. Çetrez Iscan, G., Nurçin, E., Fazlioglu, Y.: Effect of most-to-least prompting procedure on dressing skill of students with autism. Educ. Res. Rev. **11** (2016)
7. Kolodziejczyk, J., Grzegorczyk-Dluciak, N., Kuliga, E.: Rule-based expert system supporting individual education-and-therapeutic program composition in SYS-ABA, vol. 207 (2022). https://doi.org/10.1016/j.procs.2022.09.517
8. Leaf, J.B., Cihon, J.H., Ferguson, J.L., Weinkauf, S.M.: An introduction to applied behavior analysis. In: Matson, J. (ed.) Handbook of Childhood Psychopathology and Developmental Disabilities Treatment. Autism and Child Psychopathology Series, pp. 25–42. Springer, Cham (2017). https://doi.org/10.1007/978-3-319-71210-9_3
9. Lerman, D.C., Dittlinger, L.H., Fentress, G., Lanagan, T.: A comparison of methods for collecting data on performance during discrete trial teaching. Behav. Anal. Pract. **4** (2011). https://doi.org/10.1007/bf03391775
10. Matson, J.L., Hattier, M.A., Belva, B.: Treating adaptive living skills of persons with autism using applied behavior analysis: a review (2012). https://doi.org/10.1016/j.rasd.2011.05.008
11. Rodgers, M., et al.: Interventions based on early intensive applied behaviour analysis for autistic children: a systematic review and cost-effectiveness analysis. Health Technol. Assess. **24** (2020). https://doi.org/10.3310/hta24350
12. Smith, T.: Discrete trial training in the treatment of autism. Focus Autism Other Dev. Disabil. **16**(2), 86–92 (2001). https://doi.org/10.1177/108835760101600204

Ensemble Learning Approach to Fake News Detection Problem

Adrianna Kozierkiewicz, Marcin Pietranik(✉), and Aleksandra Stawarz

Faculty of Information and Communication Technology, Wroclaw University of Science and Technology, Wybrzeze Wyspianskiego 27, 50-370 Wroclaw, Poland
{adrianna.kozierkiewicz,marcin.pietranik,aleksandra.stawarz}@pwr.edu.pl

Abstract. This article is devoted to developing a method for fake news detection. Due to the immense consequences of fake news spreading, this problem has recently gained importance and popularity. However, the currently available methods' effectiveness still leaves much room for improvement - a detailed analysis of recent literature allowed for identifying current trends. Thus, in this paper, we propose a novel approach to the given task based on ensemble learning methods. Such an approach allows combining several different strategies to detect fake news detections into a unified, cohesive methodology. The performed experiments that used the widely known "Liar, Liar Pants on Fire" benchmark dataset yielded promising results which surpassed outcomes described in the literature.

Keywords: Ensemble learning · deep learning · fake news

1 Introduction

"Leonardo DiCaprio has donated \$10 m in support of Ukraine." "An Italian priest declined a respirator treatment to save the young woman's life.". Such headers appear before our eyes in a daily manner. Whether we use Facebook, Twitter, or any news service, this kind of message is nothing unusual. Moreover, at first glance, they can sound plausible. However, the quoted sentences have one thing in common - they are both fake.

In the case of the first sentence, it is a well-known fact that Leonardo DiCaprio is a well-known actor who often gets involved in various campaigns and charity collections, donating considerable sums of money to them. It is also a fact that when the news was made public, the Ukrainian state needed financial support from other countries. This information is in line with the current state of world politics. The situation is similar in the case of the second news item, as it is easy to imagine a situation in which a priest decides to undertake a noble act and save the life of a woman who still has her whole life ahead of her.

As readily seen, the news is often communicated in short statements, and its analysis is based mainly on general knowledge of the world and verification of the authors with the source of information. On this basis, it is generally possible

M. Hernes and J. Wątróbski (Eds.): ECAI 2023, LNNS 1079, pp. 15–26, 2024.
https://doi.org/10.1007/978-3-031-66761-9_2

to decide whether or not the information contained in a statement is consistent. However, as recent years have shown, even this can prove unreliable because consistency does not entail trustworthiness.

In this context, it is evident that identifying fake news is a highly complex problem, as no clear rules make it possible to distinguish clearly between fake and real news. For example, we cannot just look for specific words that could indicate the veracity of a given statement. We found many different approaches to the task in the literature, some being better in some aspects but worse in others.

The considered problem can be formally described as follows: *For a given set D of all possible documents and a set $K = \{c_1, c_2, ...c_k\}$ of classes describing different levels of truthfulness, one should determine a function λ which for a selected document $d \in D$ returns a classification of this document into one of the classes from K.*

The thorough analysis of available publications led us towards ensemble learning methods which allowed us to combine several strategies. Therefore, this paper aims to present a novel method for automatically detecting fake news, which combines outcomes collected from random forests, bidirectional neural network, and multilayered perceptron.

What is worth emphasizing is that the developed model is based on an analysis of a given document's content and context. By the context of the document, we understand a set of structured metadata which describes the considered news. Such data may include, among others, information about the document's origin, the time of its creation, its author, etc.

The developed method was experimentally verified using the widely known "Liar, Liar Pants on Fire" benchmark dataset [21]. The conducted tests ended with encouraging results. Therefore, we claim that including the document's context and its content in the truth evaluation highly increased the quality of the proposed method.

The remainder of the article is organized as follows. Section 2 contains an overview of methods devoted to natural language processing and machine learning methods in the context of the fake news detection problem. Section 3 presents our hybrid approach to the considered tasks. The results of conducted experiments can be found in Sect. 4. The last Sect. 5 summarizes the paper and briefly describes upcoming research plans and method development.

2 Related Works

According to [3], the most intuitive approach to recognize fake news automatically is to analyze a statement's content based on its linguistic aspects. It is at the heart of the problem and is what most of the methods analyzed are based on. The use of text requires the use of Natural Language Processing (NLP), which aims to represent text in the form of numerical vectors. Such an approach usually starts with the initial processing of an analyzed document, including tokenization, stemming, or lemmatization [20]. The final purpose is transforming tokenized text into a vector of numbers. The most commonly used approach is incorporating the so-called Bag of Words [23]. However, many other means exist to achieve this goal, e.g., n-grams or tf-idf [10].

Using a measurable text representation of a document makes it possible to use machine learning methods [15]. In [14], two machine learning algorithms - a naive Bayes classifier and a passive-aggressive classifier - were compared against each other in the task of fake news detection. Both models achieved good results, however the acquired models were entirely focuses on linguistic aspects of processed news.

The recent development of deep learning techniques entailed incorporating them into many aspects of the fake news detection problem. Their applications spread from text sequence processing [22] to language modeling [9].

Several newer articles focused on comparing several algorithms' performance separately or using ensemble learning techniques [4]. The ensemble method uses several machine learning models to obtain better predictions. The models are trained independently of each other and make their predictions. These predictions are combined to produce a new result, usually in a voting process. For example, in [11], the authors used a support vector machine, linear regression, a naive Bayes classifier, and a random forest method. A voting process was used to select the best result.

The ensemble learning methods allow the development of hybrid approaches that incorporate many different base tools [1]. For example, in [8], authors proposed using different models which differed in the architecture of the neural network and the natural language processing method. The first model was based on a Long Short-Term Memory (LSTM) network that used the word2vec embedded layer, while the second differed only in adding another LSTM network layer. Two further models were based on convolutional networks; one used n-grams, while the other used the Linguistic Inquiry and Word Count (LIWC) method. The ensemble learning methods have been a straight inspiration for the model we proposed in this article as promising for good results.

It is worth noticing that not only a sole analysis of the message itself is can produce good results in identifying fake news. Other aspects related to information spreading are also worth investigating. One such aspect is news author reputation [16]. As commonly known, a reputation has a significant impact on how a person is perceived in society.

The authors of [18] tried to create different sets of characteristics of users more likely to share fake and real news. They took into account inferred elements, i.e., the number of true and false tweets published in the past, age, personality traits indicated on the profile, location, profile picture, and political preferences. The analysis has shown that enriching the content of messages with this type of information highly increases the effectiveness of fake news detection.

As readily seen, most recent publications describe methods of fake news detection that focus on some selected aspects of processed documents. To the best of our knowledge, no holistic approach to the considered task has been developed. This article intends to describe the novel method of fake news detection based on a combination of several different, well-known tools using ensemble learning methods. Partial elements processed in the proposed solution involve the content of the analyzed document and the context of its creation. Such approach differs our method from the ones found in the literature, and as shown in the following sections, the obtained results are auspicious.

3 Method Overview

3.1 Basic Notions

In our research, time is expressed using a notion of a timeline, which can be treated as an ordered set of discrete moments. Formally, it is defined as $TL = \{t_n | n \in \mathbb{N}\}$. A pair $<TL, \prec>$ forms a strict partial order, where $\prec$ is an ordering relation. Having two moments of time $t_1, t_2 \in TL$ such that $t_1 \prec t_2$ we can say that t_1 is earlier that t_2.

Let $\widetilde{T}$ denote a set of words from some alphabet. By D we will denote a finite set (with repetitions) of documents (e.g. tweets, Facebook posts, email). Each document d from the set D is defined as a tuple of the following structure:

$$d = (content, context, timestamp) \tag{1}$$

where:

- $content \in 2^{\widetilde{T}}$ is a string representing a document's actual content
- $context \subset \widetilde{T}$ is a set of words describing a document in terms of its meta-information
- $timestamp \in TL$ is a moment of time in which a document has been created

Truthfulness of documents from D is not always a binary property. Therefore, in our research we use a finite set K which groups classes that describe different levels of considered truthfulness. For example, K may contain three classes $K = \{true, almost\_true, false\}$. The sought function λ has a signature $\lambda : D \rightarrow K$.

3.2 General Architecture of the Proposed Solution

In our work, we propose an ensemble learning approach which combines outcomes several different techniques using selected voting method. The general idea of our method is presented in Fig. 1.

The method's workflow is divided into three stages. The first stage extracts the document's content and context. It performs data preparation for each using natural language processing methods for preprocessing the document's content and value reduction with categorization to preprocess the document's context.

In the second stage of the workflow (described further in Sect. 3.3), selected machine learning models are used. We selected random forest and bidirectional neural network to classify preprocessed document content. To classify the document's context, we also chose random forests and multilayer perceptron.

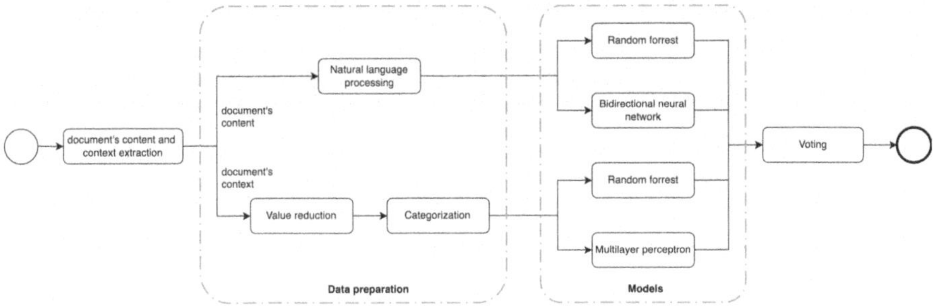

Fig. 1. The general idea for an ensemble learning fake news detection method

The third and final stage is responsible for conducting the voting procedure. Recent research has shown that combining multiple models allows for better predictive performance than could be obtained from any of the constituent learning algorithms alone. In our work we used a weighted majority voting to combine classifications from individual models. It takes as input outputs of classifications performed by partial methods in the second stage. By *RF*, we denote the output of classification of random forests, *ML* denotes the output of multilayer perceptron, and *BNN* denotes a bidirectional neural network. Then it creates a vector of partial classifications - it includes three times the classification of the document's context given by the multilayer perceptron, two times the classification of the document's context done by random forests, a single occurrence of classification of the document's content given by random forest and a single classification of document's content from the bidirectional neural network. These repetitions can be identified with weights. Finally, the algorithm calculates numbers of occurrences of classes from K in the created vector. The output is the class which most frequently occurred.

The preprocessing stage is used to split the documents into their context and content and perform the cleaning, which removes redundant or unnecessary data. The training data has been preprocessed in two ways. The context of the document (*context*) is split into its atomic values and then stored in separate columns. Then missing values are unified, unnecessary spaces are removed, and capital letters are changed to lowercase. After these operations, the data values are eventually reduced and categorized.

Natural language processing is applied to documents *content*. Uppercase letters to lowercase are converted for each document, and punctuation, special characters, and stop-words are removed. Subsequently, the text is tokenized and stemmed. Finally, the tf-idf and Glove transformations are applied to conduct word embedding. The resulting vectors are then fed to random forests and recurrent neural network.

3.3 Model Training

In order to train the selected models, we needed to choose a dataset. Our research is based on the benchmark dataset called LIAR [21]. It consists of 12811 short statements from www.politifact.com. Each statement published on this site has been manually flagged for fraud by its administrators. These statements can be treated as documents from Eq. 1. They include a *content*, but more importantly, the LIAR dataset is also provided with additional information about the described statements, i.e., the topic (e.g., politics, health, taxes), medium (e.g., interview, Twitter post, TV commercial), and the number of other statements created by the same person (based on which it is possible to estimate the author's reputation). Our method treats this additional information as a document's *context*.

Each statement is assigned to one of six categories *1. pants of fire, 2. false, 3. barely true, 4. half true, 5. mostly true, 6. true* which can be treated as elements of the set K. Thus, $|K| = 6$. This division seems more consistent with reality than the binary classification into completely true or false. Very often, it is possible to find news containing different shades of grey. However, the classification into six classes is a more complex problem and requires better models.

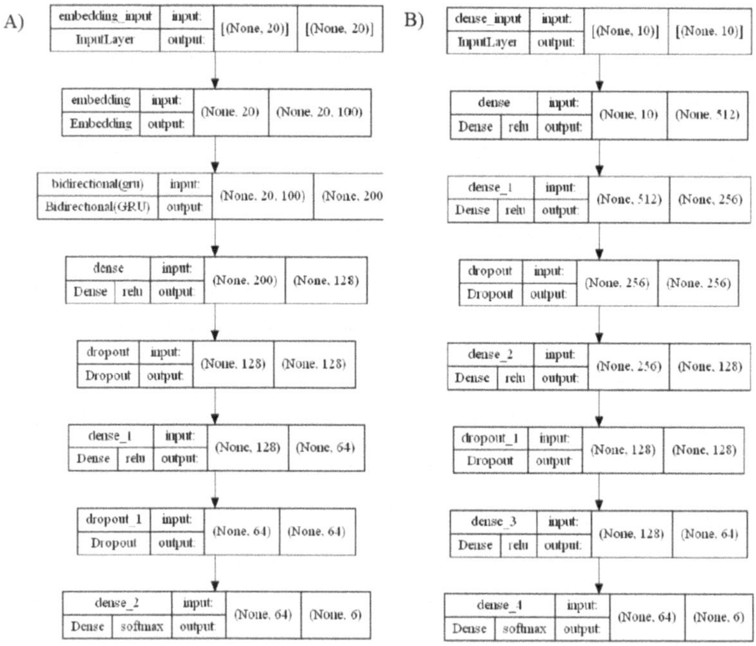

Fig. 2. The neural network architecture A) Bidirectional network for statement processing, B) Multilayer perceptron for context processing

The statement's context has been separately used for training a random forest-based model (which includes 200 trees) and multilayer perceptron with the architecture presented in Fig. 2 B. The multilayer perceptron is one of the simplest types of neural networks. The developed model consists of six dense layers: input, output, and four hidden layers. We have assumed that the input layer consisted of 10 neurons and the output layer consisted of 6 neurons. Hidden layers have 512, 256, 128, and 64 neurons, respectively. Dropout has also been added after the third and fourth hidden layers with a dropout rate of 0.1. The ReLU activation function has been selected for hidden layers, and the soft-max function has been used for the output layer. Before the training process, the data was normalized using the min-max scaling method. The Adam optimizer has been used to compile the model. Sparse categorical cross-entropy has been assumed as the loss function. Training took place over 40 epochs with a batch size equal to 64.

The statements' contents have been used to train a model based on random forests, which included 200 trees and a bidirectional neural network presented on the left side of Fig. 2 A. The developed model has a few layers, including the input and output layers. The embedding layer has mapped the words to their embedding vectors based on the Glove embedding matrix. For the embedding layer size of the vocabulary equal to 11217, the vector length for each word is equal to 20, and the maximum length of a sequence equal to 100 has been assumed. The next layer has been based on 100 GRUs. Dropout layers separated the last three dense layers with a dropout probability of 0.1. The first dense layer contains 128 neurons, the second 64, and the third 6. As an activation function, ReLu and soft-max have been chosen.

4 Experimental Results

As mentioned in Sect. 3.3, the research was based on the benchmark dataset called LIAR [21]. Its authors divide the dataset into training, testing, and validation. For evaluating the developed solution, we used classic measures: Accuracy, Precision, Recall, and F-measure.

The first research aimed to show how our ensemble method deals with the problem of detecting fake news in comparison to single methods. The Accuracy, Precision, Recall, and F-measure has been separately calculated for each model presented in Sect. 3.2. Table 1 presents the obtained results. The research confirmed that the developed solution based on the ensemble method improves the quality of fake news detection. However, interesting conclusions can be drawn by comparing the individual methods. Research has shown that the metric values were twice as high for Multilayer Perceptron and Random Forest (context) than for Bidirectional Neural Network and Random Forest (content). This means that the detection of fake news is significantly influenced not by the statement itself but by its context.

We also analyzed how the quality measures for individual classes are distributed. Based on Table 2, it can be concluded that the proposed method is

Table 1. Quality measures for individual methods

	Accuracy	Precision	Recall	F1-measure
Bidirectional Neural Network	0.2324	0.2077	0.2084	0.2014
Multilayer Perceptron	0.4562	0.5308	0.4635	0.4544
Random Forest- content	0.2394	0.2511	0.2153	0.2075
Random Forest - context	0.4155	0.4838	0.4126	0.4454
Proposed method	**0.4609**	**0.5383**	**0.4657**	**0.4579**

best at recognizing completely false messages - the values of both Precision and Recall are about 60%. For news from the *true* class, a very high Precision (86.84%) can be observed. Unfortunately, at the same time, very low Recall (only 15.79%) was obtained. This means that our method cannot correctly detect all true news. However, in the problem of fake news detection, detecting whether a given statement is false is essential because disinformation is the most harmful to society. Hence, high values of quality measures are more desirable for fake news, and for true news, they are not so important.

Table 2. Quality measures for individual classes

Class	Precision	Recall	F1-measure
pants of fire	0.6044	0.5978	0.6011
false	0.4222	0.5640	0.4829
barely true	0.4563	0.4393	0.4476
half true	0.4203	0.5431	0.4739
mostly true	0.4583	0.4929	0.4795
true	0.8684	0.1579	0.2672

The results of conducted experiments have been collected and presented as a confusion matrix in Fig. 3. The confusion matrix shows the combination of actual and predicted classes. It is a good measure of whether models can account for overlapping class properties and understand which classes are most easily confused. The method is very good at recognizing fake news. The vast majority of entirely false news was correctly classified. The most incorrect predictions concern the *false* class (23 cases) and the *barely true* class (7 cases). However, this is not a significant error since both categories contain more false than true.

The proposed method incorrectly predicted the news from the *true* class. This explains the low Recall value in Table 2. The proposed method most often made mistakes within the adjacent categories, which is not so important from

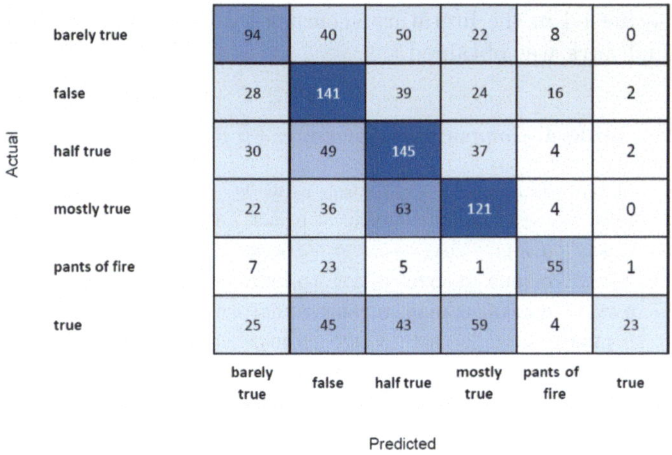

Fig. 3. Confusion matrix

the point of view of fake news detection. In many situations, we do not need the clarification of whether the news is half or mostly true or false or barely true.

The results were compared with those obtained by the authors of other works, which also used the LIAR set. The first column of Table 3 contains a brief summary of the models and the subsequent values of the Accuracy measure. Unfortunately, in many works, the authors focused only on Accuracy, omitting other equally essential performance indicators.

Table 3. Comparison of results for classifications including 6 classes

Model	Accuracy
CNN + LSTM [21]	0.2740
LSTM [13]	0.4150
Random forests + sentiment analysis [2]	0.3560
Triple BERT model [24]	0.3740
Double BERT model [12]	0.4058
CNN + LSTM + word tagging [7]	0.4200
CNN + BiLSTM [19]	0.3510
capsule network [5]	0.3950
CNN + margin loss function [6]	0.4160
Ensemble learning - machine learning methods [17]	0.4415
Proposed model	**0.4609**

To the best of our knowledge, our method has achieved the best result so far. The combination of deep learning techniques proposed in this paper and classical machine learning allowed using the potential of both methods. The improvement of the best-achieved in the literature accuracy by 1.94% and 18.69% compared to the original work was obtained.

Table 4. Comparison of measures for individual classes

Class	Margin loss function			Capsule network			Proposed model		
	Prec.	Recall	F1	Prec.	Recall	F1	Prec.	Recall	F1
pants on fire	0.2857	0.0243	0.0449	0.4737	0.0978	0.1622	**0.6044**	**0.5978**	**0.6011**
false	0.3787	**0.7669**	**0.5070**	0.3592	0.6867	0.4717	**0.4222**	0.5640	0.4829
barely true	0.4554	**0.4718**	**0.4635**	**0.7000**	0.1651	0.2672	0.4563	0.4393	0.4476
half true	**0.4554**	0.3834	0.4163	0.4477	0.4007	0.4229	0.4203	**0.5431**	**0.4739**
mostly true	**0.4959**	0.2552	0.3370	0.3801	0.4137	0.3962	0.4583	**0.4929**	**0.4795**
true	0.3729	0.3385	0.3548	0.3556	**0.3791**	**0.3670**	**0.8684**	0.1579	0.2672

Analyzing Table 4, we can conclude that the proposed model recognizes much better completely false news - each of the received values is higher than the one found in the literature. It also performs better at recognizing statements labeled as *half true* and *mostly true* where the highest values of Recall and F1 were obtained for both.

The Capsule network is better at recognizing true news, and our method is highly unstable in the case of this class. Nevertheless, the proposed model returns the highest values most often, allowing us to conclude that our solution is the best of all known models.

5 Summary

The problem of identifying fake news is a challenging and complex problem. There are a number of aspects to consider when designing a solution for this task. Careful analysis of the literature showed that currently available approaches are frequently based on a single aspect of fake news, such as its content, origin, or authors' reputation. Only a handful of publications address the issue holistically.

This paper presents our approach to fake news detection. It is a novel solution that incorporates an ensemble learning method. It allows to combine several different, independent partial strategies (such as random forests or multilayer perceptron) into a unified, cohesive methodology. More importantly, the selected strategies process different aspects of fake news, focusing on their content (like a plethora of previously developed approaches) and their context (understood as a set of meta-data describing fake news).

The most significant advantage of such a hybrid method is improving previous solutions found in the literature. The solution presented in the original work [21] describing the widely known "Liar, Liar Pants on Fire" benchmark dataset authors obtained the Accuracy equal to 27.4%. The best solution found in the literature obtained the Accuracy equal to 44.15%. As shown in Sect. 4 the method proposed in this article obtained 1.94% better Accuracy.

Our upcoming research plans include two main directions. The first incorporates the fact that the proposed solution is a hybrid approach to the considered problem. It is, therefore, possible to experiment with different models and their combinations, e.g., using neural networks of higher complexity or other machine learning algorithms.

The second direction of considered development includes testing the developed ideas using different datasets and expanding the proposed model with elements that could process additional information from independent sources, such as various social media.

References

1. Aslam, N., Ullah Khan, I., Alotaibi, F.S., Aldaej, L.A., Aldubaikil, A.K.: Fake detect: a deep learning ensemble model for fake news detection. Complexity **2021** (2021)
2. Bhutani, B., Rastogi, N., Sehgal, P., Purwar, A.: Fake news detection using sentiment analysis. In 2019 Twelfth International Conference on Contemporary Computing (IC3), pp. 1–5. IEEE (2019)
3. Choraś, M., et al.: Advanced Machine Learning techniques for fake news (online disinformation) detection: a systematic mapping study. Appl. Soft Comput. **101**, 107050 (2021)
4. Dong, X., Yu, Z., Cao, W., Shi, Y., Ma, Q.: A survey on ensemble learning. Front. Comput. Sci. **14**(2), 241–258 (2020)
5. Goldani, M.H., Momtazi, S., Safabakhsh, R.: Detecting fake news with capsule neural networks. Appl. Soft Comput. **101**, 106991 (2021)
6. Goldani, M.H., Safabakhsh, R., Momtazi, S.: Convolutional neural network with margin loss for fake news detection. Inf. Process. Manage. **58**(1), 102418 (2021)
7. Hakak, S., Alazab, M., Khan, S., Gadekallu, T.R., Maddikunta, P.K.R., Khan, W.Z.: An ensemble machine learning approach through effective feature extraction to classify fake news. Futur. Gener. Comput. Syst. **117**, 47–58 (2021)
8. Huang, Y.F., Chen, P.H.: Fake news detection using an ensemble learning model based on self-adaptive harmony search algorithms. Expert Syst. Appl. **159**, 113584 (2020)
9. Kaliyar, R.K., Goswami, A., Narang, P.: FakeBERT: fake news detection in social media with a BERT-based deep learning approach. Multimed. Tools Appl. **80**(8), 11765–11788 (2021)
10. Khan, J.Y., Khondaker, M.T.I., Afroz, S., Uddin, G., Iqbal, A.: A benchmark study of machine learning models for online fake news detection. Mach. Learn. Appl. **4**, 100032 (2021)
11. Lasotte, Y.B., Garba, E.J., Malgwi, Y.M., Buhari, M.A.: An ensemble machine learning approach for fake news detection and classification using a soft voting classifier. Eur. J. Electri. Eng. Comput. Sci. **6**(2), 1–7 (2022)
12. Liu, C., Wu, X., Yu, M., Li, G., Jiang, J., Huang, W., Lu, X.: A two-stage model based on BERT for short fake news detection. In: Douligeris, C., Karagiannis, D., Apostolou, D. (eds.) KSEM 2019. LNCS, vol. 11776, pp. 172–183. Springer, Cham (2019). https://doi.org/10.1007/978-3-030-29563-9_17
13. Long, Y., Lu, Q., Xiang, R., Li, M., Huang, C.R.: Fake news detection through multi-perspective speaker profiles. In: Proceedings of the Eighth International Joint Conference on Natural Language Processing (Volume 2: Short Papers), pp. 252–256 (2017)

14. Mandical, R.R., Mamatha, N., Shivakumar, N., Monica, R., Krishna, A.N.: Identification of fake news using machine learning. In: 2020 IEEE International Conference on Electronics, Computing and Communication Technologies (CONECCT), pp. 1–6. IEEE (2020)
15. Manzoor, S.I., Singla, J.: Fake news detection using machine learning approaches: a systematic review. In: 2019 3rd International Conference on Trends in Electronics and Informatics (ICOEI), pp. 230–234. IEEE (2019)
16. Rahmatulloh A., Shofa R.N., Darmawan I.: Sentiment analysis of ojek online user satisfaction based on the naïve bayes and net brand reputation method. In: 2021 9th International Conference on Information and Communication Technology (ICoICT), pp. 337–341. IEEE (2021)
17. Rathod, S.: Exploring author profiling for fake news detection. In: 2022 IEEE 46th Annual Computers, Software, and Applications Conference (COMPSAC), pp. 1614–1619. IEEE (2022)
18. Shu, K., Zhou, X., Wang, S., Zafarani, R., Liu, H.: The role of user profiles for fake news detection. In: Proceedings of the 2019 IEEE/ACM International Conference on Advances in Social Networks Analysis and Mining, pp. 436–439 (2019)
19. Trueman, T.E., Kumar, A., Narayanasamy, P., Vidya, J.: Attention-based C-BiLSTM for fake news detection. Appl. Soft Comput. **110**, 107600 (2021)
20. Umer, M., Imtiaz, Z., Ullah, S., Mehmood, A., Choi, G.S., On, B.W.: Fake news stance detection using deep learning architecture (CNN-LSTM). IEEE Access **8**, 156695–156706 (2020)
21. Wang, W.Y.: "Liar, liar pants on fire": a new benchmark dataset for fake news detection. arXiv preprint arXiv:1705.00648 (2017)
22. Widiastuti, N.I.: Deep learning-now and next in text mining and natural language processing. In: IOP Conference Series: Materials Science and Engineering, vol. 407, no. 1, p. 012114. IOP Publishing (2018)
23. Zhang, Y., Jin, R., Zhou, Z.H.: Understanding bag-of-words model: a statistical framework. Int. J. Mach. Learn. Cybernet. **1**(1), 43–52 (2010)
24. Triple Branch BERT Siamese Network for fake news classification on LIAR-PLUS dataset. https://github.com/manideep2510/siamese-BERT-fake-news-detection-LIAR

SMAP: Smart Personalized Medication Assistant to Help Nurses in Medication Adherence

Michele Persiani[1], Madeleine Bluise[2], Helena Norberg[2], and Juan Carlos Nieves[1(✉)]

[1] Department of Computing Science, Umeå University, Umeå, Sweden
{michelep,jcnieves}@cs.umu.se
[2] Department of Medicine, Umeå University, Umeå, Sweden

Abstract. We here report on the current development of an intelligent assistant supporting health professionals in the task of preparing pillboxes. This task is challenging for different reasons: it requires to search medications in the hospital's physical warehouse, and when pills are unavailable to search online resources for equivalents, to then search again the warehouse for availability of those. We therefore propose to support this process through an assistant residing on a holographic device that can be wore during practices. In this way, nurse can rely on an online holographic layer supporting the task showing availability of drugs and online information in a comfortable way. In addition, we plan to furnish this agent with features for explainability through an Abstract Argumentation Framework. We here introduce the argumentative framework and provide an initial formulation for deciding which drugs are equivalent and where they are in the warehouse. In the future we plan to extend the framework with more types of argumentative structures such as when suggesting drugs based on online resources.

Keywords: Intelligent Assistant · Ontology · Probabilistic Logic Argumentation Framework · Augmented Reality

1 Introduction

Preparing pillboxes for patients is a common task in nurses daily job which requires to fill pillboxes according to given prescriptions. The procedure includes taking pills from their original containers and distributing the loose pills in the compartments of the pillbox. Using pillboxes is a common stragety to improve and facilitate medication management for patients with multiple medicines, so called polypharmacy. Instead of patients taking pills from multiple packages several times throughout the day, the nurse prepare the pillbox for a week at the time [8,18]. Prescribed drugs may not always be available, and as such nurses are required to substitute the original prescribed drugs with equivalent generics from the market. In clinical nursing practice substitutions occur often. The main reason is generic drug substitution, where the prescribed drug gets substituted with

M. Hernes and J. Wątróbski (Eds.): ECAI 2023, LNNS 1079, pp. 27–38, 2024.
https://doi.org/10.1007/978-3-031-66761-9_3

one having the same chemical composition and intake administration, but selling at a cheaper price. Generic drug substitution policies have been introduced by governments in both developed and developing countries as a way to decrease pharmaceutical expenditures [10] In several European countries substitution is mandatory (European GM se ref list) and in the USA up to 84% of drug prescriptions were substituted for cheaper generic equivalents [6]. In Sweden 2019, the top 20 most used substances were together available in 1391 varieties (on average 70 varieties/substance, range 7–241) (Socialstyrelsen, 2020, FASS 2021 - these 2 are not in ref list, I will add).

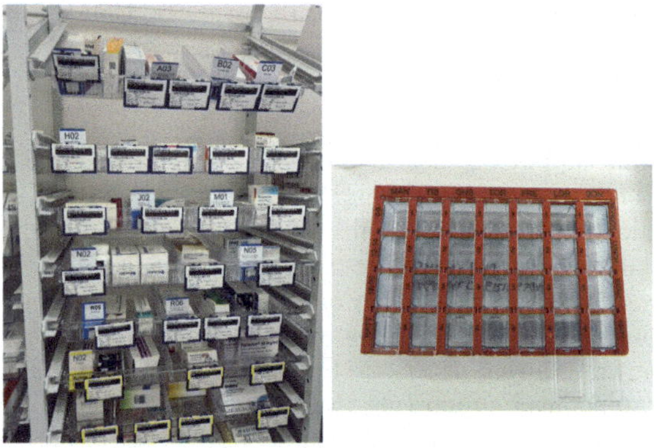

Fig. 1. Pillbox filling operation.

However, generic medicine substitution is a challenging task from the professionals perspective. Health professionals such as nurses often simultaneously manage prescriptions of multiple patients, and for each of them are required to search for generic alternatives in large catalogues. Due to the large availability of drugs nurses rely on online information systems to search for correct generic substitution drugs when the prescribed medicine is not on the shelf. These systems have become powerful tools for this practice [2,13]. Online databases (e.g. FASS for the Swedish country) contains detailed description of several thousands of drugs, including generic substitution information, indications, contraindications, physical aspect, history of medical trials, and more. Not all this information is needed when selecting generic drug alternatives; however, due to their vast amount of data being presented, which may not always be relevant for their current case, health professionals can be easily overwhelmed with information. Also, searching for the generic information in the databases is time consuming for nurses, thereby contribution to adding strain to an already busy workday. It is therefore becoming relevant to develop tools to support health professionals investigations on these vast information systems, by for example being able to suggest generic substitutions, to check compatibilities with other drugs, and so on.

Recommender systems based on argumentative frameworks are showing to be powerful tools for this type of scenarios. Since argumentation frameworks decompose interactions with the recommender system as an argumentative process, it has several advantages over traditional statistical-based methods in terms of personalization and explainability. For instance, through argumentation frameworks we can better structure the reasoning process of both patients health professionals, thus allowing more personalized and context-specific recommendations. Additionally, argumentation frameworks can handle uncertainty and conflicting information more effectively, which is often present in medical decision-making.

In this paper we report initial results on the development of an AI system to assist nurses in their daily work with preparing pillboxes for patients, with a focus on generic substitutions when prescriptions' pills are not available. We called this assistant system the Smart Personalized Medication Assistant (SMAP). SMAP aims at supporting nurses in real time during their job by providing information on available drugs, searching for alternatives and accessing related information on online resources. As we propose, this assistant is realizable through Augmented Reality (AR) technologies that are supported by devices such as the Hololens 2, with which we can create an holographic layer of the task with which nurses can interact in a quick and intuitive way. This project has 4 main research questions:

- What challenges do health professionals encounter when handling medicines for patients?
- How does smart interactive technology need to be designed and developed in order to support health professionals when managing generic medicines?
- Which factors are key for usability of such supportive technology?
- To what degree can such technology be trustworthy?

While the aim of the project is broad, in this paper we will report only initial results on the prototype. In particular we will focus on the task of preparing pillboxes (see Fig. 1), that is an important sub-task of handling patients' prescriptions.

This paper reports on the development stage of this prototype of agent. In particular, we implemented two of its main functions that are the management of an ontology of drugs and equivalents, and the management of drugs availability in the warehouse. All this information is made available on a Hololens 2. We also introduce here the Abstract Argumentation Framework (AF) that will be implemented for the reasoning part of the agent, that is concerned in enhancing the agent with features for explainability on its decisions. This part of the work is however mostly left for future immediate work, and at the moment we only provide arguments for finding drugs inside the warehouse.

2 Background

The importance of information systems (ISs) for the health-care has been known for quite some time [11]. On ISs different type of information is stored such as patients and doctors data, medications, and so on, and it would be nowadays strange to think of a health facility, such as an hospital, without an IS as a central repository of information. Usually, system decision are supported by data or knowledge bases and offer different ways of presenting and interacting with the data. For drugs a common way of storing information is through databases targeted at capturing all relevant information on drugs and potential interactions. For the task of preparing pillboxes a rich variety of databases maintained by official institutions is present on the Internet. Nurses already rely on these resources and information systems can be built by either importing the data locally or creating remote interfaces. For the nordic countries we have many freely available National Medical Information Systems (NMIS) that are information systems complying with national directives. For example, the Dental and Pharmaceutical Benefits Agency (tlv.se) provides a database of national-wide availability of drugs and relative prices, Läkemedelsverket provides a database on which drugs can be exchanged because of their therapeutic equivalence (including generics), or fass.se providing an extensive database with drugs and relevant medical trials.

New technologies are changing the way in which this information can be accessed, and enable new possibilities for users. For example, AR technologies such as smart-glasses enable an array of functionality previously not accessible, such as the possibility of continuously displaying information to their users, or to "see" what the user is seeing, thus enabling real time interaction. [3] illustrates different ways in which AR can support professionals in health-care by proposing agent-driven paradigms where an agent resides on the smart devices, and on one hand can support its user in real time in its task, for example through an AR interface, on the other enables access to local or remote ISs for consulting the relevant information.

Previous work in [9] explored the utilization of holographic technologies for medication adherence, where the authors propose an ontology schema to store information about patients' medicine schedules. In addition, they showed how AR glasses (in this case Hololens) can comfortably show information regarding medicines to the patients, by showing an overlay of information over the physical medicine boxes. This becomes especially important for medicine adherence because of the confusion that taking many medicines can induce to patients. Also, it is shown how AR technologies can illustrate in simple and intuitive ways information regarding to medicines. Results of a user study showed how patients positively perceived the technology: 50% of patients above 70 years said they would use it in the future, while for patients below 70 this number was 91%. Factors such as age and expertise with technology affected these preferences. As concluded in [14] AR technologies have potential in the health-care domain for doctors (eg. surgeons) but also patients, because of features such as being hands-free, the capacity of taking photos or videos, and their possibility of allowing quick access to online resources.

When moving towards advanced AI agents that can autonomously recommend decisions impacting patients' health, such as on their pills schedule, properties such as competence (e.g. accuracy of recommendations) and explainability (e.g. explaining provided recommendations) become of prime importance. Prior work [12] shows how this type of systems can be built on the so called evidence-based medicine, that is a systematic approach in which health care professionals use the available scientific evidence from past clinical research to help make decisions about an individual patient's case. An example of such recommender systems is proposed in [4], where deep models are utilized to extract information on drug interactions from medical records, or in [5], where the authors propose different ways to leverage data from medical trials to then predict new outcomes.

It is however debated how data mining large texts to then provide a statistical-based decision may not be the best solution for medical practices targeting individual patients. This is because there is a gap between generally valid statistical-based data and the specific conditions of patients and decisions of doctors, and it is important that doctor's prescriptions are targeted towards the specific cases rather than just being generally supported by data. Targeting this behavior, various alternatives have been proposed that are more focused on complementing or extending doctors reasoning and not replacing it with statistical-based predictions [7,17].

In this, Argumentative Frameworks represents a sweet spot between the statistical reliance offered by data mining techniques through the process of argument mining, robustness and predictability because of its inherently formal framework, and explainability and transparency, because allows to define argumentative structures that best fit the reasoning process of both patients and health professionals, thus allowing more personalized and context-specific recommendations.

Argumentative approaches for medical applications can be categorized into three main categories [1]: for medical decision making, for medical explanations and for medical dialogues. A main distinctive factor among the categories is the intended goal of the system (respectively recommending diagnosis, explaining system diagnosis, or to provide the system a conversational type of interface). Each type of system category could require different types of argumentative structures and focus on different aspects of the recommendation process such as context, usage of natural language, or type of underlying dataset or knowledge base.

3 Method

We envision SMAP as an information system for assisting health professionals in their daily duties of handling patients' medications. As backbone of the system we created an ontology merging information publicly available from NMISs with other regarding the local context of prescriptions, such as drug availability inside the ambulatory's warehouse. If a drug is not available, by automated reasoning we can search if there are generic substitutes. This allows to capture to a relevant degree the information that nurses rely on when deciding for generic substitutions while preparing pillboxes. Presenting this information through augmented reality devices makes information retrieval possible during practices, in an holographic form that allows an easy an intuitive interaction with information. For this functionality we realized an initial prototype of this system on Hololens 2 as we will describe in the next sections.

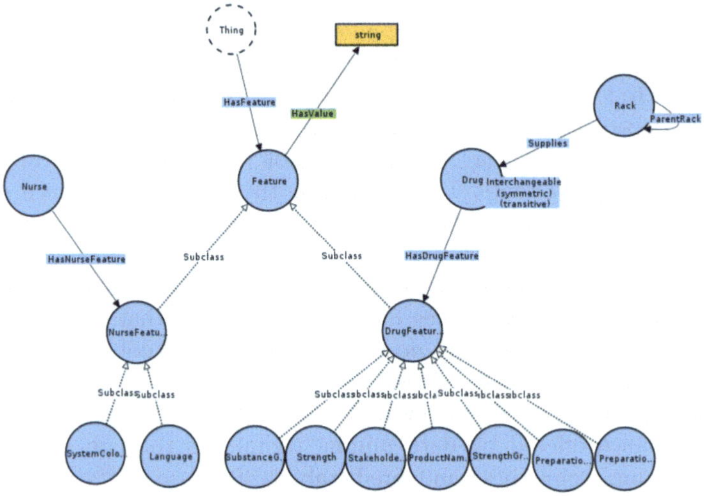

Fig. 2. Developed ontology. See text for details.

3.1 Ontology

A drug ontology provides access to information of drugs and substitutions. We designed an ontology in OWL using Protégé [15] (see Fig. 2), in which we imported the database on generic substitutions from the Swedish Medical Product Agency[1]. This database groups drugs by their therapeutic equivalence and is updated eight times a year. We also created classes and properties regarding the availability of drugs inside the warehouse, as well as a set of classes to store use preferences such as language. The resulting ontology has 16 classes and 7 properties.

Classes. The ontology classes can be divided into three main groups, modeling drugs features, user preferences, and warehouse availability. For the first and second group we are mainly concerned about modeling the features related to drugs and users, and as such this part of the ontology represent an hierarchy of types features. For warehouse availability we instead implemented a pattern of containers (racks) and objects (drugs). Table 1 contains a details description of the classes.

[1] https://www.lakemedelsverket.se/sv/behandling-och-forskrivning/utbytbara-lakemedel/utbytbarhetslista.

Table 1. Ontology classes

Class name	Description	Class individuals	Super class
Feature	Objects features that have an associated value	All features values associated to objects such as drugs and nurses	Feature ⊑ Thing
Drug Feature	Features exclusive to drugs	Feature values for drugs such as name, strength, stakeholder etc.	Drug Feature ⊑ Feature
Substance Group	Groups of drugs that are considerable as interchangeable	Names of the groups	Substance Group ⊑ Drug Feature
Preparation Form Group	Groups of drugs preparation forms	Names of preparation form groups such as "tablet", "injection", or "capsule"	Prep Form Group ⊑ Drug Feature
Preparation form	Specific drugs preparation form	Specific preparation forms such as "oral", "pulverized", "injected", etc.	Prep form ⊑ Drug Feature
Strength group	Strength (quantity) group in which the drug belongs to	All possible strength groups such as "1mg", "50ml", etc.	Strength Group ⊑ Drug Feature
Strength	Specific strength of individual products	All possible individual strengths	Strength ⊑ Drug Feature
Product Name	Drug names	All possible drug names	Product Name ⊑ Drug Feature
Product Stakeholder	Companies producing drugs	Individual companies manufacturing drugs	Product Stakeholder ⊑ Drug Feature
Drug	Drugs in the catalogue	All available drugs	Drug ⊑ Thing
Rack	Warehouse racks	All racks inside the warehouse	Rack ⊑ Thing
Nurse	Nurses utilizing the system	Individual nurses	Nurse ⊑ Thing
Nurse feature	Features related to nurses	Features such as preferred language, views, colors, etc.	Nurse Feature ⊑ Feature
Name	Nurses names	All available names	Name ⊑ Nurse Feature
Language	Selected language	All available languages	Language ⊑ Nurse Feature
System Color	Selected system color	All available system color	System Color ⊑ Nurse Feature

Properties. Properties mainly connect drugs and users to their corresponding features, and racks to their contained drugs. In addition, a property of drug interchangeability connects drugs to their equivalents. Table 2 contains a details description of the properties.

Table 2. Ontology properties

Class name	Description	⟨Domain, Range⟩	Super property
Interchangeable	Drug interchangeability relation	⟨Drug, Drug⟩	Interchangeable ⊑ Top Object Property
Has Value	Relates features to their values	⟨Feature, string⟩	Has Value ⊑ Top Object Property
Has Feature	Relates objects to their features	⟨Thing, Feature⟩	Has Feature ⊑ Top Object Property
Has Drug Feature	Relates drugs to their features	⟨Drug, Drug Feature⟩	Has Drug Feature ⊑ Has Feature
Has Nurse Feature	Relates nurses to their features	⟨Nurse, Nurse Feature⟩	Has Nurse Feature ⊑ Has Feature
Supplies	Specifies which drugs are supplied by racks	⟨Rack, Drug⟩	Supplies ⊑ Top Object Property
Parent	Specifies which rack contains other racks	⟨Rack, Rack⟩	Parent ⊑ Top Object Property

Querying this ontology for substituting a certain drug is rather simple, and we just have to query for all drugs connected to it by the *interchangeable* property.

3.2 Data Modeling for Automated Reasoning

There are two main types of entities that we need to model to perform basic automated reasoning on drug equivalence and availability: the drugs and the warehouse where drugs are stored. When combined, these will allow to define inference rules capturing which drug is available inside the warehouse, and, in the negative case, if there are generic substitutions available.

Drug. Drugs are modeled with features taken from the OWL ontology which defines drugs by a set of attributes such as name, strength, preparation form, etc. as well as equivalence groups for which each drug in the same group can be considered equivalent to all other inside the same group.

A drug is a tuple

$$d = \langle id, N, S, S_g, P, P_g, E_g, T \rangle \tag{1}$$

where id: unique drug identifier, N: drug name, S_g*: substance group dictated by the drugs active principle, S: strength i.e. quantity,* S_g*: class of strength, P: preparation form,* P_g*: class of preparation form,* E_g*: equivalence group for which drugs in the same group are equivalent, T: product stakeholder.*

Rack. Racks are warehouse units of storage (e.g. see Fig. 1). For simplicity we will model as racks warehouses, drugs shelves and the individual shelf slots where the drugs are kept. In this way we can create a recursive structure for drugs storage where racks contains other racks. In our specific case we will have that the warehouse contains shelves that contains shelf slots. Warehouse, shelves, and shelf slots are all rack objects.

A Rack is a tuple

$$r = \langle id_r, P_r, D_r, R_r \rangle \tag{2}$$

where id: unique rack id, P: parent rack or null if it doesn't have a parent (i.e. is a warehouse), D: set of drugs contained inside the rack, R: set of sub-rack of the rack[2]

We then define two base predicates that specify how racks are organized in an hierarchical structure, and how racks supplies the drugs they contain.

$$\text{parent}(x, y) \iff y \in R_x \tag{3}$$

$$\text{supplies}(x, a) \iff a \in D_x \lor \exists y \, \text{parent}(x, y) \land \text{supplies}(y, a) \tag{4}$$

specifying that a rack x is parent of another y if it contains it in one of its sub-racks. And that a rack x supplies a drug a if it is either directly supplied or if it is in one of its sub-rack. We will use these predicates for inference of equivalent drugs supply using an argumentation framework, as we will later describe.

[2] For example if we consider Fig. 1 we would a rack for every specific drug slot, plus a parent rack for the whole shelf defining all drug slots as sub-racks.

3.3 Argumentation Framework for Drug Availability

To augment the system with automated reasoning we define a logic-based probabilistic argumentation framework (LPAF) for two common queries that nurses perform when searching the warehouse for pills: checking if a specific drug has an available equivalent substitutions, and which rack contains the equivalent substitutions. For this, we developed a LPAF capturing availability of drugs inside the warehouse; that contains 8 arguments templates $A = \{a_0, a_1, a_2, a_3, a_4, a_5, a_6, a_7\}$. These templates are then instantiated for all possible combinations of their variables. The resulting arguments with corresponding probabilities are shown in Table 3. For all templates x, y are racks and a, b drugs. We can then fulfill these two queries by searching for specific extensions of the AF. For example, for the first type of query, where we want to know whether a drug equivalent to another drug a is present in the warehouse, we can search for extensions containing the predicate $\text{supplies}_{\text{eq}}(x, b)$. We will here use the complete extension; even though also others could be viable. For example, we could use the preferred extension to obtain argumentative structures of minimum size.

$$\textbf{equivalents}(AF, a) = \{\forall Y \subseteq A,\ \textbf{complete}\ Y \wedge \exists x\ \text{supplies}_{\text{eq}}(x, a) \in Y\} \quad (5)$$

Table 3. Arguments modeling drug availability and equivalence

Arg_{name}	Arg_{body}	$P(\text{Arg})$
a_0	$\langle E^g_a = E^g_b, \text{interchangeable}(a, b)\rangle$	$\mathbf{1}((E^g_a = E^g_b) \in \text{Ontology})$
a_1	$\langle E^g_a \neq E^g_b, \neg\text{interchangeable}(a, b)\rangle$	$1 - P(a_0)$
a_2	$\langle \text{supplies}(x, a) \wedge \text{interchangeable}(a, b), \text{supplies}_{\text{eq}}(\text{x,b})\rangle$	1
a_3	$\langle y \in R_x, \text{parent}(x, y)\rangle$	$\mathbf{1}((y \in R_x) \in \text{Ontology})$
a_4	$\langle y \notin R_x, \neg\text{parent}(x, y)\rangle$	$1 - P(a_3)$
a_5	$\langle a \in D_x, \text{supplies}(x, a)\rangle$	$\mathbf{1}((a \in D_x) \in \text{Ontology})$
a_6	$\langle a \notin D_x, \neg\text{supplies}(x, a)\rangle$	$1 - P(a_5)$
a_7	$\langle \exists y\ \text{parent}(x, y) \wedge \text{supplies}(y, a), \text{supplies}(x, a)\rangle$	1

4 Implementation on Hololens

We created an holographic platform to interact with the ontology using Hololens 2 [9]. The platform's development was performed on Unity with the Mixed Reality Toolkit (MRTK) [16]. The MRTK already provides multiple libraries for interacting with holograms and what is left to do is to design Unity entities to access and interact with the developed ontology. This early version of prototype allows to search the ontology for drugs and to show alternatives on an holographic tab. Instead of manually inserting the medicine's name users can also scan packets of medicines from the Hololens's camera, and through Optical Character Recognition (OCR) the medicine's name is extracted. Users can also manage the warehouse ontology by a set of dedicated holograms for racks. Ideally, these virtual racks are superimposed with the real ones such that the holographic system corresponds with the actual warehouse. The prototype fully resides on the Hololens and is composed of three main modules: the Ontology Module, the Reasoning Module and the Holographic Module (see Fig. 3). We will next describe these modules in details.

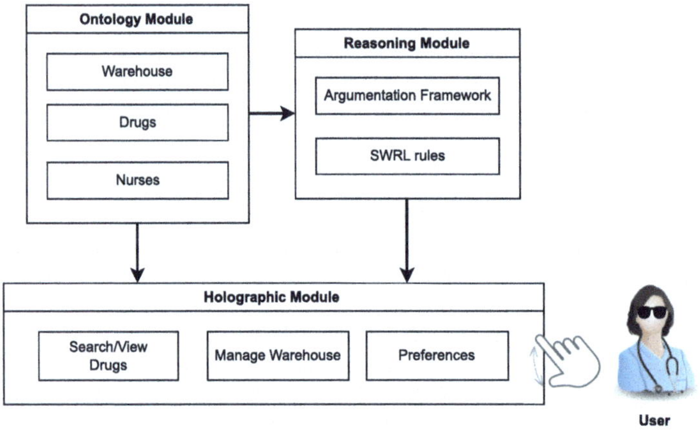

Fig. 3. System architecture running on Hololens. The main modules are the OCR module to scan drugs packages, Ontology module to store and query the ontology, and holographic module allowing nurses interaction.

Ontology Module. This module is mostly responsible of providing access to the previously described ontology, for example by searching for drugs using their features, search where to find a specific drug, or get a user preference.

Reasoning Module. The reasoning module perform inference on where to find drugs equivalents inside the warehouse, by combining information on drugs with that of the warehouse. This can be done using an argumentation framework as we previously described, but other means are also available such as through SWRL rules. We expect for this module to be extended in the future with other types of inferences.

Holographic Module. The holographic modules handles the interaction with the system components and offers a user interface to the users. At the moment we have three main sets of holograms: to search/view drugs with either manual input or through OCR, to manage the warehouse by adding/removing racks or putting drugs into them, and for managing user preferences.

5 Conclusions

In this paper we illustrated an initial version of a prototype of holographic system to support nurses in the task of preparing pillboxes. This system accorpates publicly available information on drugs with that of local availability inside the warehouse. This allows to reason on whether a specific drug is available and where it is, and in the negative case whether an equivalent is available. This reasoning is here implemented using a logic probabilistic argumentative framework. At the moment we don't leverage probabilities that much, but this could be useful in the future if we'll need to handle noisy inputs for the system queries.

With the goal of encompassing the task of managing pillboxes future work will comprehend the addition of new functionalities. A first extension will be on the argumentative framework with the goal of supporting explanations based on online resources. This would allow nurses to know in more detail why the assistant suggested a given drug. Then, we could add inside the ontology information about patients. Knowing which drugs that the current patient needs allows to get altogether all availability inside the warehouse for that specific patient, and the warehouse's supply can be obtained at a glance. Future work will also comprehend user studies where we will gather health professionals opinions on the developed system. This can be performed through standard questionnaires such as the godspeed questionnaire.

References

1. Caroprese, L., Vocaturo, E., Zumpano, E.: Argumentation approaches for explanaible AI in medical informatics. Intell. Syst. Appl. **16**, 200109 (2022)
2. Clauson, K.A., Fass, J.A., Seamon, M.J.: Legal requirements for drug information resources maintained by pharmacies. Drug Inf. J.: DIJ/Drug Inf. Assoc. **42**, 569–582 (2008)
3. Croatti, A., Montagna, S., Ricci, A.: A personal medical digital assistant agent for supporting human operators in emergency scenarios. In: Montagna, S., Abreu, P., Giroux, S., Schumacher, M. (eds.) A2HC AHEALTH 2017. Lecture Notes in Computer Science, vol. 10685, pp. 59–75. Springer, Cham (2017). https://doi.org/10.1007/978-3-319-70887-4_4
4. Fan, B., Fan, W., Smith, C., et al.: Adverse drug event detection and extraction from open data: a deep learning approach. Inf. Process. Manage. **57**(1), 102131 (2020)
5. Harrer, S., Shah, P., Antony, B., Hu, J.: Artificial intelligence for clinical trial design. Trends Pharmacol. Sci. **40**(8), 577–591 (2019)
6. Hassali, M.A., et al.: The experiences of implementing generic medicine policy in eight countries: a review and recommendations for a successful promotion of generic medicine use. Saudi Pharmac. J. **22**(6), 491–503 (2014)
7. Helgason, C.M.: The application of fuzzy logic to the prescription of antithrombotic agents in the elderly. Drugs Aging **21**, 731–736 (2004)
8. Huygens, M.W., Vermeulen, J., Swinkels, I.C., Friele, R.D., Van Schayck, O.C., De Witte, L.P.: Expectations and needs of patients with a chronic disease toward self-management and ehealth for self-management purposes. BMC Health Serv. Res. **16**, 1–11 (2016)

9. Ingeson, M., Blusi, M., Nieves, J.C.: Smart augmented reality mHealth for medication adherence. In: AIH@ IJCAI, pp. 157–168 (2018)
10. Kaplan, R.S., Porter, M.E., et al.: How to solve the cost crisis in health care. Harv. Bus. Rev. **89**(9), 46–52 (2011)
11. Kuhn, K.A., Giuse, D.A.: From hospital information systems to health information systems. Methods Inf. Med. **40**(04), 275–287 (2001)
12. Landes, J.: An evidence-hierarchical decision aid for ranking in evidence-based medicine. Uncertainty Pharmacol.: Epistemol. Methods Decis. 231–259 (2020)
13. Mayer, M.A., Rodríguez Blanco, O., Torrejon, A.: Use of health apps by nurses for professional purposes: web-based survey study. JMIR Mhealth Uhealth **7**(11), e15195 (2019)
14. Mitrasinovic, S., et al.: Clinical and surgical applications of smart glasses. Technol. Health Care **23**(4), 381–401 (2015)
15. Musen, M.A.: The protégé project: a look back and a look forward. AI Matt. **1**(4), 4–12 (2015)
16. Ong, S., Siddaraju, V.K., Ong, S., Siddaraju, V.K.: Introduction to the mixed reality toolkit. In: Ong, S., Siddaraju, V.K. (eds.) Beginning Windows Mixed Reality Programming: For HoloLens and Mixed Reality Headsets, pp. 85–110. Apress, Berkeley (2021). https://doi.org/10.1007/978-1-4842-7104-9_4
17. Sturmberg, J.P.: Evidence-based medicine–not a panacea for the problems of a complex adaptive world. J. Eval. Clin. Pract. **25**(5), 706–716 (2019)
18. Tellier, M., Auger, C., Bier, N., Demers, L.: Use of an electronic pillbox by older adults with mild Alzheimer's disease: impact on medication administration and adherence. Gerontechnology **19**(1), 66–76 (2020)

The Use of Modern Technologies in Decision-Making by Polish Farmers

Agnieszka Biernat-Jarka(✉), Monika Gębska, Wiesław Świderek, and Sławomir Jarka

Warsaw University of Life Sciences, Nowoursynowska 166, 02-787 Warsaw, Poland
agnieszka_biernat_jarka@sggw.edu.pl

Abstract. Modern technological solutions, such as information and communication technologies may significantly contribute to the proper decision-making process in farming. Using data and intelligent decision-making support systems can improve the productivity and profitability of farming and stimulate sustainable development. With modern technology, agricultural data, such as weather information, soil conditions, land use, and animal diseases, can be collected, analysed and processed to help farmers make the right decisions and achieve better production results. Therefore, the question should be: What kind of farmers' decisions are influenced by modern technologies? Is it possible to distinguish groups of farmers similar in implementing modern technologies? The research covered 86 farm owners located in various regions of Poland. These farms varied in terms of area, and the average area of a farm was about 107 ha. The research distinguished three segments of farmers: firstly, enthusiasts of modern technologies, who had no problems using them; the second group were sceptics, who did not use modern technologies; and the third one was the farmers, who used modern technologies, but often had problems with their use.

Keywords: Smart Farming · Precision Agriculture · Decision Support

1 Introduction

One of the current challenges related to agriculture is how to feed the growing world population, with limited external inputs and limited environmental impact, taking into account the increasingly variable and extreme climatic conditions [1]. That is why the sustainable use of existing natural resources is so important. Agriculture, as a sector dedicated to food production, given the growth of the population, has to cope with the increasing demand for food, considering social, environmental and economic issues. More than ever, it is necessary to find innovative, modern technologies to improve the functioning and development of the agricultural sector [2]. In addition, one of the biggest challenges currently facing agriculture in many countries is the workforce shortage for fieldwork. This trend is assumed to continue in the coming years [3]. In this light, the use of modern technologies in agriculture makes it possible to deal with this problem, on the one hand, through the use of technological solutions in the production process and, on the

M. Hernes and J. Wątróbski (Eds.): ECAI 2023, LNNS 1079, pp. 39–49, 2024.
https://doi.org/10.1007/978-3-031-66761-9_4

other hand, in making management decisions on farms. Smart farming offers the ability to analyse agricultural data on an unprecedented scale. Using multiple data and intelligent decision-making support can improve the productivity and profitability of framing and optimise the consumption of scarce natural resources. However, this process requires technological progress and a detailed understanding of the factors affecting the acceptance of technologies in smart farming [4]. Even small changes in the production process, appropriate use of plant protection products, fertilisation, and appropriate prevention and early action in case of animal diseases can significantly impact agricultural production's profitability. The foundation of efficiency is the effective acquisition, processing and management of multiple sources of information to enable making economically viable and environmentally friendly decisions [5].

2 The Impact of Modern Technologies on the Decision-Making Process

Smart farming is based on using information and communication technologies to enable more efficient, productive and profitable agricultural enterprises [6]. The ongoing development of information and communication technologies offers considerable potential for information management on the farm. Sensor technologies allow farmers to monitor their farms in unprecedented detail, in multiple aspects and near real-time [6]. Modern farm management strategies use sensors, data and communication networks in cropping systems [7, 8] to increase yields and reduce production losses [9]. Agriculture 4.0, as the fourth evolution in agricultural technology, poses four basic requirements: increased productivity, reasonable allocation of resources, adaptation to climate change and prevention of food waste [10].

With the adoption of advanced information systems and Internet technologies in agriculture 4.0, agricultural data, such as weather information, soil conditions or land use, can be collected, analysed and processed in order to help farmers in making the right decisions and achieving better production results [10]. Jawad [8] found that wireless sensor networks can be used in agriculture to provide farmers with much information. Using wireless sensor technologies and management tools by farmers can result in more efficient and sustainable agriculture. With modern technology, the right nutrients for crops and the proper use of pesticides can be ensured to control weeds, pests and diseases effectively.

With the advent of smart machines and sensors in farming and the increase in the amount and scope of farm data, farming processes will become increasingly data-driven and data-enabled. The rapid development of the Internet of Things and cloud computing drives smart farming [11]. The smart farm is based on autonomous decision-making aimed at improving product quality, product safety and environmental sustainability, reducing production costs, reducing time to delivery of products to consumers, increasing market share and retaining the workforce [12]. Smart farms, like smart factories, should use the following modern technologies: artificial intelligence, cyber-physical systems (CPS), Internet of Things (IoT), Big Data and cloud computing. Smart farms should also reduce the traditional delay in using new technologies concerning their use in industry [12].

The emergence of intelligent machines and sensors on farms, as well as the ability to use more and more information, allows for better management and more efficient decision-making. While precision agriculture considers the variability in the field, smart farming goes much further, basing the management process not only on the specific location of the farm but also on many other real-time data [13]. Machines increasingly assist people in analysis and planning, so the cyber-physical cycle becomes almost autonomous. People will always be involved in the entire process, but increasingly at a much higher level of decision-making, leaving most operating activities to machines [13]. The survey's main objective was to determine whether farmers were interested in implementing modern technological solutions in their farms and whether they use them in the decision-making process.

3 Research Method

Research on the scope of using modern technologies on farms in the decision-making process was carried out in the fourth quarter of 2022. Data were collected through a survey. The sample was selected deliberately, and the survey participants were people with knowledge about the use of modern technologies in agriculture. Respondents were people living in Western, Central and Eastern Poland, from Wielkopolskie, Mazowieckie, Warmińsko-Mazurskie and Podlaskie voivodeships.

The survey's main objective was to determine whether farmers were interested in implementing modern technological solutions in their farms and whether they use them in the decision-making process. Based on the conducted research, farmers' differences in the approach to the use of modern technological solutions were also determined, and, on that basis, farmers were classified into three groups, which differed in the approach to the use of modern technologies in their farms.

The survey was conducted on a sample of 86 farm owners located in various regions of Poland. These farms varied in terms of area; namely, the area of farms ranged from 10 to 2000 ha. The average area of a farm was about 107 ha, of which the (own) average agricultural area was 54 ha. Notably, a large amount of agricultural land was leased in the analysed farms – on average, about 66 ha. The average age of the respondent was 44.3, with the largest group being young farmers (up to 30 years old – 23.5%) and farmers aged 50–60 (22.4%). Most farms had an area ranging from 50 to 100 ha of agricultural land (34.1%), which, together with farms of 100 to 300 ha, accounted for over 50% of the surveyed population. The smallest farms with an area of up to 30 ha accounted for 23.5%, and medium-sized farms (30–50 ha) 18.8% of the sample. The least numerous were the largest farms, with an area of more than 300 ha with a share only of 3.5%.

Among the respondents, secondary education (49%) and higher education (28%) prevailed. More than half of the surveyed farms (61.6%) carried out mixed production: crop and livestock production, and about 27% of farms had only livestock production. Farms with plant production were the least numerous (11.6%).

The results of the surveys were statistically processed in the IBM SPSS Statistics 27 program. The internal consistency of the question scale was assessed using the Cronbach alpha reliability coefficient. A cluster analysis was performed using the k-means method to identify homogeneous groups of respondents. The significance of differences between the identified groups was assessed by one-way analysis of variance.

4 Results

According to the research, half of the respondents (43) declared using modern technologies in crop production (Fig. 1). Most of them (24 people, 28% of all respondents and 56% of those using technologies used them to assess the condition and content of soils) and to organise supply processes (25.6% of all respondents, 22 people, 51% of those using technologies). Modern irrigation control and navigation technologies in driving machines are least willingly used (2 people, 2.3% of all respondents, 4.6% of those using technologies).

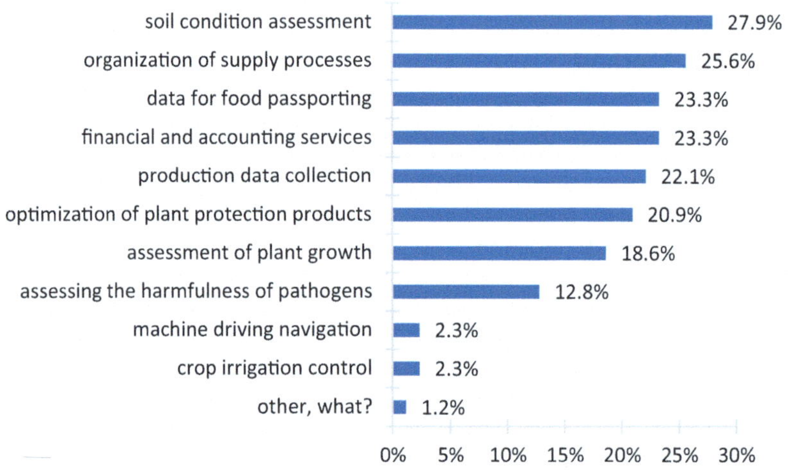

Fig. 1. The use of modern technologies by Polish farmers in crop production

Thirty-seven people (43% of the sample) applied modern technologies in livestock production. These farmers most often used modern technologies to make decisions regarding the organisation of supply processes (23 people, 62% using technologies) and for financial and accounting services (22 people, 26% of all respondents, 60% using technologies). Other usages include using technology to monitor animal behavior (15 people, 23% of the total, 54% using technologies) and animal feeding (15 people, 23% of all respondents, 54% using technologies). Technologies dedicated to the animal watering system control and data collection for Food passports were used the least often. However, those answers were still selected by 32% of those using technologies (14% of the total) (Fig. 2).

The frequency of using telephones or laptops with internet access for farm management was also determined (a five-point scale was used). Smartphones or laptops with Internet access were most often used to check the weather (average m = 4.22), make decisions about fieldwork, and search for purchase offers (m = 3.94). The most rarely indicated uses are e-mail communication (3.27) and sales (m = 3.24).

Another research aspect was getting to know farmers' opinions on modern technologies. The analysis of the averages for the answers to individual questions indicating a

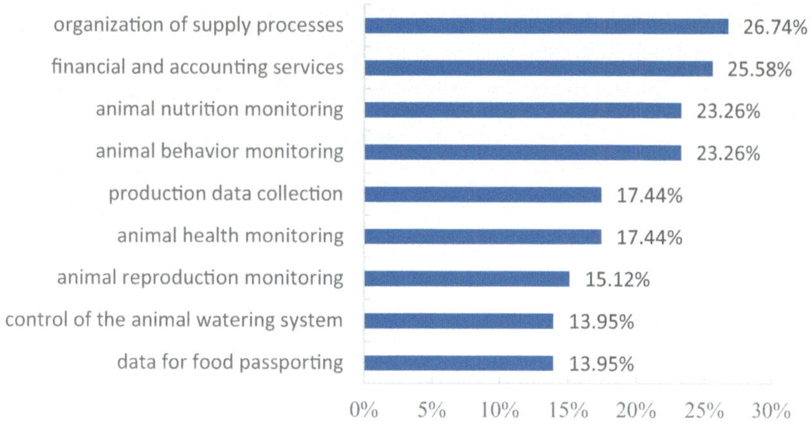

Fig. 2. The scope of using modern technologies in livestock production.

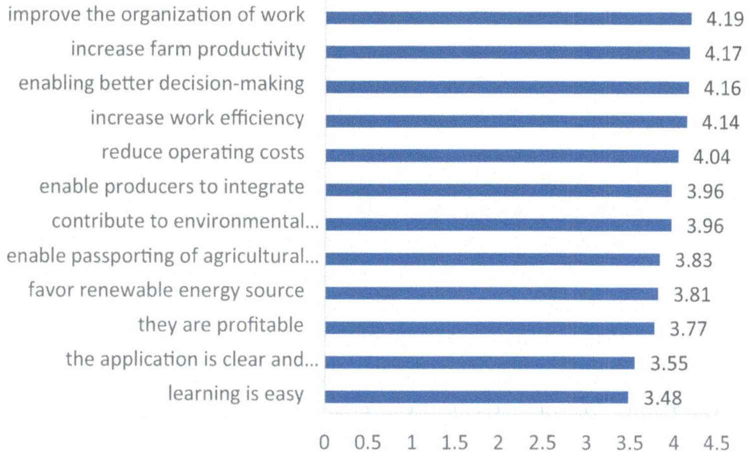

Fig. 3. Positive attitude towards the implementation of modern technologies – technophilia (scale 0–5 was used)

positive attitude towards modern technologies showed that the respondents assessed all the examined aspects of new technologies similarly.

To a lower extent, they agreed that learning new technologies was accessible (m = 3.48) and that using modern technologies was understandable (m = 3.55). The respondents were mainly convinced that new technologies improved the quality and organisation of work (m = 4.19) and provided the information to make better decisions (m = 4.16) (Fig. 3).

Concerning barriers in the implementation of modern technologies, the respondents pointed primarily to the issues with the cost of new technologies (m = 3.43) and lack of compatibility (m = 3.19). Additionally, the farmers stated that technologies are too complex and too complicated for them (m = 2.92). The minor issues were related to their inadequacy (m = 2.56) and insufficient precision (m = 2.73) (Fig. 4).

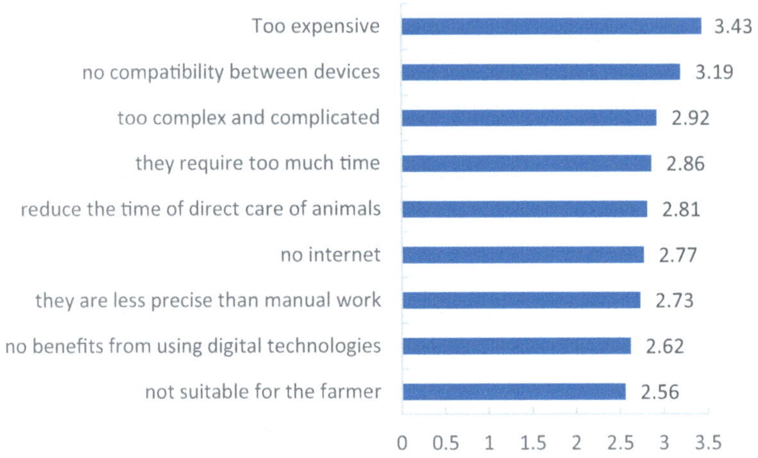

Fig. 4. Barriers related to the implementation of modern technologies (technophobia).

The conducted cluster analysis made it possible to distinguish three groups of farmers. Group 1 – farmers declaring a very low intensity of problems and obstacles in using modern technologies, showing high technophilia and frequent, voluntary use of modern technologies – 21 respondents, 28% of the sample. They can be called enthusiasts of modern technologies. Group 2– farmers declaring a very high technophobia (they did not believe in the benefits of modern technologies) and their occasional use.

They did not indicate significant issues with its use (possibly because they rarely use technology). They constitute 17% of the sample (15 people). We can call them sceptical and not using modern technologies. Group 3 – moderate technophiles farmers struggling with technology usage, often showing slight technophilia and substantial problems and obstacles to its use. They constitute as much as 55% of the sample (47 people) and are distinguished primarily by problems with using modern technologies.

The obtained ANOVA results (Table 1) indicate significant differences between the identified three groups of farmers (technophiles, technophobes, and moderate technophiles). It confirms the right quality of the segmentation made by the cluster method.

The Table 1 presents the average values in the groups and the results of the ANOVA tests (F statistics and the corresponding significance level). The results showed a significant correlation between the age of farmers and for all variables P11 except P11_2 (decisions regarding incurring inputs for production).

The presented data show that the age of the surveyed farmers in different groups differs significantly. Farmers from group 1 are the youngest ones (m = 41.83), respondents from group 3 are not much older (m = 42.93), and those from group 2 are much older (m = 52.07). On the other hand, farmers from group 1 are most convinced that new

Table 1. ANOVA results, characteristics of the groups of farmers

Segment	Cluster Mean Square	df	Error Mean Square	df	F	Sig
Moderate technophiles	8.200	2	0.826	83	9.922	<0.001
Technophiles	20.468	2	0.531	83	38.554	<0.001
Technophobes	24.287	2	0.439	83	55.339	<0.001

technologies contribute to making decisions regarding the joint purchase of farm inputs (m = 4.25). The respondents from group 2 are the most sceptical (m = 3.47). Notably, farmers from group 1 are most convinced that new technologies contribute to decisions on the joint marketing of agricultural products (m = 4.21) (Table 2).

Table 2. Characteristics of the groups of farmers in relation to decision-making attitudes

Segment	Statistics	Area	Age	Experience	P11_1	P11_2	P11_3	P11_4	P11_5	P11_6
1.	Mean	180.7	41.83	18.46	4.25	4.00	4.21	3.88	4.08	4.17
N = 24	SD	400.2	15.93	16.69	0.68	0.88	0.78	0.80	0.65	0.76
2.	Mean	50.55	52.07	27.73	3.47	3.40	3.33	3.07	3.20	3.20
N = 15	SD	32.8	9.34	10.05	0.83	0.74	0.62	1.03	0.94	0.68
3.	Mean	87.89	42.93	19.78	3.72	3.66	3.77	3.60	3.91	3.82
N = 47	SD	88.8	13.70	13.74	0.85	0.87	0.79	0.90	0.78	0.78
Total	Mean	107.2	44.24	20.82	3.83	3.71	3.81	3.58	3.84	3.81
N = 86	SD	223.8	14.08	14.32	0.84	0.87	0.80	0.93	0.82	0.81
F		1.9	3.005	2.26	5.22	2.469	6.354	3.756	6.511	7.557
Sig.		0.142	0.055	0.111	0.007	0.091	0.003	0.027	0.002	0.001

Where: P11_1 – decisions regarding the purchase of input materials for production, P11_2 – decisions regarding agricultural inputs, P11_3 – decisions regarding the marketing of agricultural products, P11_4– decisions regarding investments in renewable energy sources, P11_5 – decisions regarding development investments, P11_6 – decisions regarding management functions

The respondents from group 2 are the most sceptical (m = 3.33). This result is interesting because joint marketing activities allow for building strong local brands and increasing the recognition of agricultural products on the market. It is the basis for a unique sales proposal for the offered products. It can be considered as a significant conclusion in the era of increasing energy prices and the energy transformation of the European economy that farmers from group 1 were most convinced that new technologies contributed to decision-making in planning and implementing joint investments in renewable energy sources (m = 3.88). The respondents from group 2 are the most sceptical (m = 3.07). The last issue taken up in the research was to determine the attitude of farmers towards making strategic decisions for their farms. Decisions of this type usually concern real estate investments (land for agricultural production, buildings and structures) and movable fixed assets in machinery, means of transport, equipment or entire technological lines. Such investments are particularly capital-intensive and can be a source of competitive advantage for farms, especially those operating within the framework of horizontal integration. Again, farmers from group 1 were most convinced that new technologies contribute to joint development investments (m = 4.08). The respondents from group 2 are the most sceptical (m = 3.2).

Summing up the results, farmers classified in the first segment use modern technologies to the greatest extent to make farming decisions, impacting not only the results obtained from the perspective of current activity but also strategic activities regarding future development directions.

5 Discussion with Results of Other Empirical Studies

The literature indicates many factors influencing modern technology acceptance by farmers. Davis [14] listed the factors influencing an individual's intentions to use new technologies, including perceived ease of use and usefulness. In our study, farmers reported that technologies are too complex and too complicated for them. That suggests that technologies supporting decision-making should be more user-friendly. This result aligns with other scientists researching this area [15]. Flett et al. [16] stated that the technology's ease of understanding and use influences its adoption level. Similarly, Borchers and Bewley [17] explained that simplicity and ease of use are the most critical factor when deciding whether to implement digital technology at dairy farms. Gittings et al. [18] showed that the practicality and usability of the new technology had a decisive impact on its implementation. That is why some Polish farmers (17%) showing high Technophobia who did not believe in the benefits of modern technologies used them occasionally. According to other research, voluntary implementation is the factor favouring technology use. Gittings et al. [18] indicated that farmers, who felt intense external pressure to adopt the technology, were less likely to implement a system for herd management. Our study results indicate that farmers who showed high technophilia are enthusiasts of modern technologies and use them frequently. The opposite attitude of the Polish farmers, classified as high Technophobia, could be explained, among others, by economic reasons. Respondents declared that this technology was too expensive. It was also confirmed by the research of Caffaro and Cavallo [19].Further barriers to implementing modern technologies listed by Polish farmers were incompatibility with other systems and poor Internet coverage. Daberkow [20], in his research, found that farmers were more likely to adopt precision technologies if they cultivated a large area of corn, had high yields, and achieved significant income from agricultural production. Our study analysed only the farmland size, partially connected with production volume, but this factor was not statistically correlated with decisions on digital technology use or any production decisions. The land factor does not play the same role in decision-making across farm types. At the same time, our study demonstrates that a farmer's age is essential in decision-making concerning modern technology use. Marescoti [21] obtained a similar result showing that farmers under 50 were more likely to adopt modern technology. Several authors indicated other factors negatively associated with the adoption level of technology at farms, such as low levels of education and working on farms alone [19, 22]. The factors were not the subject of our research.

6 Study Limitations

Limitations of this study include but are not limited to the method of collecting data, that is, the survey, which is related to the reliability and truthfulness of the answers provided. Therefore, respondents were informed about using information for scientific purposes only. The second limitation is the research sample. Purposeful selection of respondents was used because they had to be people running farms. Obtaining information from people, who were not engaged in agricultural activities, would be useless and could lead to falsification of research results. Due to the respondents' selection method, the results do not apply to the entire community but only to the sample considered. In the future, the research sample should be expanded, which could show better and more comprehensive attitudes contributing to the implementation of modern technologies in farming.

7 Conclusions

In the conducted research, the respondents were mostly convinced that new technologies improved the quality and organisation of work and provided the information to make better decisions. The ability to collect and process data and draw the correct conclusions allows the use of modern technologies in the decision-making process.

The research shows that farmers classified in the first group use modern technologies to the greatest extent to make operational and strategic decisions regarding future development directions. This group of farmers declared a very low intensity of problems and obstacles in using modern technologies. They also showed high technophilia and frequent use of modern technologies.

In turn, farmers classified in second group did not believe in the benefits of using modern technologies on farms. Interestingly, they did not point to significant problems with modern technologies, probably due to their lack of experience in this area. They should have used the potential of using modern technologies in agriculture to optimise the decision-making process I pn their farms. Therefore, they did not achieve synergy in making decisions regarding their farms' production and development (investment) activities. This group of farmers were declared sceptics and even opponents of modern technologies.

The largest share in the surveyed sample was made up of farmers showing slight technophilia, vital problems, and obstacles in using modern technologies supporting the decision-making process. These farmers used modern technologies to make decisions regarding development investments and management. They used these technologies relatively less for making decisions regarding agricultural inputs.

The results of this study indicate the need for further research to determine whether farm productivity and financial situation depend on the use of technologies supporting management decision-making on the farm.

References

1. Pittelkow, C.M., et al.: Productivity limits and potentials of the principles of conservation agriculture. Nature **517**(7534), 365–368 (2014). https://doi.org/10.1038/nature13809

2. Couliably, S., Kamsu-Foguem, B., Kamissoko, D., Traore, D. a. K.: Deep learning for precision agriculture: a bibliometric analysis. Intell. Syst. Appl. **16**, 200102 (2022). https://doi.org/10.1016/j.iswa.2022.200102
3. Yang, Q., Du, X., Wang, Z., Meng, Z., Ma, Z., Zhang, Q.: A review of core agricultural robot technologies for crop productions. Comput. Electron. Agric. **206**, 107701 (2023). https://doi.org/10.1016/j.compag.2023.107701
4. Thomas, R.J., O'Hare, G.M.P., Coyle, D.: Understanding technology acceptance in smart agriculture: a systematic review of empirical research in crop production. Technol. Forecast. Soc. Chang. **189**, 122374 (2023). https://doi.org/10.1016/j.techfore.2023.122374
5. Bewley, J., Russell, R., Dolecheck, K., Borchers. M.R.: 1.1. Precision dairy monitoring: what have we learned? Wageningen Academic Publishers eBooks, pp. 13–24 (2015). https://doi.org/10.3920/978-90-8686-815-5_1
6. O'Grady M.J., O'Hare G.M.P.: Modelling the smart farm. Inf. Process. Agric. **4**(3), pp. 179–187 (2017). ISSN 2214-3173, https://doi.org/10.1016/j.inpa.2017.05.001
7. Ferrandez-Pastor, F., García-Chamizo, J.M., Nieto-Hidalgo, M., Mora-Pascual, J., Mora-Martínez, J.M.: Developing ubiquitous sensor network platform using internet of things: application in precision agriculture. Sensors **16**(7), 1141 (2016). https://doi.org/10.3390/s16071141
8. Jawad, H.M., Nordin, R., Gharghan, S.K., Jawad, A.M., Ismail, M.: Energy-efficient wireless sensor networks for precision agriculture: a review. Sensors (2017). https://doi.org/10.3390/s17081781
9. Ponraj, A.S., Vigneswaran, T.: Machine learning approach for agricultural IoT. Int. J. Recent Technol. Eng. **7**(6), 383–392 (2019)
10. Zhai, Z., Martínez, J., Beltran, V., Martínez, N.B.: Decision support systems for agriculture 4.0: survey and challenges. Comput. Electron. Agric. **170**, 105256 (2020). https://doi.org/10.1016/j.compag.2020.105256
11. Sundmaeker, H., Verdouw, C., Wolfert, S., Freire, L.P.: Internet of food and farm 2020. In: Digitising the Industry Internet of Things Connecting the Physical, Digital and Virtual Worlds, pp. 129–151. River Publishers (2016)
12. Gonzalez-De-Santos, P., Fernández, R., Sepúlveda, D., Navas, E., Emmi, L., Armada, M.: Field robots for intelligent farms—inhering features from industry. Agronomy **10**(11), 1638 (2020). https://doi.org/10.3390/agronomy10111638
13. Wolfert, S., Goense, D., Sørensen, C.G.: A Future Internet collaboration platform for safe and healthy food from Farm to Fork. In: Wolfert, J., Sørensen, C.G., Goense, D., (eds.) Global Conference 2014. Annual SRII, pp. 266–273. IEEE, San Jose, CA San Jose, CA, USA (2014). https://doi.org/10.1109/srii.2014.47
14. Davis, F.D.: Perceived usefulness, perceived ease of use, and user acceptance of information technology. Manag. Inf. Syst. Q. **13**(3), 319 (1989). https://doi.org/10.2307/249008
15. Giua, C., Materia, V.C., Camanzi, L.: Smart farming technologies adoption: which factors play a role in the digital transition? Technol. Soc. **68**, 101869 (2022). https://doi.org/10.1016/j.techsoc.2022.101869
16. Flett, R., Alpass, F., Humphries, S.E., Massey, C., Morriss, S., Long, N.: The technology acceptance model and use of technology in New Zealand dairy farming. Agric. Syst. **80**(2), 199–211 (2004). https://doi.org/10.1016/j.agsy.2003.08.002
17. Borchers, M.R., Bewley, J.M.: An assessment of producer precision dairy farming technology use, prepurchase considerations, and usefulness. J. Dairy Sci. **98**(6), 4198–4205 (2015). https://doi.org/10.3168/jds.2014-8963
18. Gittins, P.J., McElwee, G., Tipi, N.S.: Discrete event simulation in livestock management. J. Rural. Stud. **78**, 387–398 (2020). https://doi.org/10.1016/j.jrurstud.2020.06.039

19. Caffaro, F., Cavallo, E.: The effects of individual variables, farming system characteristics and perceived barriers on actual use of smart farming technologies: evidence from the Piedmont Region Northwestern Italy. Agriculture **9**(5), 111 (2019). https://doi.org/10.3390/agriculture9050111
20. Daberkow, S., McBride, W.: Socioeconomic profiles of early adopters of precision agriculture technologies. J. Agribus. **16**(2), 151–168 (1998). https://doi.org/10.22004/ag.econ.90442
21. Marescotti, M.C., Demartini, E., Filippini, R., Gaviglio, A.: Smart farming in mountain areas: investigating livestock farmers' technophobia and technophilia and their perception of innovation. J. Rural. Stud. **86**, 463–472 (2021). https://doi.org/10.1016/j.jrurstud.2021.07.015
22. Carrer, M.J., De Souza Filho, H.M., Batalha, M.O.: Factors influencing the adoption of Farm Management Information Systems (FMIS) by Brazilian citrus farmers. Comput. Electron. Agric. **138**, 11–19 (2017). https://doi.org/10.1016/j.compag.2017.04.004

Decision Support in Production Environment Using Artificial Intelligence

Artur Skoczylas[1(✉)], Bogdan Franczyk[2], Wiesława Gryncewicz[1], and Ryszard Zygała[1]

[1] Faculty of Management, Wroclaw University of Economics and Business, Komandorska 118/120, 53-345 Wroclaw, Poland
artur.skoczylas@ue.wroc.pl
[2] Leipzig University, Grimmaische Straße 12, 04109 Leipzig, Germany

Abstract. Operational decisions are the backbone of every production focused company. Their optimization is the key to obtaining and maintaining an efficient enterprise. This task can be achieved with information systems, especially ones called Decision Support Systems. The current data-driven world means that the DSS subcategory – Intelligent Decision Support Systems are of increasing popularity. This trend can be observed in both the literature and real-life implementations. Unfortunately, the increase in popularity visible in the literature is also causing the terminological boundaries to become blurred. The goal of this article is to examine the essence and scale of this trend with regard to solutions dedicated to operational decisions. The research also examines other aspects of the solutions, and presents some insights into areas of application and the methods used to build such solutions.

Keywords: Artificial Intelligence · Production Environment · Operational Decisions · Decision-Making · Decision Support System · Decision Support

1 Introduction

Nowadays, the task of effective management of an production enterprise poses a difficult challenge. The digital revolution has brought some fundamental changes that have forced companies into a dynamic and radical evolution towards new technologies and operating models that are mostly focused on the internet [1]. These challenging times were further enhanced dramatically by the COVID-19 pandemic [6], and new challenges are continually arising, mainly due to the current geopolitical situation [17]. All of these factors, along with others not listed here, have added up to creating a harsh environment in which companies must be more efficient and optimized on a scale that has been seen never before, not only to thrive but simply to survive. For this reason, each and every decision that is made in an organization should be the best possible one. This is especially the case when it comes to decisions regarding the company's day-to-day operations where, as Lee [18] showed, there is a clear link between operational variables and the financial performance of a given industry.

M. Hernes and J. Wątróbski (Eds.): ECAI 2023, LNNS 1079, pp. 50–61, 2024.
https://doi.org/10.1007/978-3-031-66761-9_5

Every decision taken can be described by the general process: problem formulation, problem modelling and problem optimization [16]. Operational decisions made in production enterprises follow this pattern, and their structure can be further divided into 3 main blocks: input data, decision logic and conclusion or action. For a decision to be effective and optimal, the quality of the input data or information provided is crucial. Every piece of information that could potentially be useful in the decision-making process should be gathered, processed and taken into account [32]. The structure of input data can take various forms depending on the decision being made. Taking into account the technological transformation of IoT (the Internet of Things), Industry 4.0 and others, the amount of information that is beneficial to the decision-making process can be enormous. Reinsel, Gantz, and Rydning [24] estimated that the global amount of data in 2018 was 33 ZB, and that by 2025 this will increase to 175 ZB, with part of this growth affecting decision-making processes. This is especially true when it comes to production environments, where thousands of sensors generate a tremendous quantity of useful data [30]. This presents a challenge, not only when extracting information from such data, but also at every other main step of the process, that is the gathering, processing, and storage of the data [31]. The role of information systems in this process is crucial, and models such as data lakes are becoming standards even for small enterprises [26].

Unfortunately, information overload is not the only element that negatively affects the decision-making process. Several other large-scale factors need to be taken into account, and their source is mainly in the decision-maker themselves. In the case of operational decisions, they usually need to be performed instantly, which creates an environment of time pressure and stress. Mumaw [21] researched the impact of stress on operational decisions using the example of nuclear plant workers. The study showed that stress reduced working memory capacity, impaired communication skills and narrowed attentional focus. However, factors affecting decisions are not limited to the environment because the person themselves adds a large degree of instability to the process. In his book, Davis [8] describes biases that are people-based, and divides them into 3 main categories: judgements regarding risks, evaluation of outcomes and bounded rationality. The overall conclusion states that when dealing with uncertainty, people can become biased by miscalculating probabilities, misevaluating possible outcomes and neglecting other factors. All these effects are further reinforced by the uncertainty that comes from the environment [19]. For all of the above-mentioned reasons, decision makers should be assisted, and the best possible way to do this is with the use of information technology support. However, one should always be cautious because the introduction of technology into the decision-making process can potentially generate other kinds of bias [7]. Accordingly, to avoid such problems and to maximize profits from investments, the systems involved should be the best possible ones.

There are many ways in which technology can be used to improve decision making. In a data-driven world there are 4 main categories of systems used, as shown by Volkova et al. [28]. These include: OLAP technologies (Online Analytical Processing), knowledge extraction technologies (such as Data Science), imitation modelling technologies and SADT (Structured Analysis and Design Technique) technologies. Currently, data-driven decision making is gaining high popularity among companies, especially with the use of Data Science methods [23]. Brynjolfsson et al. [3] found a connection between data-driven decision making and better productivity and market value, along with a possible further connection to measures of profitability. In this particular field of data usage, artificial intelligence (AI) methods, especially neural networks, tend to gain better results compared to more traditional ones [9]. Since the birth of AI in 1956, there have been 3 periods of excitement separated by two periods of stagnation [29]. Currently, we are in the third and as yet largest period of popularity, initiated by deep learning and big data, therefore now is the best time to take advantage of AI.

In the literature there is a vast amount of research concerning decision support and the AI methods used to perform it. Over the years, a strong terminology has also developed. However, in recent times, there has been a lack of precision in its application, and the boundaries of some concepts are blurring. In the authors' opinion, this is especially the case when it comes to the two most basic concepts: Decision Support Systems (DSS) and Intelligent Decision Support Systems (IDSS). Therefore, the goal of this article is to assess the current state of the terminology applied, along with the AI based methods involved. This will increase knowledge in a decision support domain, especially when comes to the correctness of terminology. This review is performed on the basis of production environment, especially the operational decisions made within it.

The structure of this article is as follows. The next chapter presents the two most used methods of AI implementation in decision support. Then, the general literature review that was carried out is presented. After the selection of interesting works, specific results are presented for the selected and manually reviewed articles describing the use of AI in operational decision making. Finally, conclusions and remarks regarding future research end the paper.

2 Idea of Decision Support Solutions Using Artificial Intelligence

Many decisions that are made within production enterprises can be reduced to the optimization problem with multiple objectives and uncertainties [20]. The most basic and commonly used solution to solve this problem is the decision support system, which is a subtype of an IT system. As was mentioned before, the term DSS is often misused, therefore the decision was taken to introduce another term – the Isolated Decision Support Model (IDSM). The DSS is a systemic and advanced approach, while the IDSM differs from this by being more a concept than a full-scale system.

When a decision is influenced by an algorithm or algorithms that possess little to no infrastructure, then this is a conceptual approach (IDSM). Such solutions are usually built with a process at least similar to the CRIPS-DM (Cross Industry Process for Data Mining) model presented by Chapman et al. [4]. This approach divides the data mining process into 6 steps: business understanding, data understanding, data preparation, modelling, evaluation and finally deployment. In this process, the implemented algorithms are usually a stand-alone product. The input data as well as the output information are forged within the solution, and can only be changed to a certain degree. Usually, the security of such products can also be significantly improved. The conceptual approach is generally the result of initial research work where the algorithms have already been created, but there was no need, time or resources to design and build the whole system.

The natural evolution of the Isolated Decision Support Model solution is the Decision Support System. DSS's are a subcategory of information systems whose main role is to influence the decision-making process so that it is more optimized. Such systems are clearly divided into layers, where the generic structure can be described as the input layer, processing layer and presentation layer. The vast amount of uses in various areas have forced DSS's to evolve in almost every way. As shown by Holsapple [14], a DSS can differ in almost every aspect, in particular: form of information storage, algorithms used for decision support, presentation layer, and the general architecture that commonly is based on the ecosystem. All of these factors affect the final system architecture, service uses or technology employed [10].

When it comes to DSS, there are no general point by point instructions for the creation process. Due to the case dependent architecture and unique features, every company can take a different approach to this problem. However, as DSS is an IT system, the agile approach is recommended as it generally performs well in similar situations [13]. This particular approach provides the means to consecutively adapt the product, which is an important feature when speaking about DSS's.

The structure of both DSS and IDSM is presented in Fig. 1. Even in this simplified state, the DSS can be distinguished by its more complex, multiple layer architecture. As opposed to IDSM, where there is nothing more, the DSS can be further divided into 7 subcategories [2], depending on its methods, tasks and environment. In this article, additional focus has intentionally been placed on intelligent systems, and authors like Tariq and Rafi [27] define these as interactive DSS's that solve complex and ill-structured problems using expert knowledge and models that incorporate artificial intelligence techniques. Artificial intelligence itself can have many possible definitions. However, when comes to AI techniques, the definition is much simpler and usually revolves around a sub-area of computer science that combines a mixture of Data Science and Big Data.

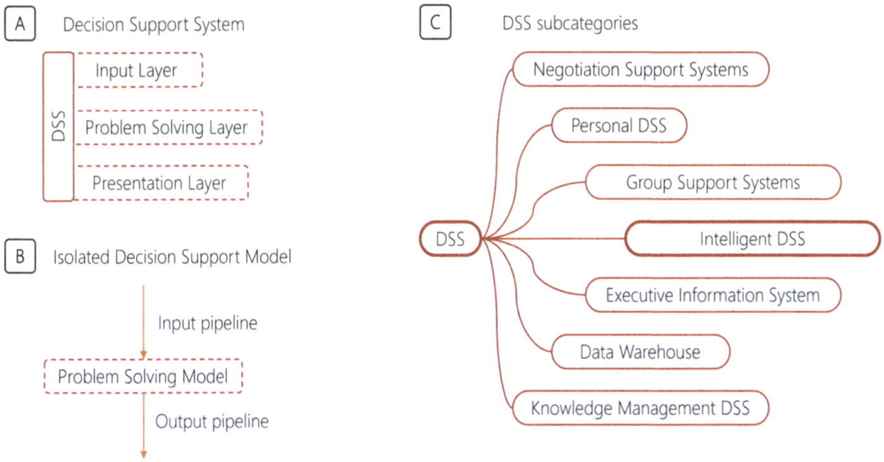

Fig. 1. Architecture of A – Decision Support System (DSS) and B – Isolated Decision Support Model (IDSM) along with C – subcategories of DSS.

Even with clarified boundaries, the AI field is vast and complex. Collins et al. [5] performed a systematic review of artificial intelligence used in information systems, and found that the top 5 subtypes used were: Machine Learning, Machine Vision, Natural Language Processing, Expert Systems and Robotics. Their study also confirmed the abundance of AI definitions (the authors found 28 definitions) and narrowed down the application to three business needs: process automation (50% – automating a business process), cognitive insight (33% – gaining insight through data analysis), cognitive engagement (17% – engaging the customer and employees).

This scale and uniqueness may cause additional difficulties in the selection of the proper method. When it comes to decision support, the task itself imposes no boundaries or general limitations on the choice of algorithm. However, each case of usage can and probably will make some particular algorithms more effective or easier to implement than others. Some important factors that affect the final choice are: the data itself (size, characteristics, behaviour and type), the required accuracy and ability for generalization, the speed of the solution and the required features that the model needs to handle. The issue of choosing the right algorithm has been researched for some time. Methods have been created such as one presented by Zhongguo et al. [33], where choosing the right classification algorithm is done based on data set characteristics. Nonetheless, to this day the issue remains unresolved, and there is no unified, general approach that can be used to solve this problem.

While the costs, trouble and effort of initial research and implementation of decision support solution may be high, the benefits are potentially larger. Holsapple and Sena [15] researched the topic of the possible results of such implementation in organizations. They gave the following list of the noted effects: better knowledge processing, better coping with large/complex problems, reduced decision time and costs, greater exploration/discovery, fresh perspective, substantiation, greater reliability, better communication and coordination, greater satisfaction, decision empowerment and competitive advantage. A similar analysis was later conducted in practice by Gholami and Salimi [12], and overall, their results were consistent. Unfortunately, the use of decision support comes with some downsides [22], the main one being possible over-reliance. The assumption should be made that for the system used, there constantly exists the possibility of failure (no matter how small). If not, a mistake may not be detected and filtered out by the expert. Schöll, Möller, and Longino [25] examined the situation when inaccurate decision support is introduced. In a clinical decision-making scenario, the authors used 3 groups of medical specialists (with correct decision support, with incorrect decision support and without decision support). The result was that the specialist with correct decision support completed the task faster and better, while the specialist with incorrect decision support had the same outcomes and required the same amount of time as participants with no decision support. This effect was to be expected as staff need time to come to trust the system to such a level that possible error can occur and not be filtered by user. In addition, as decision support is usually implemented as a system, the security of the said solution should be inspected and maintained [11].

3 Artificial Intelligence-Based Decision Support in the Literature

For the assumed goals to be achieved, a literature review was performed based on the Scopus database. The query process started with selecting all research articles (or conference papers) that had phrase "Decision Support System" in the title. This phrase and condition were chosen because, in the authors' experience and opinion, they should place emphasize on solutions that involve real-life implemented systems. This process resulted in 14 981 literature sources, which was later narrowed down by specifying the use of the next two phrases: "artificial intelligence" and "machine learning" in the title, abstract or keywords. Although machine learning is a part of artificial intelligence, both keywords are sometimes used interchangeably. At this point, the total amount of papers meeting the criteria was 6 180. To further narrow it down, articles (or conference papers) released after 2017 and written in English were selected, giving a total of 1 662 documents. Finally, the type of decision was filtered for by selecting only papers that included the word "Operational" in one of all the possible fields. In addition, all reviews were also removed. The final number of documents accepted for this review was 355, and was achieved using the following Scopus query:

TITLE ("Decision Support System") AND TITLE-ABS-KEY ("Artificial Intelligence") OR TITLE-ABS-KEY ("Machine Learning") AND PUBYEAR AFT 2017 AND DOCTYPE (ar) OR DOCTYPE (cp) AND LANGUAGE (English) AND ALL ("Operational") AND NOT TITLE ("Review").

Given the above query, the authors made the assumption that all of the results should describe Intelligent Decision Support Systems. The selected publications were first compared by publishing date and number of citations. This analysis was performed in the form of histograms, as presented in Fig. 2. During the selected period of time (from 2018 to 2023), an increase in the number of relevant research works can be observed until 2020, when the yearly number of publications drops by around 30. After this point, the number of publications continues to increase, but at a much slower rate. As for the citation count, the majority of the articles (around 114) have been cited less than 5 times. The overall citation count of the papers has decreasing exponential distribution.

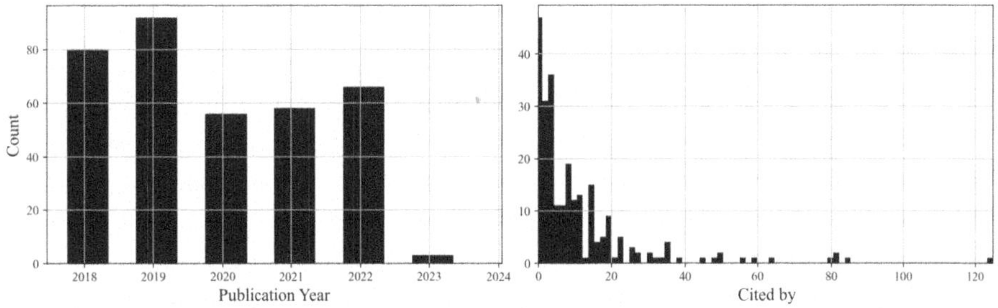

Fig. 2. Year and citation number distributions of the selected works.

By analyzing the subject area of all the selected works, some main fields can be established for the use of decision support. This analysis is shown in Fig. 3, with the possible fields of implementation including: Business, Management, Accounting, Environmental Sciences, Energy, Social Sciences, Physics and Astronomy, Psychology, Materials Science, Art and Humanities, Earth and Planetary Sciences, and Agricultural and Biological Sciences. The fields of Computer Science, Engineering, Business Sciences and Mathematics should be excluded as they cover the construction and use of technology-based decision support solutions.

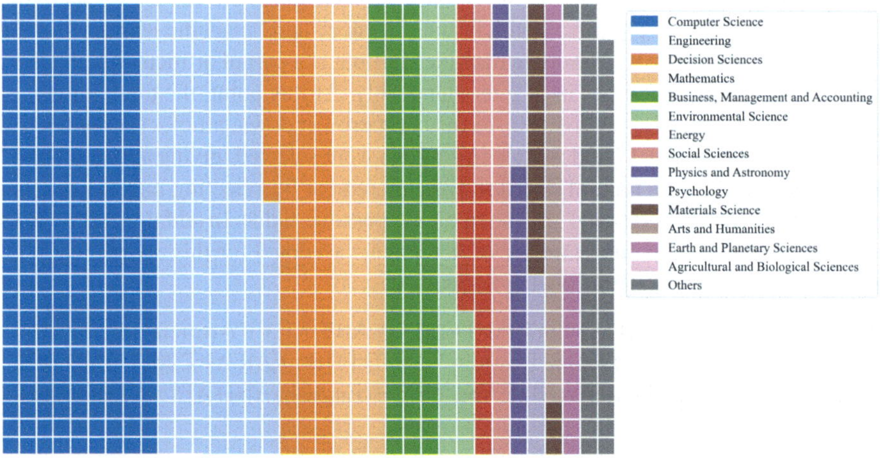

Fig. 3. Division of selected research papers based on subject area.

Finally, the general analysis concludes with an examination of keywords. To perform this action, the word cloud method was applied, but before that keyword cleaning needed to be applied. All of the keywords were reduced to lower cases, and had any spaces replaced with '_' to merge multiple words into one. The keywords: artificial intelligence, decision support system and machine learning were removed as they were part of the Scopus query.

The word cloud generated from the author-specified keywords is presented in Fig. 4. The majority of the space is taken up by the words: *optimization, simulation, decision making* and *multi criteria decision making*. This represents the principal problems that AI algorithms are being applied to. However, several specific problems are also visible, such as: *supply chain management, renewable energy, industry 4.0, supplier selection, sensitivity analysis, risk management, scheduling* and *healthcare*. As for the algorithms commonly used, a few can be found here, including: *decision trees, genetic algorithms, fuzzy logic, neural networks* and *logistic regression*.

Fig. 4. Word cloud made from keywords specified by the authors.

The above analysis shows the general direction that AI-based decision support systems take. Unfortunately, they are also burdened with a significant degree of error because of diversity of the keyword usage. For this reason, much of the information is still hidden and cannot be acquired automatically. Therefore, the authors performed a thorough review manually.

4 Advanced Review of DSS Solutions

The initial sample consisted of 355 publications. In order to perform a thorough analysis, the authors analyzed 125 open access publications (35% of the whole sample), on which this review is based. Each research work has been validated by the authors of this article and 3 types of information have been extracted:

- Type of Solution – whether it is a Decision Support System or an Isolated Decision Support Model. In addition, some of the selected literature samples described different topics, and for those a separate "Other" category was created.
- Business Area – the area to which the solution is dedicated. In most cases, the main area of research was concluded as being the area of the company using the DSS. However, if the solution was intended as multi-domain, it was labelled with the "Universal" category.
- Algorithms – the algorithms used to build the solution. The article is focused on AI-based methods, thus only these were of interest. For others not corresponding to the topic, a more generic label was created "Mathematical Models or/with Logic Rules", as they were the most commonly exploited techniques.

For the solutions to be classified as systems, the paper needed to at least mention the existence of larger infrastructure or give some remarks about a layered structure. Both of the requirements could also be mentioned indirectly, for example by a screenshot of the GUI (Graphical User Interface), and the paper would still be classified as a system. With such criteria, half of the total research works correctly described the researched effects as a Decision Support System. In the case of 36%, the solution described by the authors or the method of description did not show enough indications, therefore the Isolated Model (IDSM) label was assigned. The rest of the literature sample (11%) did not refer to the correct subject. These articles consisted mostly of as yet unfiltered reviews, or more theoretical works with remarks on how to build Decision Support Systems. This demonstrates that based on the analyzed sample, the term Decision Support System appears to be quite often misused.

When it comes to Intelligent Decision Support Systems, the main criterion was to incorporate at least one standard artificial intelligence technique. The results of this analysis are presented in Fig. 5A. Once more, a substantial number of papers described solutions that incorporated standard mathematical models with some logic rules (in general). In the authors' opinion, these methods should not be classified as originating in the field of artificial intelligence. Consequently, the solutions that used them should not be called Intelligent Decision Support Systems. In terms of techniques from the field of AI, methods based on Fuzzy Logic are most often used. Commonly in use are also: Decision Trees, the Analytical Hierarchy Process, Genetic Algorithms, Random Forests and Neural Networks. This was to be expected as every single one of them appeared earlier in the keyword-based word clouds. It is also worth mentioning that several literature sources described an AI-based system as a concept without details, and as such the concept category was assigned. Another large proportion (26%) is covered by the "Other" category. The main sources for this category are a large quantity of highly varied solutions. All of them can still be classified as from the domain of AI.

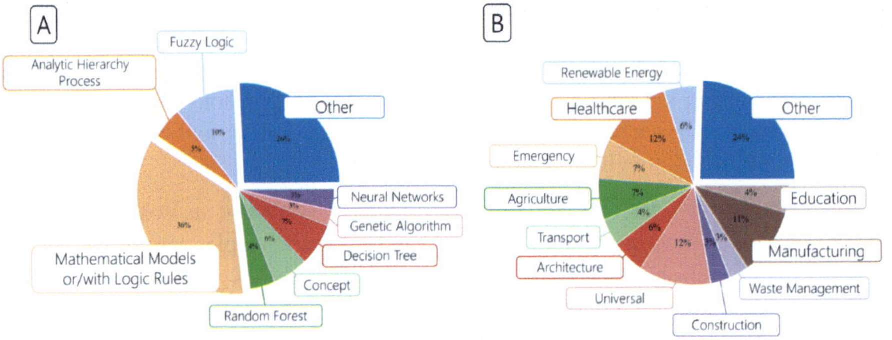

Fig. 5. A – Main techniques incorporated into the described solutions (DSS or IDSM), B – Areas of application found in the research papers.

The last analysis regarded the possible areas of application, the results of which are shown in Fig. 5B. It can be seen that the top 3 areas of possible DSS implementation are healthcare, manufacturing and agriculture, along with emergency. Healthcare is especially interesting as it comprises a large proportion of the sample, but did not show up earlier in the subject area analysis. As for the others, only construction, waste management and education were not referred to in the word clouds by closely correlated keywords. Some solutions have also been found that are not strictly specialized in one particular area but are rather generic (e.g., employee time scheduling). For such solutions, the universal label was assigned.

The results obtained from an advanced review performed on a selected portion of the whole sample correlate strongly with the initial insights given by general analysis. Therefore, it can be assumed that they can be extrapolated to the whole field, providing it is understood that the terms Decision Support System and Intelligent Decision Support System are frequently misused.

5 Conclusions

Operational decisions are an integral part of the operating pattern of any company. Such operations should be as optimized and effective as possible for the company to survive and expand in current world market conditions. To achieve this, technology-based decision support solutions are more and more often introduced into decision making processes. This is facilitated by the fact that every day, the world is becoming more data-driven. In such an environment, AI solutions perform exceptionally well. Therefore, the best way to implement decision support is to use intelligent decision support systems that make intensive use of AI technologies.

Unfortunately, some studies in the literature misidentify the solutions they describe as DSS. Such works focus mainly on the conceptual side of their results (algorithmic solution of the described problem) and do not have the characteristics of an information system (data, algorithms and user interface). In this paper, the authors propose to distinguish these from DSS and call such solutions isolated decision support models (IDSM).

Based on cases described in the literature, the authors attempted to indicate how and where the terms Decision Support System and Intelligent Decision Support System are used in operational decision-making. In addition, some insights are provided into possible application areas and the methods incorporated into the solutions in question. To achieve this, a Scopus query was prepared so as to select a literature sample which was later analyzed using the terms: publication year, citation count, area of research, authors keywords and Scopus keywords. The authors decided to further deepen the analysis by preparing a manual review of part of the initial literature sample. Identified features included type of solution, area of use and algorithms powering the solution. The results of this detailed review showed that about 36% of the papers described an isolated decision support model instead of the claimed decision support system. It was also found that 36% of solutions declared as intelligent did not in fact use any of AI technologies. It should also be stated that the overwhelming majority of solutions used rather simple and general AI methods. Those most commonly used include fuzzy logic, decision tree, the analytical hierarchy process, random forest, genetic algorithms and neural networks. In addition, the three most common uses were found to be in healthcare, manufacturing and agriculture, along with emergency. A large proportion of the sample also described universal solutions that do not fit into one particular field.

Due to the prevalence of the researched phenomenon in operational decisions, there is a distinct likelihood that it may be found in other decision-making fields. Therefore, exploring other areas of application would seem to be the most appropriate direction for future research. In addition, there is also the possibility that other fundamental concepts such as "data warehouse" may also be affected, and this should therefore be verified.

References

1. Angeloska-Dichovska, M., Petkovska-Mirchevska, T.: Challenges of the company in the new economy and development of e-business strategy. Strat. Manag. **22**(2), 27–35 (2017)
2. Arnott, D., Pervan, G., O'Donnell, P., Dodson, G.: An analysis of decision support systems research: Preliminary results. In: Proceedings of the 2004 IFIP International Conference on Decision Support Systems, pp. 25–37 (2004)
3. Brynjolfsson, E., Hitt, L.M., Kim, H.H.: Strength in numbers: how does data-driven decision making affect firm performance? Available at SSRN 1819486 (2011)
4. Chapman, P., et al.: CRISP-DM 1.0: step-by-step data mining guide. SPSS **9**(13), 1–73 (2000)
5. Collins, C., Dennehy, D., Conboy, K., Mikalef, P.: Artificial intelligence in information systems research: a systematic literature review and research agenda. Int. J. Inf. Manage. **60**, 102383 (2021)
6. Craven, M., Liu, L., Mysore, M., Wilson, M.: COVID-19: Implications for Business. McKinsey & Company, 8 (2020)
7. Darioshi, R., Lahav, E.: The impact of technology on the human decision-making process. Hum. Emerg. Technol. **3**(3), 391–400 (2021)
8. Davis, A.M.: Biases in individual decision-making. In: Donohue, K., Katok, E., Leider, S. (eds.) The Handbook of Behavioral Operations, pp. 149–198. Wiley (2018). https://doi.org/10.1002/9781119138341.ch5
9. Delen, D., Sharda, R.: Artificial neural networks in decision support systems. In: Burstein, F., Holsapple, C.W. (eds.) Handbook on Decision Support Systems 1, pp. 557–580. Springer Berlin Heidelberg, Berlin, Heidelberg (2008). https://doi.org/10.1007/978-3-540-48713-5_26
10. Demirkan, H., Delen, D.: Leveraging the capabilities of service-oriented decision support systems: putting analytics and big data in cloud. Decis. Support Syst. **55**(1), 412–421 (2013)
11. Finne, T.: What are the information security risks in decision support systems and data warehousing? Comput. Secur. **16**(3), 197–204 (1997)
12. Gholami, A., Salimi, Y.: Investigate implementation decision support system (dss)'s impact on predicting and reducing the costs of organizations. Acad. J. Res. Econ. Manag. **2**(3), 50–56 (2014). https://doi.org/10.12816/0006543
13. Gryncewicz, W., Sitarska-Buba, M., Zygała, R.: Agile approach to develop data lake based systems. In: Hernes, M., Rot, A., Jelonek, D. (eds.) Towards Industry 4.0 — Current Challenges in Information Systems, pp. 201–216. Springer International Publishing, Cham (2020). https://doi.org/10.1007/978-3-030-40417-8_12
14. Holsapple, C.W.: DSS architecture and types. In: Burstein, F., Holsapple, C.W. (eds.) Handbook on Decision Support Systems 1, pp. 163–189. Springer Berlin Heidelberg, Berlin, Heidelberg (2008). https://doi.org/10.1007/978-3-540-48713-5_9
15. Holsapple, C.W., Sena, M.P.: ERP plans and decision-support benefits. Decis. Support Syst. **38**(4), 575–590 (2005)
16. Koopialipoor, M., Noorbakhsh, A.: Applications of artificial intelligence techniques in optimizing drilling. In: Azizi, A. (ed.) Emerging Trends in Mechatronics. IntechOpen (2020). https://doi.org/10.5772/intechopen.85398
17. Korhonen, I., Simola, H., Solanko, L.: Sanctions, counter-sanctions and Russia: effects on economy, trade and finance (No. 4/2018). BOFIT Policy Brief (2018)
18. Lee, J.: Effects of operational performance on financial performance. Manag. Sci. Lett. **9**(1), 25–32 (2019)
19. Lin, Z.: Environmental determination or organizational design: an exploration of organizational decision making under environmental uncertainty. Simul. Model. Pract. Theory **14**(4), 438–453 (2006)

20. Matasniemi, T. (2008). Operational decision making in the process industry. Multidisciplinary Approach, 143
21. Mumaw, R.J.: The effects of stress on nuclear power plant operational decision making and training approaches to reduce stress effects (No. NUREG/CR–6127). Nuclear Regulatory Commission (1994)
22. Power, D.J.: Decision Support Basics. Business Expert Press (2009)
23. Provost, F., Fawcett, T.: Data science and its relationship to big data and data-driven decision making. Big Data **1**(1), 51–59 (2013)
24. Reinsel, D., Gantz, J., Rydning, J.: The digitization of the world from edge to core. IDC white paper, Nov. (2018)
25. Schöll, C., Möller, S., Longino, A.: Assessing the impact of inaccurate decision support systems on experts' behavior and decisions. In: 2017 Ninth International Conference on Quality of Multimedia Experience (QoMEX), pp. 1–3. IEEE (2017)
26. Sitarska-Buba, M., Zygała, R.: Data lake: strategic challenges for small and medium sized enterprises. In: Towards Industry 4.0—Current Challenges in Information Systems, pp. 183–200. Springer, Cham (2020)
27. Tariq, A., Rafi, K.: Intelligent decision support systems-A framework. Inf. Knowl. Manag. **2**(6), 12–20 (2012)
28. Volkova, V.N., Vasiliev, A.Y., Efremov, A.A., Loginova, A.V.: Information technologies to support decision-making in the engineering and control. In: 2017 XX IEEE International Conference on Soft Computing and Measurements (SCM), pp. 727–730). IEEE (2017)
29. Wang, L., Liu, Z., Liu, A., Tao, F.: Artificial intelligence in product lifecycle management. Int. J. Adv. Manuf. Technol. **114**(3), 771–796 (2021)
30. Wu, X., Zhu, X., Wu, G.Q., Ding, W.: Data mining with big data. IEEE Trans. Knowl. Data Eng. **26**(1), 97–107 (2013)
31. Yan, J., Meng, Y., Lu, L., Li, L.: Industrial big data in an industry 4.0 environment: challenges, schemes, and applications for predictive maintenance. IEEE Access **5**, 23484–23491 (2017)
32. Zager, K., Zager, L.: The role of financial information in decision making process. Innov. Mark. **2**(3), 35–40 (2006)
33. Zhongguo, Y., Hongqi, L., Ali, S., Yile, A.: Choosing classification algorithms and its optimum parameters based on data set characteristics. J. Comput. **28**(5), 26–38 (2017)

Identification of Barriers to AI Implementation in Medical Information Systems

Agnieszka Pilch[1], Ryszard Zygała[1], Wiesława Gryncewicz[1](✉), Mykola Dyvak[2], and Andriy Melnyk[2]

[1] Wroclaw University of Economics and Business, Wroclaw, Poland
wieslawa.gryncewicz@ue.wroc.pl
[2] West Ukrainian National University, Ternopil, Ukraine

Abstract. Based on a literature review between 2014 and 2022, barriers to implementing artificial intelligence in medical information systems were identified. It was noted that these barriers are independent of the sub-discipline of medicine (e.g., oncology, ophthalmology, dermatology) and are recurrent in nature. The most important ones were described and grouped. The following four subgroups of barriers were identified: organizational and legal, technical, economic, and human factor-related barriers. Awareness and consideration of these barriers to implementing AI in health systems can help eliminate (or minimize) the risk of project failure. The study can be useful both for decision-makers who want to implement an AI-based medical information system (e.g., in a hospital), and for researchers and managers responsible for organizing medical facilities.

Keywords: Artificial Intelligence · Medical Information Systems · Implementation · Barriers

1 Introduction

Artificial intelligence (AI) is being used in various sectors and is impacting many areas of life, including healthcare, finance, marketing, logistics, retail, education, customer service, defense, and even sports. In particular, it allows for streamlining business processes, optimizing decisions, reducing costs, saving time, and eliminating routine tasks. AI is also leading to lifestyle changes through smart buildings, self-shop stores, integrated living enhancements, and IoT devices [1]. Despite the aforementioned benefits of artificial intelligence in every field, there are also challenges and barriers that need to be addressed. Identifying and then defining barriers to implementation will help eliminate (or minimize) the risk of failure for future AI-based projects.

Due to the high intensity of AI use in the medical field, the authors considered this area attractive from a research perspective. The application of AI in this field brings many benefits, especially in predictive analytics and image recognition [2]. It has the potential to improve the way medical data is processed, diagnoses are made, treatments are developed, and diseases are prevented [3]. Through analysis of the use of AI in medical systems, the authors have sought to fill a research gap related to the feasibility

M. Hernes and J. Wątróbski (Eds.): ECAI 2023, LNNS 1079, pp. 62–73, 2024.
https://doi.org/10.1007/978-3-031-66761-9_6

of implementing AI-based decision support systems in medicine. This will contribute to the existing literature on identifying and classifying barriers.

The authors hope that the article will be helpful not only for decision-makers who want to implement AI-based systems but also for researchers who can use this knowledge in their future work. The key contributions of this study are: identification and categorization of barriers to AI implementation in medical information systems based on literature analysis and classification of barriers.

The paper is organized as follows. Section 2 outlines related works that inspired the research. Section 3 presents the scientific research method, describes the databases, and gives the two stages of barrier identification, i.e., (1) Search terms, (2) Descriptions and classification of terms. Section 4 provides a summary of the study and recommendations of the authors as they relate to future research.

2 Related Work

In medical practice, there are systems such as Clinical Decision Support Systems (CDSS) and Medical Information Systems (MIS) that are used to improve decision-making and analysis. An important part of developing these types of systems, especially those based on artificial intelligence, is exploring the barriers associated with implementation. Radhakrishnan et al. [4] created an initial overview of the determinants and barriers associated with AI adoption in areas such as autonomous vehicles, big data analytics, cognitive engagement, robots, and medicine at both the organization and individual level. The authors [4] identified the following barriers: regulatory, transparency, and reproducibility, differences between clinical and experimental data, and potential risk to life. Krian Grant et al. [5] discuss the barriers associated with AI in emergency medicine. Their researchers identified three types of barriers from this perspective: technical challenges (the „black box" phenomenon, data input, ambiguity, and scientific evidence), regulatory and legislative barriers, and integration into routine clinical work. In addition, there are a lot of publications that address the issue of identifying barriers to the implementation of AI, but only for specific sub-disciplines of medicine, such as radiology [6], oncology [7], and gastroenterology [8]. Another aspect to be taken into consideration is that the use of AI in medicine involves analysing the following forms of information concerning the patient: textual (e.g., unstructured text of medical records, patient feedback and comments [9]), image-based (e.g., radiological imaging [10]) and data stream – various types of signals (e.g., ECG analysis [11]).

The authors of this paper have identified a research gap related to the potential of comprehensive research aimed at identifying barriers to the implementation of AI in medicine. The literature reviewed lacks a holistic classification of barriers across the medical sector, without narrowing down only to the specifics of individual sub-disciplines.

3 Research Method

The scope of the study focused on a search strategy to identify and analyse implementation barriers associated with AI-based systems and algorithms in the medical sector. The authors primarily collected articles from the Scopus database. In addition, searches were conducted on other search engines using the snowball method to look for relevant articles and identify barriers.

The search strategy: a Scopus database query was focused on the medical domain. The use of the Scopus database allowed access to the bibliographic metadata of these articles, which made it possible to exclude articles that were inconsistent with the research area. Our main exclusion criteria were papers related to drug discovery and development. Articles were required to include: (1) topics related to medical issues, (2) topics related to an algorithm or decision support system based on AI, and (3) identification of barriers. Articles from 2014 to 2022 were selected. The titles of the articles were broadly related to medicine and AI in various sub-disciplines of medicine. Our study consists of three parts: search terms, description of terms, and classification of terms.

First, the search terms – barriers encountered by the researchers were extracted from the articles and named. Based on them, a table was prepared in which each term related to barriers was placed along with the relevant sources. In the next step, descriptions of terms were created. Using an analysis of terms found in the literature, a definition of each barrier and each threat was presented. Taking into account the concepts and their definitions, the terms were classified into four categories that provide a high-level view of the barriers to implementing AI-based information systems in the medical field.

4 Results

4.1 Search and Analysis

The identification of the discussed barriers was carried out on the basis of selected research articles. These papers were related to various sub-disciplines of medicine. On this basis, the following list of various barriers encountered when implementing AI-based DSS was created. The barriers discussed in Table 1 have been grouped into four groups: (1) organizational and legal, (2) technical, (3) economic, and (4) human factors. Table 1 presents all the barriers found in the search.

4.2 Descriptions of Terms

This part of the article focuses on the characteristics of the barriers listed in Table 1.

Organizational and Legal

1. *Regulatory.* Country-specific – refers to various regulations on data ownership, privacy, consent and data use standards, including HIPAA, GDPR [6] and FDA [8]. Nair et al., [6] noted that if the software is used to generate profits, then attention should be paid to conflicts of interest and intellectual property rights. Papadimitroulas et al. [7] found it difficult to gather data from multiple sources. For this barrier, they also highlighted that the regulatory approval process may have a lengthy waiting period [12].

Table 1. Identified barriers based on the research. In the first column: O&L – organizational and legal, Tech – technical, Eco – economic and HF – human factors

	Barriers	References
O&L	regulatory	[5–8, 12–26]
	responsibility	[5, 19, 21, 27–29]
	sensitive data	[5, 6, 15–17, 19, 20, 28, 30]
	data privacy	[5, 6, 17, 19, 22, 23, 28, 30–33]
Tech	data quality and labelling	[7, 8, 14, 16, 20, 31, 32, 34–39]
	erroneous or falsified annotation	[6, 18, 19, 28]
	small datasets	[3, 7, 8, 14, 16, 19, 20, 24, 31, 32, 34, 36, 37]
	lack of diverse data set	[7, 8, 16, 31, 36–38, 40, 41]
	data drift	[37]
	lack of standardization	[5, 7, 14, 21, 22, 37, 39, 42]
	lack of structured data	[35, 42]
	customization of the user interface	[20]
	data storage	[15, 43]
Eco	education	[15, 19, 21, 22, 29, 31, 33, 44]
	implementation costs	[17, 18, 21–23, 26, 29, 33, 43, 45]
	integration with existing systems	[5, 25, 32, 46]
	risk of system failure	[22, 29, 47]
	data labelling is time-consuming	[23, 34]
HF	fear of technology	[30]
	lack of interpretability and explainability	[5–7, 13, 15, 16, 18, 19, 21, 25, 27, 28, 36, 37, 39, 48, 49]
	lack of confidence in the algorithm	[8, 13, 15, 16, 21, 26, 29, 50]
	human resources	[19, 20, 22]

(*continued*)

Table 1. (*continued*)

	Barriers	References
	lack of people with strategic business culture	[2]
	technology	[12, 14, 19, 20, 22, 26, 32, 51]
	overreliance on the system	[15]
	ethics	[6, 13, 14, 27, 28, 30]
	lack of reproducibility of results	[27]
	lack of confidence in modern solutions	[2, 40]

2. *Responsibility.* The researchers pointed out a very high priority aspect related to the responsibility of the AI algorithm for decision-making and asked, "Who is responsible for errors?" [5]. Therefore, there are justifications for arguing that AI should be a job-support tool rather than a diagnostic tool [21].
3. *Access to Private and Sensitive Data.* AI systems collect large amounts of sensitive data that are processed in a specific way, hence every element of the data lifecycle should be included in discussions on data privacy [5]. According to O. F. Ahmad et al. [15], patients may want to know about the processing of their data, and have access to their data as well as information regarding the extent to which data is anonymized. This is due to the fear of data disclosure, such as medical diagnoses.

Technical

1. *Data Problem (Data Quality and Labelling).* Several problems have been identified that can be described as "data problems", including a lack of data annotation, erroneous or falsified data annotation, data cleansing, and erroneous data. In particular, research [15], above all, that data used to train AI algorithms may contain inherent errors or mispresent the population as a whole. Human error can occur during data collection and entry, which will later cause errors and inconsistencies [35]. According to Wang et al. [34], data cleaning consumes 80% of the time spent on AI development, and the labels that are used in AI systems are frequently very simplistic – usually involving binary classifications - "disease" and "no disease", with intermediate states such as disease progression or regression being ignored.
2. *Data Problem (Small Datasets).* Many research currently employ tiny datasets, which restricts the generalizability of AI models [14]. Each doctor or medical institution maintains its own patient information, which presents a significant challenge to an AI system because one individual database may not correspond with another patient

group, either demographically or geographically [8]. Reproducing medical data is also difficult [34] as it is diverse and not all transmitted digitally and it should be appropriately structured and labelled. Reddy et al. [14] have highlighted that models trained on selected datasets fail when applied to real-world instances, resulting in poor generalization performance. Some of authors argue that gathering data from several sources is difficult for regulatory, ethical and technical reasons. An further barrier, according to Brennan et al. [19], is long-term data storage.

3. *Data Problem (Lack of a Diverse Dataset).* The best solution for an AI algorithm is to use data from different sources during training so that the algorithm can generalize to diverse scenarios and work with a wide range of target populations [7]. The data should be representative of the populations for whom the AI system is designed [17]. According to Wang et al. [34], treatment data only includes what has been provided, but does not include information on options under consideration, such as other possible treatment. For example, the documentation includes information only on treatment, without information about lifestyle.
4. *Data Problem (Data Drift).* This occurs when the input data has changed, and the decision boundary remains unchanged. According to Wang et al. [34], molecular markers evolve, so an AI system trained on outdated data may make erroneous predictions.
5. *Lack of Standardization.* Not having a specific standard can cause instability in the input data, i.e., a change in the character of the input data is largely dependent on physical factors, e.g., device manufacturer, version, model, settings – [7], so it is important to ensure that the data is comparable.
6. *Software.* The biggest impediment is the lack of a modular approach to software development in medical institutions, making integration with contemporary clinical practice and other systems difficult [5].
7. *Risk of System Failure.* A system failure can disrupt the clinical workflow (of the facility and the doctors). In such a case, it is necessary to plan the system flow in advance so that it is reliable [47].

Economic

1. *Implementation Costs and Data Storage Costs.* An important barrier to the implementation of AI-based solutions is the issue of ensuring that benefits outweigh the costs of their implementation, which in the case of costs applies to all activities related to this implementation. Yoshida et al. [20] found a disparity between the overall expense of implementing the system and the advantage of identifying only breast cancer metastasis to lymph nodes in actual pathology practice. When deciding whether to implement an AI system, it is important to consider the structure of the organization and communication between departments, as the benefits and costs of using the system may be unevenly divided between departments, and this complicates funding decisions [21]. Data storage costs require specific consideration. The cost of data storage is mostly related to regulatory limits, but it is also vital to consider how much data will be gathered[15] because data storage is expensive, and it is critical to focus on processing efficiency as well as security. It should be noted that data quality assurance is costly

since, preparing data for modeling (e.g. data labeling) is time-consuming - it requires specialist knowledge of personnel, finances and time resources [18].

Human Factors

1. *Lack of Experience.* Physicians' lack of expertise in innovative AI-based technologies may be a barrier to the implementation of AI systems [20]. In the first instance, the focus should be on a GUI, where there is an opportunity for physicians to register their uncertainty or disagreement with the AI results in the medical record [15]. This raises another barrier – the fear of technology, which will be described in the next section.
2. *Fear of Technology.* According to Oprescu et al. [30], some employees fear technology due to a lack of autonomy and fear of losing their authoritative position (especially when the doctor disagrees with the algorithm) [15].
3. *Lack of Interpretability and Explainability.* AI algorithms are treated as "black boxes", and so the decision-making process in the context of AI is not understandable. Therefore, the issue of interpretability and explainability is important, *especially* when the model should work as expected [7]. Explainability is essential when making clinical decisions. According to Henry et al. [36], interpretability should be inherent in creating a model for clinical use. The lack of these two aspects creates a barrier of sorts that makes patients, doctors and other stakeholders sceptical of AI solutions.
4. *Lack of Trust in the Algorithm.* The use of sensitive data creates various problems that can have a significant impact on artificial intelligence algorithms – not everyone wants to *share* their data, which generates additional problems such as data scarcity, lack of data diversity, etc. The ability to explain and interpret also has a significant impact on the problem of trust in the algorithm - the more the tool can show and explain the diagnosis, the more people believe the algorithm.
5. *Lack of Reproducibility of Experiments.* Reproducibility of experiments is a very important element not only in terms of systems but also with regard to further research on *systems* using AI. Reproducibility of results requires the appropriate environment, source code, and raw data to be used in experiments, moreover, for open-source projects, sharing data and code is a recommended approach [27].
6. *Education and Training.* According to Mugabe [44], the safe adoption of AI depends on combining the discipline (e.g., clinical oncology) with data science, as the workforce is currently untrained in data science, and has a limited awareness of the system. With the emergence of new technologies, a need has arisen to update education programmes to include knowledge of AI methodologies and limitations [31]. In *order* to teach new technologies, experienced teaching staff are necessary, which also represents a certain barrier. According to Brennan et al. [19], there is a need to develop an educational programme, special device training, courses on AI, and various support systems in order to be able to effectively implement AI in medical practice. When training employees, it is important that they are prepared to critically evaluate diagnoses made by the AI [15].

7. *Lack of Belief in Modern Solutions.* An important aspect is that it is the physician who must agree to bring an innovative approach to the conventional workflow. In most cases in healthcare, technical skills and tasks are outsourced [2].
8. *Ethical Problems.* These include accountability, transparency, permission, and privacy [28]. Above all, this concerns the way information is communicated and the attitude among users that the system makes mistakes.

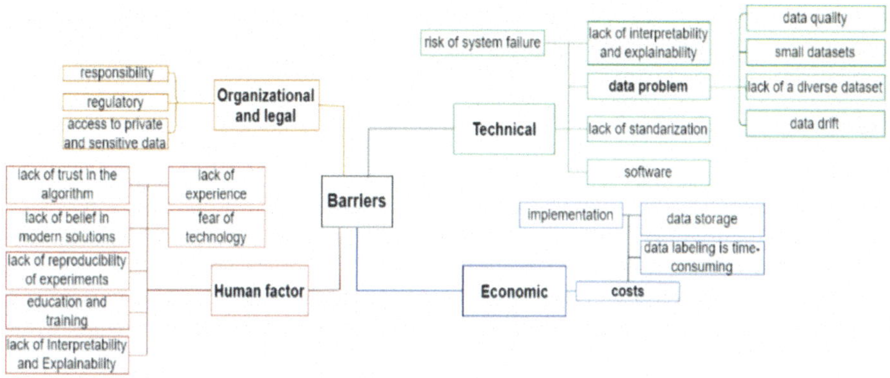

Fig. 1. Classification of barriers to AI system implementation in medicine

Figure 1 presents a synthetic picture of the discussed barriers. The above examples of barriers can serve as an introduction to risk analysis for implementing a decision support system project using AI. The analysis can be used at any stage of a project, depending on the needs of the organization.

Based on the listed barriers, the strengths and weaknesses of a project or its idea can be assessed.

5 Conclusion

In this article, the authors have focused on identifying, defining, and classifying barriers to implementing AI systems in medicine. They noted that the problems identified by the researchers appear regardless of the sub-discipline of medicine (e.g., radiology and oncology). Currently, there are a relatively limited number of articles that deal with AI decision support systems in medicine, which affects the possible scope of research. At the same time, the authors express the belief that research is needed to make decision-makers aware of the problems they may encounter during planned and ongoing AI implementation projects in medicine.

As a purpose of the future work, the authors intend to conduct surveys at several medical institutions. They will include questions related to the barriers identified in this paper, as well as open-ended questions about new barriers noted by those working in the medical environment. This approach will enable an assessment of the relevance of the identified barriers.

The identification of the barriers (organizational and legal, technical, economic, and human factors) will allow stakeholders to assess the risk of project failure at each stage of projects involving AI in medicine. Moreover, it should eliminate such problems as: over-fitting [36], bias in the model [15], unreliability of the system (erroneous decisions of the system, especially in the case of false-negative responses, as then it does not detect the disease in a patient who is actually ill [52]), re-identification [6], generation of inaccurate results [6], and technical problems with the removal of data (once they are part of the system [15]). Given the large amount of literature related to "regulatory constraints", it remains an open question: how to find a compromise between the use of patient data in AI systems and patient privacy?

References

1. Wamba-Taguimdje, S.-L., Fosso Wamba, S., Kala Kamdjoug, J.R., Tchatchouang Wanko, C.E.: Influence of artificial intelligence (AI) on firm performance: the business value of AI-based transformation projects. Bus. Process. Manag. J. **26**, 1893–1924 (2020). https://doi.org/10.1108/BPMJ-10-2019-0411
2. Dicuonzo, G., Donofrio, F., Fusco, A., Shini, M.: Healthcare system: moving forward with artificial intelligence. Technovation **120**, 102510 (2023). https://doi.org/10.1016/j.technovation.2022.102510
3. Timiliotis, J., et al.: A novel diagnostic decision support system for medical professionals: prospective feasibility study. JMIR Format. Res. **6**, e29943 (2022). https://doi.org/10.2196/29943
4. Radhakrishnan, J., Chattopadhyay, M.: Determinants and barriers of artificial intelligence adoption – a literature review. In: Sharma, S.K., Dwivedi, Y.K., Metri, B., Rana, N.P. (eds.) Re-imagining Diffusion and Adoption of Information Technology and Systems: A Continuing Conversation, pp. 89–99. Springer International Publishing, Cham (2020)
5. Grant, K., McParland, A., Mehta, S., Ackery, A.D.: Artificial intelligence in emergency medicine: surmountable barriers with revolutionary potential. Ann. Emerg. Med. **75**, 721–726 (2020). https://doi.org/10.1016/j.annemergmed.2019.12.024
6. Nair, A.V., Ramanathan, S., Sathiadoss, P., Jajodia, A., Blair Macdonald, D.: Barriers to artificial intelligence implementation in radiology practice: what the radiologist needs to know. Radiología (Eng. Edn.) **64**, 324–332 (2022). https://doi.org/10.1016/j.rxeng.2022.04.001
7. Papadimitroulas, P., et al.: Artificial intelligence: deep learning in oncological radiomics and challenges of interpretability and data harmonization. Physica Med. **83**, 108–121 (2021). https://doi.org/10.1016/j.ejmp.2021.03.009
8. Sutton, R.A., Sharma, P.: Overcoming barriers to implementation of artificial intelligence in gastroenterology. Best Pract. Res. Clin. Gastroenterol. **52–53**, 101732 (2021). https://doi.org/10.1016/j.bpg.2021.101732
9. Harrison, C.J., Sidey-Gibbons, C.J.: Machine learning in medicine: a practical introduction to natural language processing. BMC Med. Res. Methodol. **21**, 158 (2021). https://doi.org/10.1186/s12874-021-01347-1
10. Ker, J., Wang, L., Rao, J., Lim, T.: Deep learning applications in medical image analysis. IEEE Access **6**, 9375–9389 (2018). https://doi.org/10.1109/ACCESS.2017.2788044
11. Lipps, C., Bergkemper, L., Schotten, H.: Distinguishing Hearts: How Machine Learning identifies People based on their Heartbeat (2021)
12. Barua, I., Mori, Y., Bretthauer, M.: Colorectal polyp characterization with endocytoscopy: ready for widespread implementation with artificial intelligence? Best Pract. Res. Clin. Gastroenterol. **52–53**, 101721 (2021). https://doi.org/10.1016/j.bpg.2020.101721

13. Loftus, T.J., et al.: Executive summary of the artificial intelligence in surgery series. Surgery. **171**, 1435–1439 (2022). https://doi.org/10.1016/j.surg.2021.10.047
14. Reddy, C.D., Van den Eynde, J., Kutty, S.: Artificial intelligence in perinatal diagnosis and management of congenital heart disease. Semin. Perinatol. **46**, 151588 (2022). https://doi.org/10.1016/j.semperi.2022.151588
15. Ahmad, O.F., Stoyanov, D., Lovat, L.B.: Barriers and pitfalls for artificial intelligence in gastroenterology: ethical and regulatory issues. Techn. Innov. Gastro. Endosc. **22**, 80–84 (2020). https://doi.org/10.1016/j.tgie.2019.150636
16. Lee, E.E., et al.: Artificial intelligence for mental health care: clinical applications, barriers, facilitators, and artificial wisdom. Biol. Psychiat. Cogn. Neurosci. Neuroimag. **6**, 856–864 (2021). https://doi.org/10.1016/j.bpsc.2021.02.001
17. Jairam, M.P., Ha, R.: A review of artificial intelligence in mammography. Clin. Imag. **88**, 36–44 (2022). https://doi.org/10.1016/j.clinimag.2022.05.005
18. Ramaswamy, A., et al.: It's the data, stupid: inflection point for artificial intelligence in Indian healthcare. Artif. Intell. Med. **128**, 102300 (2022). https://doi.org/10.1016/j.artmed.2022.102300
19. Brennan, H.L., Kirby, S.D.: Barriers of artificial intelligence implementation in the diagnosis of obstructive sleep apnea. J. Otolaryngol. Head Neck Surg. **51**, 16 (2022). https://doi.org/10.1186/s40463-022-00566-w
20. Yoshida, H., Kiyuna, T.: Requirements for implementation of artificial intelligence in the practice of gastrointestinal pathology. World J. Gastroenterol. **27**, 2818 (2021). https://doi.org/10.3748/wjg.v27.i21.2818
21. Gomolin, A., Netchiporouk, E., Gniadecki, R., Litvinov, I.V.: Artificial intelligence applications in dermatology: where do we stand? Front. Med. **7**, (2020)
22. Turnquist, C., et al.: On the edge of a digital pathology transformation: views from a cellular pathology laboratory focus group. J. Pathol. Inform. **10**, 37 (2019). https://doi.org/10.4103/jpi.jpi_38_19
23. Krajcer, Z.: Artificial intelligence for education, proctoring, and credentialing in cardiovascular medicine. Tex Heart Inst. J. **49**, e217572 (2022). https://doi.org/10.14503/THIJ-21-7572
24. Fellous, J.-M., Sapiro, G., Rossi, A., Mayberg, H., Ferrante, M.: Explainable artificial intelligence for neuroscience: behavioral neurostimulation. Front. Neurosci. **13** (2019)
25. Bukowski, M., et al.: Implementation of eHealth and AI integrated diagnostics with multidisciplinary digitized data: are we ready from an international perspective? Eur. Radiol. **30**, 5510–5524 (2020). https://doi.org/10.1007/s00330-020-06874-x
26. Sarwar, S., et al.: Physician perspectives on integration of artificial intelligence into diagnostic pathology. npj Digit. Med. **2**, 28 (2019). https://doi.org/10.1038/s41746-019-0106-0
27. Toh, T.S., Dondelinger, F., Wang, D.: Looking beyond the hype: applied AI and machine learning in translational medicine. EBioMedicine **47**, 607–615 (2019). https://doi.org/10.1016/j.ebiom.2019.08.027
28. Davenport, T., Kalakota, R.: The potential for artificial intelligence in healthcare. Future Healthc J. **6**, 94–98 (2019). https://doi.org/10.7861/futurehosp.6-2-94
29. Parikh, D., Armstrong, G., Liou, V., Husain, D.: Advances in telemedicine in ophthalmology. Semi. Ophthalmol. **35**, 1–6 (2020). https://doi.org/10.1080/08820538.2020.1789675
30. Oprescu, A.M., et al.: Towards a data collection methodology for Responsible Artificial Intelligence in health: a prospective and qualitative study in pregnancy. Inform. Fusion **83–84**, 53–78 (2022). https://doi.org/10.1016/j.inffus.2022.03.011
31. Briganti, G., Le Moine, O.: Artificial intelligence in medicine: today and tomorrow. Front. Med. **7** (2020)

32. Martinez, D.S.-L., et al.: Artificial intelligence opportunities in cardio-oncology: Overview with spotlight on electrocardiography. Am. Heart J. Plus Cardiol. Res. Pract. **15**, 100129 (2022). https://doi.org/10.1016/j.ahjo.2022.100129
33. Paranjape, K., et al.: The value of artificial intelligence in laboratory medicine: current opinions and barriers to implementation. Am. J. Clin. Pathol. **155**, 823–831 (2021). https://doi.org/10.1093/ajcp/aqaa170
34. Wang, J.W., Williams, M.: Registries, databases and repositories for developing artificial intelligence in cancer care. Clin. Oncol. **34**, e97–e103 (2022). https://doi.org/10.1016/j.clon.2021.11.040
35. Akosman, I., et al.: The emerging role of artificial intelligence in adult spinal deformity. Semi. Spine Surg. **34**, 100986 (2022). https://doi.org/10.1016/j.semss.2022.100986
36. Kwok, T.C., et al.: Application and potential of artificial intelligence in neonatal medicine. Semin. Fetal Neonatal. Med. **27**, 101346 (2022). https://doi.org/10.1016/j.siny.2022.101346
37. Chong, J.H., Abdulkareem, M., Petersen, S.E., Khanji, M.Y.: Artificial intelligence and cardiovascular magnetic resonance imaging in myocardial infarction patients. Curr. Probl. Cardiol. **47**, 101330 (2022). https://doi.org/10.1016/j.cpcardiol.2022.101330
38. Kim, K.-J., Tagkopoulos, I.: Application of machine learning in rheumatic disease research. Korean J. Intern. Med. **34**, 708–722 (2019). https://doi.org/10.3904/kjim.2018.349
39. Oliveira, S.P., et al.: CAD systems for colorectal cancer from WSI are still not ready for clinical acceptance. Sci. Rep. **11**, 14358 (2021). https://doi.org/10.1038/s41598-021-93746-z
40. Peng, G.C.Y., et al.: Multiscale modeling meets machine learning: what can we learn? Arch. Computat. Meth. Eng. **28**, 1017–1037 (2021). https://doi.org/10.1007/s11831-020-09405-5
41. Murphy, D., Saleh, D.: Artificial Intelligence in plastic surgery: what is it? where are we now? what is on the horizon? Annals **102**, 577–580 (2020). https://doi.org/10.1308/rcsann.2020.0158
42. Van den Eynde, J., Lachmann, M., Laugwitz, K.-L., Manlhiot, C., Kutty, S.: Successfully implemented artificial intelligence and machine learning applications in cardiology: State-of-the-art review. Trends Cardiovasc. Med. **S1050–1738**(22), 00012–00013 (2022). https://doi.org/10.1016/j.tcm.2022.01.010
43. Xie, L., Yang, S., Squirrell, D., Vaghefi, E.: Towards implementation of AI in New Zealand national diabetic screening program: cloud-based, robust, and bespoke. PLoS ONE **15**, e0225015 (2020). https://doi.org/10.1371/journal.pone.0225015
44. Victor Mugabe, K.: Barriers and facilitators to the adoption of artificial intelligence in radiation oncology: A New Zealand study. Techn. Innov. Patient Sup. Radiation Oncol. **18**, 16–21 (2021). https://doi.org/10.1016/j.tipsro.2021.03.004
45. Scherer, J.D.S., Pereira, J.S., Debastiani, M.S., Bica, C.G.: Beyond technology: can artificial intelligence support clinical decisions in the prediction of sepsis? Rev. Bras. Enferm. **75**, e20210586 (2022). https://doi.org/10.1590/0034-7167-2021-0586
46. Seyam, M., Weikert, T., Sauter, A., Brehm, A., Psychogios, M.-N., Blackham, K.A.: Utilization of artificial intelligence–based intracranial hemorrhage detection on emergent noncontrast CT images in clinical workflow. Radiol. Artif. Intell. **4**, e210168 (2022). https://doi.org/10.1148/ryai.210168
47. Fujimori, R., et al.: Acceptance, barriers, and facilitators to implementing artificial intelligence-based decision support systems in emergency departments: quantitative and qualitative evaluation. JMIR Form Res. **6**, e36501 (2022). https://doi.org/10.2196/36501
48. Alhussain, G., Kelly, A., O'Flaherty, E.I., Quinn, D.P., Flaherty, G.T.: Emerging role of artificial intelligence in global health care. Health Pol. Technol. **11**, 100661 (2022). https://doi.org/10.1016/j.hlpt.2022.100661
49. Milam, M.E., Koo, C.W.: The current status and future of FDA-approved artificial intelligence tools in chest radiology in the United States. Clin. Radiol. **78**, 115–122 (2023). https://doi.org/10.1016/j.crad.2022.08.135

50. Nam, S., et al.: Introduction to digital pathology and computer-aided pathology. J. Pathol. Transl. Med. **54**, 125–134 (2020). https://doi.org/10.4132/jptm.2019.12.31
51. Mathenge, W., et al.: Impact of artificial intelligence assessment of diabetic retinopathy on referral service uptake in a low-resource setting: the RAIDERS randomized trial. Ophthalmol. Sci. **2**, 100168 (2022). https://doi.org/10.1016/j.xops.2022.100168
52. Strohm, L., Hehakaya, C., Ranschaert, E.R., Boon, W.P.C., Moors, E.H.M.: Implementation of artificial intelligence (AI) applications in radiology: hindering and facilitating factors. Eur. Radiol. **30**, 5525–5532 (2020). https://doi.org/10.1007/s00330-020-06946-y

Exploring Cause-and-Effect Relationships Between Public Company Press Releases and Their Stock Prices

Filip Grzonkowski[1] and Paweł Weichbroth[2](✉)

[1] Faculty of Electronics, Telecommunications and Informatics, Gdansk University of Technology, Gdańsk, Poland
[2] Faculty of Electronics, Telecommunications and Informatics, Department of Software Engineering, Gdansk University of Technology, Gdańsk, Poland
pawel.weichbroth@pg.edu.pl,
https://mostwiedzy.pl/en/pawel-weichbroth,49696-1

Abstract. The aim of the work is to design and implement a method of exploring the cause-and-effect relationships between company announcements and the stock prices on NASDAQ stock exchange, followed by a brief discussion. For this purpose, it was necessary to download the stock quotes of selected companies from the NASDAQ market from public web sources. Additionally, media messages related to selected companies had to be downloaded, and then a news sentiment analysis mechanism had to be prepared. The mechanism of sentiment analysis was prepared based on the supervised machine learning approach. The implemented method was used to analyze the sentiment of a set of stock exchange announcements and it was correlated with changes in the share prices of selected companies. Based on the collected data, the association rules were extracted using the Apriori algorithm. While the obtained results are very promising, however, one should also estimate the rate of return for the interrelated transactions to determine the true and ultimate value of discovered relationships.

Keywords: Sentiment Analysis · Stock · NASDAQ · Association Rules · Apriori

1 Introduction

The stock exchange is an institution that allows anyone to buy or sell shares in public companies such as Apple, VISA or Exxon Mobil, to name but a few. However, it is not the mere possession of shares that is an end in itself, but the obtaining of benefits from them. These benefits include the ability to sell the shares at a higher price, of course, but also the right to share in the company's profits in the form of dividends. Every day, millions of investors around the world receive information and rumours from the media as well as corporate

M. Hernes and J. Wątróbski (Eds.): ECAI 2023, LNNS 1079, pp. 74–82, 2024.
https://doi.org/10.1007/978-3-031-66761-9_7

information, including official press releases (news and announcements), quarterly and annual reports, proxy statements and additional disclosures [6]. Public disclosure is designed to protect investors by providing timely and meaningful information.

Among many investors' efforts and interests, the overriding concern is to assess the impact of current information on future share prices [3]. However, human judgement is inherently influenced by personality traits [13], as evidenced by patterns of behaviour, emotion and thought that are relatively consistent over time and across situations [5]. Investors and traders therefore rely not only on self-directed qualitative analysis, but also on computer-based quantitative (technical) analysis to make informed decisions [10]. In addition, recent advances in machine learning (ML) [2], particularly in sentiment analysis, have attracted considerable attention to the application of ML methods to determine the emotional tone behind a body of textual stock information [11,12,14].

Among the many freely available sources of stock information of particular interest to a large and independent group of investors and traders, official corporate press releases are still considered credible and viable [1,7,9]. On the other hand, their value could still be questioned due to numerous cases of market manipulation, including the release of some form of market-moving news that is not true or has no legitimate basis [15], such as false claims of a significant deal [18], fake product sales [17] or manipulation of financial data [16], to name a few.

Nevertheless, in our study we decided, with deliberate caution and hope, to investigate the impact of public press releases on the corresponding stock prices by integrating technical analysis with machine learning methods. More specifically, by adopting technical indicators and combining them with the techniques of sentiment analysis and association rule induction, our research objective is to extract significant and meaningful cause-and-effect relationships. While the impact of the broad category of financial news has been extensively studied in previous literature [4,8,20], to the best of our knowledge, exploring both quantitative and qualitative datasets by incorporating methods at the intersection of statistics and computational intelligence seems to be a novel approach in academic research on this topic.

The rest of the paper is organized as follows. In Sect. 2 we provide the theoretical foundations for each of the methods applied. In Sect. 3, we outline the methodological assumptions assumed for the study performed. In Sect. 4, we analyze and interpret the results obtained. In Sect. 5, we briefly discuss the relevance of the results and suggest future research directions.

2 Theoretical Background

2.1 Sentiment Analysis

Sentiment is the emotional overtone of a given text. Typically, the sentiment of review text is categorised into positive, negative and neutral [19]. There are different ways of classifying sentiment, but we can divide them into two groups:

statistical and dictionary. Statistical methods are characterized by describing the text with quantitative data, including all machine learning models, neural networks, decision trees. Dictionary methods, are based on dictionaries and usually classify sentiment based on the occurrence of keywords and phrases.

2.2 Technical Analysis

In order to determine as accurately as possible when an impulse can be expected and when not, four indicators known from technical analysis were used: (*a*) Simple Moving Average, (*b*) Exponential Moving Average, (*c*) Relative Strength Index (RSI), and (*d*) Moving Average Convergence/Divergence (MACD).

Moving Averages usually help to determine what the trend is and when it changes. Oscillators, on the other hand, help to assess the strength of the trend and help predict a correction in the trend or when it will end. The closing price is used as the average value in most cases. An uptrend starts when the price breaks above the moving average, and a downtrend begins when the price falls below the moving average.

Choosing the length of the moving average is not an easy task, because the longer the period of the average, the less indications to buy/sell we will find, but at the same time they will be more reliable. At the same time, the delay with which we will notice a change in the trend will also be greater. Typically, averages of 5 to 45 sessions are used for short- and medium-term trends. Choosing the right type of average is also important, e.g. for long averages such as 200 period, the choice of exponential average is pointless, because the last values are of such little importance that they are almost negligible. In the further part of the work, I most often adopted 9-period moving averages in accordance with the recommendations of Appel, who found that such a period works very well, because it used to be 1.5 trading weeks, today, despite the fact that the trading week has shortened from 6 to 5 days still this medium length works very well.

Relative Strength Index (RSI) is a momentum indicator used in technical analysis that measures the speed and magnitude of a security's recent price changes to evaluate overvalued or undervalued conditions in the price of that security.

$$RSI = 100 - (\frac{100}{1 + \frac{U}{D}}) \tag{1}$$

where:

- U is sum of positive price changes from n sessions;
- D is sum of negative price changes from n sessions.

The value of n can be chosen arbitrarily, but the most popular is 14, recommended by Wilder The oscillator takes values from 0–100 and it is assumed that a value below 30 means that the market is oversold, i.e. there are less and less people willing to sell and more and more people willing to buy, so you can expect at least temporary increases in value. On the other hand, values from 70 to 100 mean that the market has been bought out, i.e. the market is saturated, there are slowly ending willing buyers, which may herald a decrease in the valuation of securities.

MACD indicator shows the relationship between two moving averages. Typically 26 and 12 hundred period exponential averages. The signal line of this indicator is the 9-period exponential average calculated from the difference mentioned above. Nothing stands in the way of using other average lengths, but these have been proposed by the author of this indicator, Gerald Appel.

2.3 Association Rules

In order to formulate the cause and effects relationships between press releases and corresponding stock prices, we used the association rule, written in the form of $(X) \rightarrow (Y)$, where left side stands for the cause, and right side represents the effect. The association rules are generated from the frequent itemsets, discovered from the input data by the adapted Apriori algorithm. By definition, a frequent itemset is one that occurs at least a minimum number of times in the transactions, defined as an input algorithm's parameter.

3 Methodology

In this study, we combined the sentimental analysis with the technical analysis. Therefore, two separate datasets were used.

Considering the first approach, a total of 3,810 articles were downloaded. Of which 500 of them were used as a training set for the sentiment analyzer. These 500 articles were assigned to one of 3 groups: positive, negative or neutral sentiment. Then, using machine learning, a model was trained to evaluate the sentiment of press release texts. The Fast Forest algorithm was used to train the model, which, together with the appropriate hyperparameters, was selected by the AutoML tool as the best fit. Then, the sentiment of each of the remaining news items was classified using the prepared model.

Considering the second approach, the quotes were downloaded for each press release with a 5-minute interval for the period from 3 days before the release to 6 days after the release. The quotes were imported, using *finnhub.io* API.

In the study, it was arbitrarily assumed that if the impulse is to occur, it should occur no more than 1 h after the publication of the message. A price movement greater than the standard deviation for the 9 previous quotes was considered an impulse (Table 1).

For each news, labels from Table 2 were assigned, which were later used in the process of obtaining association rules. Then, frequent sets were searched for, with a minimum support of 5% and a certainty of 55%, using the Apriori algorithm (Fig. 1).

Association rules were generated from frequent sets, so that there was a sentiment among the premises and an impulse as a conclusion. 5 rules with the

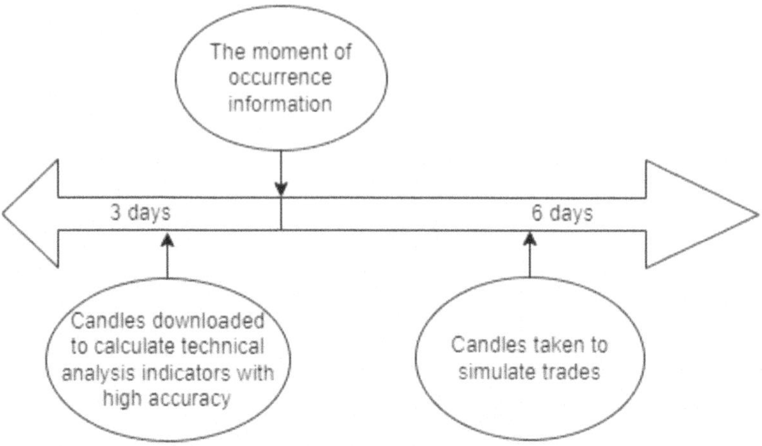

Fig. 1. The range of candles (the period from which the candles were taken).

highest level of support and confidence were selected for the upward impulse and the downward impulse, respectively, and trades executed using these rules were simulated and the rate of return was calculated for them. Rules allow you to determine whether positions should be opened. The moment of closing the position has been adopted the same for all rules and the transaction is closed when one of the following situations occurs:

1. There was no impulse or correction within an hour.
2. There was an impulse when a correction was expected.
3. There was a correction when the impulse was expected.
4. The price closed below the minimum price in the previous period (5 min).

4 Results

In total, 20 rules were received. However, due to the paper limitations, for further analysis only 5 rules were selected with the highest level of support (see Table 3). The minimum support level was set at the level of 5%, while the confidence at a level of 55%.

The best results are achieved by rules 1, 2, 3, they can be understood as:

1. If the news has a positive sentiment and additionally the MACD line is below the signal line and the exponential 50-period average is above the course, then an impulse is expected.
2. If the news has a positive sentiment and, in addition, the MACD line is below the signal line and the 50-period simple average is above the course, then an impulse is expected.
3. If the news has a positive sentiment and additionally the MACD line is below the signal line and the 50-period simple and exponential average is above the rate, an impulse is expected (Figs. 2 and 3).

Table 1. List of tested companies

Ticker	Name	Description
ATVI	Activision Blizzard	American video game holding company
AMD	Advanced Micro Devices	Hardware manufacturer
PYPL	PayPal	Fin-tech offering payment services
CMCSA	Comcast	Media corporation
MRNA	Moderna	Pharmaceutical company
SBUX	Starbucks	Coffee shop chain
ABNB	Airbnb	An online marketplace for homestays
PEP	PepsiCo	Manufacturer of soft drinks and food products
AVGO	Broadcom	Manufacturer and supplier of semiconductors
WDAY	Workday	Software vendor for HR departments
AZN	AstraZeneca	Pharmaceutical company
ADBE	Adobe	Graphics software company
TEAM	Atlassian	Software vendor
ADSK	Autodesk	Software vendor
EBAY	eBay	Auction company
PANW	Palo Alto Networks	Multinational cybersecurity company
NET	Cloudflare	Content delivery solution company
SNPS	Synopsys	Electronic design automation company

Table 2. Labels list

Label	Description
IMPULSE_UP	A price increase is expected
IMPULSE_DOWN	A price decrease is expected
SENTIMENT_POSITIVE	Positive sentiment detected
SENTIMENT_NEUTRAL	Neutral sentiment detected
SENTIMENT_NEGATIVE	Negative sentiment detected
SMA9_UP	The share price is above the 9-period simple average
SMA9_DOWN	The share price is below the 9-period simple average
SMA50_UP	The share price is above the 50-period simple average
SMA50_DOWN	The share price is below the 50-period simple average
EMA9_UP	The share price is above the 9-period exponential average
EMA9_DOWN	The share price is below the 9-period exponential average
EMA50_UP	The share price is above the 50-period exponential average
EMA50_DOWN	The share price is below the 50-period exponential average
RSI_OVERSOLD	RSI indicates oversold (<30)
RSI_NORMAL	RSI without strict direction $<30; 70>$
RSI_OVERBUY	RSI points to overbought (>70)
MACD_UP	The MACD line is above the signal line
MACD_DOWN	The MACD line is below the signal line

Table 3. A list of rules with levels of support and confidence

No.	X (Causes)	Y (Effects)	Support [%]	Confidence [%]
R1	MACD_DOWN	IMPULSE_UP	10,54%	56,16%
	SENTIMENT_Positive			
	EMA50_UP			
R2	SMA50_UP	IMPULSE_UP	10,42%	56,78%
	MACD_DOWN			
	SENTIMENT_Positive			
R3	EMA50_UP	IMPULSE_UP	9,86%	56,51%
	MACD_DOWN			
	SENTIMENT_Positive			
	SMA50_UP			
R4	EMA50_DOWN	IMPULSE_UP	9,17%	56,76%
	SENTIMENT_Negative			
R5	SMA50_DOWN	IMPULSE_UP	9,04%	56,12%
	SENTIMENT_Negative			

5 Discussion

This paper shows that both sentiment and the history of company valuations (in the form of technical analysis indicators) have an impact on price formation and that we are able to generate a positive return based on the rules generated. However, it is important to note that the final return is also affected by the accuracy of the sentiment analysis model and ideally a separate model could be used for each company or sector, as what is a growth driver for one sector may be a bearish factor for another.

In addition, the rules only apply to the time period for which they have been studied and should not be extrapolated to other time periods. It is also possible to check which set of rules is generated for each company or industry separately, but at the same time it is necessary to collect enough data only for a given scope so that the credibility of the rules is at a satisfactory level, which is not easy.

This study has provided quantitative evidence of the impact of the sentiment of individual press releases on share prices. Out of 3104 releases, 2333 were verified, i.e. in 75% of the cases there was a significant price change within one hour of the release. Sentiment analysis combined with technical analysis can aid traders and investors to determine the direction of price movement.

Our future research plans include the rate of return checking that can be obtained from a set of generated association rules.

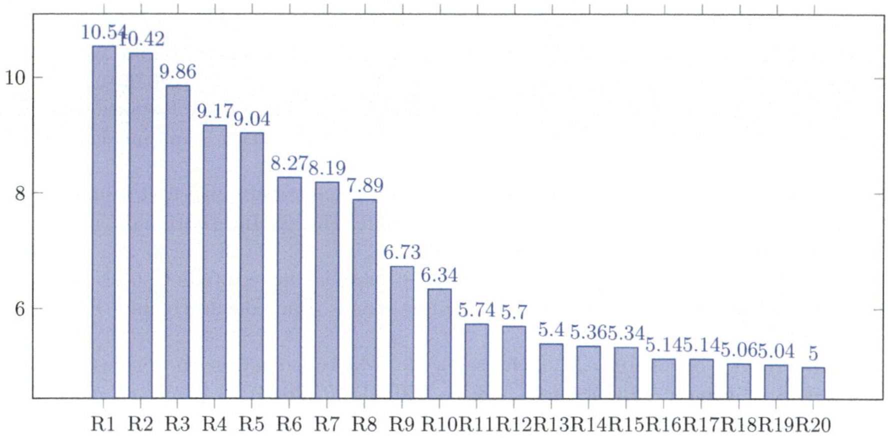

Fig. 2. Rules support histogram

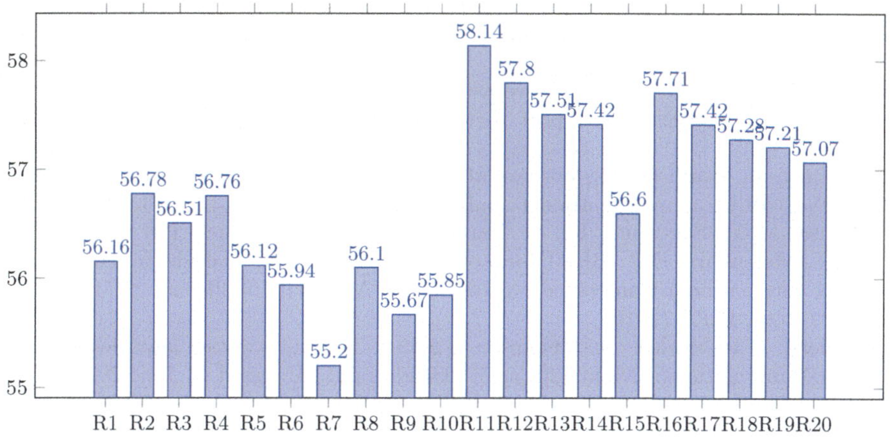

Fig. 3. Rules confidence histogram

References

1. Dyck, A., Zingales, L.: The media and asset prices. Technical report, working paper. Harvard Business School (2003)
2. Khang, P.Q., Hernes, M., Kuziak, K., Rot, A., Gryncewicz, W.: Liquidity prediction on Vietnamese stock market using deep learning. Procedia Comput. Sci. **176**, 2050–2058 (2020)
3. Kobiela, D., Krefta, D., Król, W., Weichbroth, P.: ARIMA vs LSTM on NASDAQ stock exchange data. Procedia Comput. Sci. **207**, 3836–3845 (2022)
4. Kollintza-Kyriakoulia, F., Maragoudakis, M., Krithara, A.: Measuring the impact of financial news and social media on stock market modeling using time series mining techniques. Algorithms **11**(11), 181 (2018)

5. Korczak, J., Fafuła, A.: A method to discover trend reversal patterns using behavioral data. In: Wrycza, S. (ed.) SIGSAND/PLAIS 2011. LNBIP, vol. 93, pp. 81–91. Springer, Heidelberg (2011). https://doi.org/10.1007/978-3-642-25676-9_7
6. Lubis, M., Kusumasari, T.F., Hakim, L.: The Indonesia public information disclosure act (UU-KIP): its challenges and responses. Int. J. Electr. Comput. Eng. **8**(1), 94 (2018)
7. Mittermayer, M.A.: Forecasting intraday stock price trends with text mining techniques. In: 2004 Proceedings of the 37th Annual Hawaii International Conference on System Sciences, pp. 10–pp. IEEE (2004)
8. Narayan, P.K., Bannigidadmath, D.: Does financial news predict stock returns? New evidence from Islamic and non-Islamic stocks. Pac. Basin Finance J. **42**, 24–45 (2017)
9. Neuhierl, A., Scherbina, A., Schlusche, B.: Market reaction to corporate press releases. J. Financ. Quant. Anal. **48**(4), 1207–1240 (2013)
10. Nti, I.K., Adekoya, A.F., Weyori, B.A.: A systematic review of fundamental and technical analysis of stock market predictions. Artif. Intell. Rev. **53**(4), 3007–3057 (2020)
11. Pagolu, V.S., Reddy, K.N., Panda, G., Majhi, B.: Sentiment analysis of Twitter data for predicting stock market movements. In: 2016 International Conference on Signal Processing, Communication, Power and Embedded System (SCOPES), pp. 1345–1350. IEEE (2016)
12. Ren, R., Wu, D.D., Liu, T.: Forecasting stock market movement direction using sentiment analysis and support vector machine. IEEE Syst. J. **13**(1), 760–770 (2018)
13. Samuel, J., et al.: Feeling positive about reopening? New normal scenarios from COVID-19 US reopen sentiment analytics. IEEE Access **8**, 142173–142190 (2020)
14. Sousa, M.G., Sakiyama, K., de Souza Rodrigues, L., Moraes, P.H., Fernandes, E.R., Matsubara, E.T.: BERT for stock market sentiment analysis. In: 2019 IEEE 31st International Conference on Tools with Artificial Intelligence (ICTAI), pp. 1597–1601. IEEE (2019)
15. Swiers, T.: Market manipulation (2019). https://www.sec.gov/files/Market%20Manipulations%20and%20Case%20Studies.pdf. Accessed 13 June 2023
16. The United States Department of Justice: Former chief financial officer at publicly traded transportation company charged with $245 million securities and accounting fraud scheme; two other defendants previously indicted charged with additional offenses (2019). https://www.justice.gov/opa/pr/former-chief-financial-officer-publicly-traded-transportation-company-charged-245-million. Accessed 14 June 2023
17. U.S. Securities and Exchange Commission: Litigation release no. 16725/September 26, 2000. Accounting and auditing enforcement 1310/September 26, 2000 (2000). https://www.sec.gov/litigation/litreleases/lr16725.htm. Accessed 13 June 2023
18. U.S. Securities and Exchange Commission: EC charges father-son duo and associate in market manipulation schemes resulting in a new jersey deli with a $100 million valuation (2022). https://www.sec.gov/news/press-release/2022-172. Accessed 13 June 2023
19. Weichbroth, P., Sroka, W.: A note on the affective computing systems and machines: a classification and appraisal. Procedia Comput. Sci. **207**, 3798–3807 (2022)
20. Zhao, X., Yang, J., Zhao, L., Li, Q.: The impact of news on stock market: quantifying the content of internet-based financial news. In: Proceedings of the 11th International DSI and 16th APDSI Joint Meeting, pp. 12–16 (2011)

Cybersecurity of the Unmanned Marine Vehicles in the Conditions of Partial or Complete Interruption Multi-GNSS Signals by Jamming and/or Spoofing

Łukasz Lemieszewski[1](✉), Piotr Borkowski[2], Aleksandra Radomska-Zalas[1], Larisa Dobryakova[3], and Evgeny Ochin[1]

[1] The Jacob of Paradies University, 25 Teatralna Street, 66-400 Gorzów Wielkopolski, Poland
llemieszewski@ajp.edu.pl

[2] Maritime University of Szczecin, 1-2 Wały Chrobrego, 70-500 Szczecin, Poland

[3] West Pomeranian University of Technology in Szczecin, 17 Piastów Street, 70-310 Szczecin, Poland

Abstract. Multi-GNSS (GPS, GLONASS, Galileo, and BejDou) are used for PNT (Positioning, Navigation, and Timing). UMVs (Unmanned Marine Vehicles) are unmanned vessels that can navigate autonomously or remotely, carry payloads, and be disposable or repairable. UMV networks are formations of autonomous vessels without a command, capable of carrying payloads. Technologies to suppress and counter UMVs have become kye areas of research. Jamming disrupts multi-channel GNSS receivers with noise, disabling UMVs' positioning. This article discusses suppressing UMVs navigation and countermeasures. The cybersecurity of UAVs under conditions of partial or complete interruption of several GNSS signals due to jamming and/or spoofing cannot be calculated by analytical methods, but the results of predicting the movement of the UMVs under extreme conditions can be obtained using simulation methods at the design stage of the UMVs. The main goal of the article is to increase the cybersecurity of unmanned marine vehicles in extreme conditions of partial or complete interruption of GNSS signals in general and especially because of jamming and/or spoofing.

Keywords: Unmanned Marine Vehicles (UMVs) · SANET · Ship Simulation · Ship Routing · Multi-GNSS · Jamming · Mikoning · Spoofing

1 Introduction

By now many countries have accumulated some experience in the design and manufacture of unmanned ships. Here is an example of a recent project. In mid-January 2023, China commissioned the world's first fully automated research vessel Zhuhaiyun [1], which can navigate autonomously in open water or be controlled by remote control. A science and technology laboratory in China's Guangdong province tested the vessel's autonomous navigation capabilities. The artificial intelligence system uses information

M. Hernes and J. Wątróbski (Eds.): ECAI 2023, LNNS 1079, pp. 83–94, 2024.
https://doi.org/10.1007/978-3-031-66761-9_8

from sensors and satellites to solve the positioning problem. Operators can remotely monitor its operation and, if necessary, take over control if it deviates from set targets. At 88.5 m long, 14 m wide, and 2,100 t displacement, the vessel sails at an average speed of 13 knots and can reach speeds of up to 18 knots if necessary.

A few terminological definitions must be introduced before anything specific can be discussed:

- The autonomous vessel does not provide for the interference of the navigator with the navigation.
- The remotely piloted vessel is connected to the outside navigation station(s), the necessary radio control and monitoring lines, and any other elements of the system that may be required during navigation.
- The outboard pilot manipulates the controls of the remotely piloted vessel during navigation.
- Remotely piloted navigation is the control of a vessel from the pilot's workstation, which is not on board the vessel (the pilot may be on board another sea, air, or space vessel).

Given these definitions, we can formulate what an unmanned ship is - a ship that operates without a crew, which uses advanced methods and algorithms, intelligent awareness, data fusion, communication, Internet of Things (IoT), control technology, and other technologies to implement automatic control, navigation, and mooring [2]. An unmanned ship is intelligently controlled based on navigation technology, computer technology, and automatic control technology during navigation, management, maintenance, and cargo transportation so that the unmanned ship has advantages in terms of safety and economy [3]. Currently, unmanned ships are hybrid and electric objects.

An Unmanned Marine Vehicles (UMVs) can navigate without a command on board [4–6], but the skipper can be on the ground or another ship remotely from another ship, for example on the UMVs. UMVs have diverse applications including search and rescue, weather monitoring, people tracking, network extension, construction, and goods transportation. SANET enables UMVs to expand their coverage. It is built upon the development of client-server architecture due to the challenges of managing wired unstructured networks. Routers, switches, and access points form the network infrastructure, enabling data routing, traffic management, and connections across segments. File servers, email servers, and web servers host and manage services. Infrastructure networks are the foundation of the Internet, enabling connectivity, information sharing, web browsing, video streaming, cloud computing, and more.

Unstructured networks, also known as networks without infrastructure or spontaneous networks, are a type of network that operates without reliance on a centralized infrastructure. In such networks, individual devices or nodes communicate with each other directly, without the need for an existing network infrastructure. Let us briefly list the most well-known solutions for building non-infrastructure networks, which also have other names: Infrastructure-less, spontaneous, temporary, for a specific purpose: Ad hoc NETworks (ANET), Sensor networks [7], Peer-to-peer (P2P) networks [8], Mesh networks [9], Opportunistic networks, Delay-tolerant networks (DTNs) [10], Mobile Ad hoc NETworks (MANET) [11], Maritime (Sea) Ad hoc NETworks.

Of great practical importance, today is the development of Maritime Ad hoc NETworks technology [12], which is considered part of the global e-Navigation technology. Without Maritime (Sea) Ad hoc NETworks innovations, the development of navigation architecture in terms of ship control systems, energy conservation, artificial intelligence on the ship's bridge, cybersecurity and privacy of navigation, localization and route planning is impossible [13].

Maritime Ad hoc NETworks technology is divided into Ship Ad hoc NETworks technology (SANET) and Underwater Ad hoc NETworks technology (UANET), including smart autonomous underwater vehicle design technology. For this article, we will focus only on SANET cybersecurity.

The main problems of this study are deepening the concept of SANET in relation to cybersecurity, systematization of UMVs driving modes using only GNSS, using only INS and using GNSS+INS, increase the cybersecurity of UMVs in extreme conditions of partial or complete interruption of GNSS signals in general and especially because of jamming and/or spoofing.

The main research methodology is modeling the dynamics of UMVs using a combination of MS Excel and Matlab, partially presented in Sect. 3.3 SANET organizes the movement of the ship using GNSS and/or INS.

2 Ship Ad Hoc NETworks (SANET)

SANETs can operate in complex marine environments, which include factors such as limited space and potential signal interference due to electromagnetic wave reflection from metal structures, as SANETs use Wi-Fi, Bluetooth or specialized marine wireless communication standards. SANET nodes include computers, sensors, control systems, navigation equipment, communication systems, and other onboard systems. Network nodes can establish connections and create a network on the fly, as vessel configurations and communication requirements change.

SANETs use ad hoc routing protocols that enable data exchange between ship nodes. Routing algorithms determine optimal data paths, considering factors such as node proximity, link quality, available bandwidth, and other network characteristics.

SANETs support voice communications, video surveillance, environmental monitoring, sensor data collection, remote equipment control, and more. These applications provide onboard monitoring, cybersecurity, and operational efficiency, which is especially important for unmanned ships and vessels, which must include security measures to protect against unauthorized access, data leakage, and cyber threats. Encryption, authentication mechanisms, access controls and intrusion detection systems are some of the security measures used to ensure the integrity and confidentiality of shipboard communications and data in SANET technology.

It should be emphasized that SANETs are not only for unmanned ships but also for ships with a crew on board. In this case, the SANET must provide normal telephone communication with the shore without the use of telecommunication satellites, if the SANET is in the coastal zone.

Various shipboard devices, including computers, sensors, control systems, and communication systems, form SANET nodes. These devices exchange data wirelessly (Wi-Fi or Bluetooth). SANET uses ad hoc routing protocols that allow shipboard devices to establish direct connections and exchange data. Ad hoc routing algorithms determine the best paths for data transmission based on factors such as proximity, link quality, and network status.

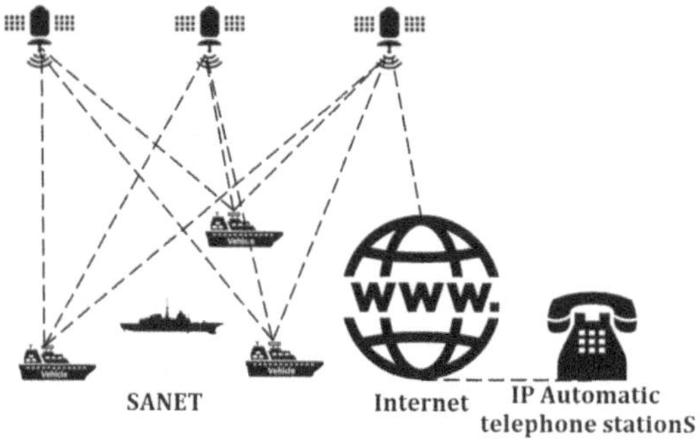

Fig. 1. An architecture of Ship Ad hoc NETworks and IP-enabled satellite network (own design).

SANETs are formed dynamically as shipboard devices find themselves in a mutual proximity zone. Nodes self-organize and establish connections on the fly, providing flexible adaptive communications in the shipboard network, supporting voice communications, data exchange, environmental monitoring, and sensor data collection while applying security measures to protect against unauthorized access and ensure shipboard communications privacy such as encryption, authentication, and access control.

Analyzing the SANET's interaction with the Internet, SANET can be thought of as an IP subnet. The transmission of packets on the network is terminated by the routing protocol on the network (a packet may have to go through several hops to reach its destination), and when a packet enters or leaves a subnet, the standard IP routing mechanism is used. This requires the gateway host to be able to run multiple routing protocols. SANET can be seen as an extension of a conventional IP network using satellite links (Fig. 1).

The network uses geostationary and/or low-Earth orbit satellites (Starlink, etc.) to provide communications between ground stations and satellite terminals. Satellites act as repeaters, transmitting data between ground stations and satellite terminals, which are installed on ships and consist of satellite dishes, satellite modems, routers, and other network equipment that manage the flow of data packets.

Network security measures such as firewalls, encryption, and authentication mechanisms are used to protect the satellite network from unauthorized access and ensure data confidentiality and integrity.

In practice, the IP-enabled satellite network and SANET can be integrated to provide ship-to-ship communications. The satellite network provides long-range communication with ground stations, while SANETs provide wireless communication and networking between shipboard devices. This integration allows ships to communicate with the outside world via satellite links and enables efficient communication and data exchange between ship systems and personnel.

UMVs on the SANET can act as nodes and routers, enabling communication between other UMVs on the network, and with other terrestrial or satellite communications systems. SANETs can be used in a variety of applications such as military surveillance, disaster surveillance, and search and rescue. They can provide a fast and efficient means of communication in areas where traditional communication infrastructure is unavailable or damaged.

However, there are several issues associated with SANET such as communication range, and the need for efficient routing protocols to ensure reliable and efficient communications. In addition, there are security issues associated with SANET such as data confidentiality, integrity, and authentication, as well as the risk of cyber-attacks on the network.

There is still a lot of work to be done to overcome the challenges and make them a practical and reliable means of communication in a variety of applications.

The cybersecurity of UMVs and SANET refers to the measures taken to protect these unmanned vehicles from hacking, which are mainly jamming, meaconing and spoofing, unauthorized access, and other forms of cyber-attacks [14, 15].

3 How to Deal with UMVs

UMVs, as a rule, combine the functions of a reconnaissance UMVs and a vehicle. It should be emphasized that UMVs can be classified as remotely operated vehicles (ROVs) and/or UMVs. When UMVs are controlled by a remote operator, they are considered ROVs. In this case, the operator uses the ground control station to remotely control the movements and actions of UMVs.

On the other hand, when UMVs are programmed to operate autonomously without the direct intervention of a skipper, they are considered UMVs. At the same time, the UMVs are equipped with advanced sensors and navigation systems that allow you to make decisions and complete missions without human intervention.

One of the advantages of such UMVs is their efficiency. Compared to other operations, the cost per UMV is relatively low, making it an attractive option to complete reconnaissance missions quickly and on budget. This makes them a suitable option for targeting high-risk targets.

The use of UMVs has some disadvantages. They can be easily intercepted or destroyed by terrorist forces. GNSS signals are susceptible to various types of interference, both unintentional and intentional. The low power of GNSS signals makes them vulnerable to interference from other devices and systems operating in the same frequency range.

In addition, there is an increasing number of intentional interferers available on the COTS (Component-off-the-shelf) market, mostly sold over the internet. Spoofing is a growing concern for GNSS security, as it allows an attacker to manipulate a GNSS receiver into estimating a fake position and/or time, without the receiver recognizing it. This type of interference can have serious implications for navigation, navigation safety, and time synchronization, as well as national security and critical infrastructure.

Therefore, it's important to be aware of the risks posed by GNSS interference and to implement measures to mitigate these risks, such as using jamming resistant GNSS receivers, using multiple GNSS systems, and regularly monitoring and analyzing GNSS signals for signs of interference.

Note that the most important technologies for combating unmanned aerial vehicles are GNSS jamming, meaconing, and spoofing.

3.1 GNSS Jamming

GNSS Jamming [16] is the deliberate interference of an external source with the GNSS signal. The Jammer sends out a signal that is stronger than the GNSS signal, causing the GNSS signal to be blocked. Jamming is already recognized as a serious problem [17, 18] (Fig. 2a). GNSS jamming is the deliberate interference with or disruption of radio signals. This can be accomplished using electronic devices that emit radio frequency signals designed to disrupt or override the desired signal. Radio jamming is illegal in many countries and is often used to prevent the transmission of information or communication by disrupting the normal functioning of radio-based communication systems. In military or security contexts, radio jamming can be used as a form of electronic warfare to deny the enemy the use of communication and navigation systems. In other cases, radio jamming may be used to block signals from cell phones, Wi-Fi networks, or other wireless devices.

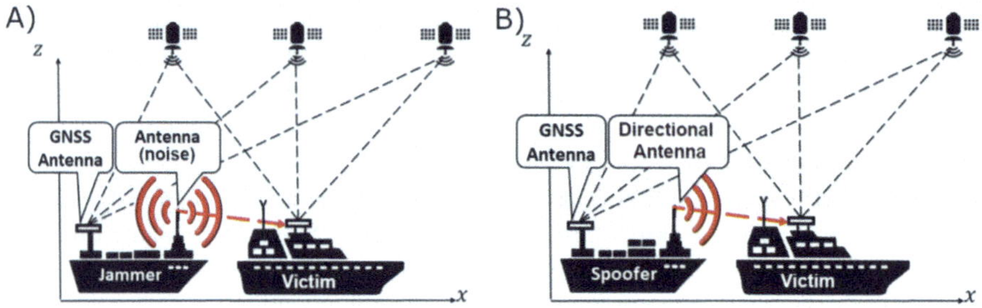

Fig. 2. GNSS Jamming (a), GNSS Spoofing (b) of the Ship as a Victim (own design).

Jamming, especially straightforward jamming, is a very disruptive method that can cause significant problems and attract attention. In addition to the GNSS receivers themselves detecting the loss or degradation of the signal, other devices, such as communication systems, navigation systems, and even cell phone networks, can also be affected.

Moreover, jamming can also disrupt the normal functioning of important communication and navigation systems, leading to potential safety hazards and economic losses.

3.2 GNSS Meaconing and Spoofing

GNSS Meaconing [19] and Spoofing [20] create a false GNSS signal to trick the drone into believing it is in a different location (Fig. 2b). Meaconing, as a special case of spoofing, is the re-transmission of received GNSS signals. Meaconing uses duplication (repetition) of the real GNSS signal, making the user believe that the beacon is a legitimate satellite. Meaconing is a technique closely related to spoofing but avoids the need to be able to actively produce the GNSS signal structure. Meaconing is subject to the same countermeasures as spoofing. However, cryptographic methods and authentication of navigation messages do not always protect against meaconing since the repeater does not need to know the structure of the relayed GNSS signal [21].

Meaconing is a system of receiving radio beacon signals and rebroadcasting them on the same frequency to confuse navigation. The meaconing stations cause inaccurate bearings to be obtained by a ship or ground station.

Spoofing is the use of a false GNSS signal to fool the receiver. Since the motivation for implementing a spoofing attack is less obvious than the motivation for using a PPD and given that the complexity of the spoofer is much higher than the PPD, it is not surprising that the number of spoofing incidents is less than that of jamming. We only note the event of December 4, 2011: Iran captured the American Lockheed Martin RQ-170 drone, claiming that it spoofed GPS [22]. This can be done by jamming the legitimate signal and sending a false signal that appears to be coming from a legitimate satellite.

3.3 SANET Organizes the Movement of the Ship Using GNSS and/or INS

GNSS and Inertial Navigation Systems (INS) are the two most used technologies for positioning and navigation w SANET, with satellites used to calculate position and INS using a system of accelerometers and gyroscopes to measure vessel motion. GNSS and INS are used together to provide more accurate navigation solutions. If GNSS is lost, the system automatically enters Dead Reckoning Mode, allowing it to sail in areas where GNSS has been forced to distort.

By combining the two technologies, errors from both systems can be minimized and more accurate navigation solutions can be obtained. For example, an integrated navigation system may use a GNSS receiver to provide the initial position of the ship and an INS system to track the motion of the ship.

A UMV trajectory plan refers to the route it follows to complete a specific task or reach a target destination. The trajectory plan must be optimized in terms of safety, efficiency, and reliability. The plan is created using specialized software that includes algorithms and simulations to determine the optimal trajectory. Formally, the trajectory plan is written as some function in 2D space ***xy***:

$$xy(t_{begin} \leq t \leq t_{end}) \tag{1}$$

Preliminary calculation of UMVs trajectory in 2D space ***xy*** can be done like this:

$$\begin{Bmatrix} x_0 = 0 \\ y_0 = 0 \end{Bmatrix} FOR_{i=1}^{i=i_{max}} DO \begin{cases} x_i = x_{i-1} + \Delta x\left(\frac{x_{i-1} - x_{GNSS}}{x_{i-1}}\right) + randbetween(-r; r) \\ y_i = y_{i-1} + \Delta y\left(\frac{y_{i-1} - y_{GNSS}}{y_{i-1}}\right) + randbetween(-r; r) \end{cases} \tag{2}$$

where x_{GNSS} i y_{GNSS} are current coordinates xy from a GNSS navigator; r is the range of values of random variable *randbetween*().

So, for example, Fig. 3a shows the plan of the UMVs trajectory in 2D space ***xy*** for conditionally dimensionless time $0 \leq t \leq 100$:

$$\left\{\begin{matrix} x_0 = 0 \\ y_0 = 0 \end{matrix}\right\} FOR_{i=1}^{i=100} DO \left\{\begin{matrix} x_i = x_{i-1} + \frac{x_{i-1} - x_{GNSS}}{x_{i-1}} + randbetween(-1;\,1) \\ y_i = y_{i-1} + \frac{y_{i-1} - y_{GNSS}}{y_{i-1}} + randbetween(-1;\,1) \end{matrix}\right. \quad (3)$$

where $\Delta x = \Delta y = 1$, $r = 1$.

UMVs missions can vary greatly depending on their design, capabilities, and intended use. Some common missions include structural inspections (bridges, buildings, and pipelines), intelligence gathering, surveillance, and situational awareness. UMVs can also be used to search for missing people or support rescue efforts in disaster areas. The actual trajectory of the UMVs is determined by its software, which will provide the flight and target.

Partial interruption of GNSS navigation signals due to interference refers to the deliberate disruption of GNSS signals by electronic interference between time $25 \leq t \leq 74$ (Fig. 3b). This can result in inaccurate or lost GNSS signals, which can cause problems for devices and systems that use GNSS for navigation and positioning. Jamming is usually done with a device that transmits strong radio signals on the same frequency as the GNSS signals, effectively jamming the real signals. In the event of partial interruption of GNSS navigation signals due to jamming, it may be necessary to temporarily switch to Inertial Navigation System (INS) navigation. INS is a backup navigation system that uses accelerometers and gyroscopes to measure the motion and orientation of a device and calculate its position, velocity, and orientation without relying on external signals.

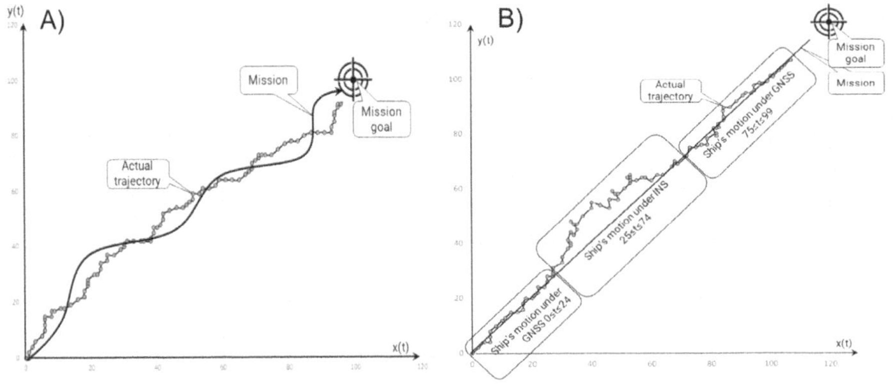

Fig. 3. UMVs under GNSS (a), UMVs under Spoofing attack (b) (own design).

During GNSS jamming, INS can provide reliable navigation information and maintain system functionality. However, INS has limitations such as error accumulation and sensitivity to initial conditions.

GNSS/INS integration is commonly used to enhance navigation reliability and accuracy. GNSS signals provide an initial position fix and periodically update the INS solution, reducing accumulated errors. In the absence of GNSS signals, INS continues to provide navigation information based on its own measurements until GNSS signals are restored.

So, for example, Fig. 4 shows the plan of the UMV trajectory for conditionally dimensionless time $0 \leq t \leq 100$ divided into two-time intervals $1 \leq t \leq 24$ and $25 \leq t \leq 100$.

$$\left\{\begin{matrix} x_0 = 0 \\ y_0 = 0 \end{matrix}\right\} FOR_{i=1}^{i=24} DO \left\{\begin{matrix} x_i = x_{i-1} + \frac{x_{i-1} - x_{GNSS}}{x_{i-1}} + RANDBETWEEN(-1;\,1) \\ y_i = y_{i-1} + \frac{y_{i-1} - y_{GNSS}}{y_{i-1}} + RANDBETWEEN(-1;\,1) \end{matrix}\right. \tag{4}$$

where $\Delta x = \Delta y = 1$,

$$FOR_{i=25}^{i=100} DO \left\{\begin{matrix} x_i = x_{i-1} + 1 + RANDBETWEEN(-1;\,1) \\ y_i = y_{i-1} - 1 + RANDBETWEEN(-1;\,1) \end{matrix}\right. \tag{5}$$

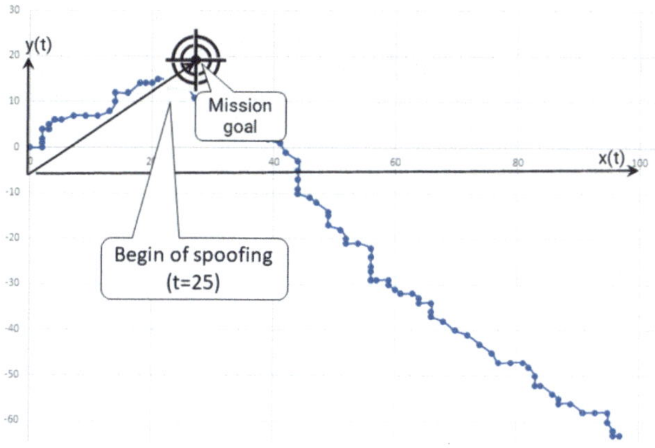

Fig. 4. The interruption of UMV under GNSS and INS by spoofing (t > 25) (own design)

GNSS spoofing refers to the intentional transmission of false GNSS signals to manipulate position, velocity, and time measurements. Unlike jamming, spoofing deceives the receiver into accepting false signals as genuine. This poses serious risks, especially in safety-critical applications relying on GNSS for navigation and positioning.

During GNSS spoofing, switching to INS as a backup may be challenging. INS relies on GNSS for initial position initialization, and without a valid GNSS position, reliable navigation information may not be available. Additionally, INS is susceptible to spoofing attacks, as false GNSS signals can manipulate the INS solution.

To mitigate GNSS spoofing risks, employing GNSS security measures like encryption, utilizing multiple GNSS constellations, and implementing anti-spoofing algorithms is crucial. GNSS/INS integration, where GNSS signals update and correct the INS solution, can also help reduce the impact of spoofing attacks.

3.4 How to Effectively Deal with UMVs

Radiofrequency signals are the main communication technology between UMVs and their control center. The jamming system paralyzes the radio channel, making it uncontrollable, using more powerful interfering RF signals, and the control center forces it to transfer control to the spoofer.

These jamming systems can be directional or omnidirectional. With directional jammers, power is limited in space, which reduces the potential for interference with co-located RF devices/equipment compared to omnidirectional jammers. It is also important to keep in mind that it is possible to install stationary jammers in a fixed location and mobile jammers operating from portable devices or mounted on a vehicle.

Modern UMVs are "floating computers" and therefore most of the attacks designed for traditional computer systems are also effective against them. And, like the basic principles of computer security, you should follow the sequence "Protect – Detect – Mitigate".

To protect against GNSS interference, it is necessary to use jamming detection systems, use secure communication and authentication protocols. These systems may use RF detection methods or software methods to detect interfering signals. Secure authentication protocols are used to ensure that only authorized users can access the navigation and positioning system. These protocols may use encryption or other methods to ensure that only authorized users can access the system.

Some of the common methods for protecting against GNSS Meaconing and Spoofing include [23]: using multiple GNSS receivers to cross-check signals and detect any anomalies, implementing anti-spoofing algorithms that can detect and reject false or manipulated signals, using encrypted GNSS signals that are more difficult to interfere with or access, installing GNSS jamming detectors to alert users to any potential interference with GNSS signals.

4 Conclusions

The SANET network's main properties include autoconfiguration, self-optimization, and self-healing, enabling automatic configuration, easy adaptation to structural changes, and automatic restoration in case of failure. Ship Ad hoc Networks simplify equipment monitoring, with SANETs being resistant to architectural changes and characterized by simplicity and fast installation.

Unfortunately, cybersecurity issues are worsening, as autonomous vessels become prime targets for control interception. However, sea drones already exist, with companies like Kongsberg, Yara, Nippon Yusen, Rolls-Royce Holdings, and Bourbon developing them. Unmanned Marine Vehicles (UMVs) is vessels capable of autonomous or remote navigation, with no human operator, and can be disposable or maintainable while carrying payloads. Research on suppressing UMVs and developing countermeasures is crucial. Jamming disrupts GNSS receivers through narrowband noise, impacting their positioning capabilities. Cybersecurity of UMV networks cannot be analytically calculated when multiple GNSS signals are interrupted by jamming or spoofing. However, modeling during the design stage provides insights into extreme condition behavior.

UMVs face cybersecurity challenges during GNSS signal interruptions caused by jamming or spoofing. Jamming interferes with GNSS signals, while spoofing transmits false signals to deceive navigation systems. Reliable navigation in such scenarios is challenging. Research focuses on modernizing navigation technologies, integrating inertial navigation systems, astronomical navigation, and computer vision-based navigation to reduce GNSS reliance. Resilient maritime communication systems, like cognitive radio with dynamic frequency switching, are crucial. Artificial intelligence and machine learning detect and mitigate cyber threats, enhancing UMVs' security by identifying anomalies and proactively responding to potential attacks or failures.

The implications of research into the security of UMVs in the face of GNSS interference may stimulate further research and technological development in cybersecurity. Innovations in detection, attack protection, and authorization of GNSS signals can significantly improve the security of UMVs and other systems that depend on satellite navigation systems.

Research on the security of UMVs in the context of GNSS interference is essential for advancing cybersecurity theories, implementing effective engineering methods, creating appropriate regulations, raising awareness, and furthering technological advances. Collaboration between researchers, practitioners, and regulators is critical to achieving a higher level of UMV security in the face of GNSS jamming threats.

Cybersecurity is an ongoing process involving continuous monitoring, threat analysis, and updates. Regular vulnerability assessments and penetration testing help identify weaknesses and implement necessary security measures. Addressing UMVs' cybersecurity challenges requires a multidisciplinary approach, incorporating technological advancements and collaboration among stakeholders. Continuous research, development, and robust cybersecurity measures improve the safety and reliability of unmanned ships and vessels.

References

1. Zhen, L.: Giant Chinese drone-carrying AI ship enters service as research vessel. https://www.scmp.com/news/china/science/article/3206781/giant-chinese-drone-carrying-ai-ship-enters-service-research-vessel. Last accessed 18 May 2023
2. Pietrzykowski, Z., et al.: The autonomous navigation system of a sea-going vessel. Ocean Eng. **261** (2022)
3. Xu, X., et al.: Performance analysis of sea unmanned ship routing protocol based on ad hoc network. In: 2019 International Conference on Information Technology and Computer Application (ITCA), IEEE, pp. 221–224 (2019)
4. Mozaffari, M., Saad, W., Bennis, M., Nam, Y.H., Debbah, M.A.: Tutorial on UAVs for wireless networks: applications, challenges, and open problems. IEEE Commun. Surv. Tutor. **21**, 2334–2360 (2019)
5. Wang, J., Xiao, Y., Li, T., Chen, C.L.P.: A survey of technologies for unmanned merchant ships. IEEE Access **8**, 224461–224486 (2020)
6. Lewicka, O., et al.: Integration data model of the bathymetric monitoring system for shallow waterbodies using UAV and USV platforms. Remote Sens. **14**(16), 4075 (2022)
7. Akyildiz, I.F., Wu, S., Sankarasubramaniam, Y., Cayirci, E.: Wireless sensor networks: a survey. Comput. Netw. **38**(4), 393–422 (2022)

8. Shen, X., et al.: Handbook of Peer-to-Peer Networking, vol. 34. Springer Science & Business Media (2010)
9. Akyildiz, I.F., Wang, X., Wang, W.: Wireless mesh networks: a survey. Comput. Netw. **47**(4), 445–487 (2005)
10. Rodrigues, J.J. (ed.): Advances in Delay-Tolerant Networks (DTNs): Architecture and Enhanced Performance. Woodhead Publishing (2020)
11. Hogie, L.: Mobile Ad Hoc Networks: Modelling, Simulation and Broadcast-based Applications. University of Le Havre (France), University of Luxembourg, Luxembourg (2007)
12. Joseph, S.: Nautical ad-hoc network application development for maritime communications. Indonesian J. Electr. Eng. Comput. Sci. **8**(3), 667–669 (2017)
13. Chauhan, E., Sirswal, M., Gupta, D., Khanna, A.: A critical review: SANET and other variants of ad hoc networks. In: International Conference on Innovative Computing and Communications, pp. 1093–1114 (2021)
14. Lemieszewski, Ł.: Cybersecurity of satellite navigation based on GNSS spoofing detection systems, AJP Gorzów Wielkopolski (2022)
15. Sałaciński, T.: Unmanned vehicles – new challenge for explosives. Sieć badawcza Łukasiewicz – Instytut Przemysłu Organicznego **11**(1), 45–55 (2019)
16. Morales Ferre, R., de la Fuente, A., Lohan, E.S.: Jammer classification in GNSS bands via machine learning algorithms. Sensors **19**(22), 4841 (2019)
17. Dobryakova, L., Lemieszewski, Ł, Ochin, E.: Protecting vehicles vulnerable to terrorist attacks, such as GNSS jamming, by electromagnetic interference shielding of antenna. Scientific J. Maritime Univ. Szczecin **50**(122), 77–83 (2017)
18. Dobryakova, L., Lemeszewski, Ł., Ochin, E.: Method, algorithm and implementation of vehicles GNSS information protection with help of anti-jamming and anti-spoong. In: REIT 2 2017 Radio Electronics & Information Technologies Hogie, L.: Mobile Ad Hoc Networks: Modelling, Simulation and Broadcast-based Applications. University of Le Havre (France), University of Luxembourg (Luxembourg) Vol-2005, pp. 31–40 (2017)
19. Zhu, X., Lu, Z., Hua, T., Yang, F., Tu, G., Chen, X.: A novel GPS meaconing spoofing detection technique based on improved ratio combined with carrier-to-noise moving variance. Electronics **11** (2022)
20. Semanjski, S., Semanjski, I., De Wilde, W., Muls, A.: Use of supervised machine learning for GNSS signal spoofing detection with validation on real-world meaconing and spoofing data-Part I. Sensors **20**(4), 1171 (2020)
21. Broumandan, A., Jahromi, A.J., Daneshmand, S., Lachapelle, G.: GNSS Vulnerability to Spoofing Threats and a Review of Anti-Spoofing Techniques. Schulich School of Engineering (2014)
22. Jones, M.: Spoofing in the black sea: what really happened. GPS World. https://www.gpsworld.com/spoofing-in-the-black-sea-what-really-happened/. Last accessed 30 May 2023
23. Dobryakova, L., Lemieszewski, L., Ochin, E.: Cloud-based GNSS navigation spoofing detection. Scientific J. Maritime Univ. Szczecin **57**(129), 29–37 (2019)

Advanced Machine Learning Methods for Support Business Processes

Applying Convolutional Neural Networks to Stock Market Forecasting – A Case Study of Stock Volume Prediction

Iga Rudawska and Grzegorz Wojarnik(✉)

Institute of Economics and Finance, University of Szczecin, Mickiewicza 64, 71-101 Szczecin, Poland
{iga.rudawska,grzegorz.wojarnik}@usz.edu.pl

Abstract. Stock market forecasting is always a challenge because of dynamic, non-linear, non-parametric and high-noise features of stock data. Researchers have been applying many statistical methods to handle complex financial data, but since DL techniques have emerged, Artificial neural networks have been successfully utilized for automatic feature extraction and prediction. Among many ANN tools, convolutional neural networks (CNN) prove to produce better learning capabilities and therefore manage big data in capital markets, enabling stock market prediction. Therefore, with this work authors seek to explore the possibility of using CNN to process graphics containing stock charts and demonstrate the potential of predicting volume changes. The aim of the study is to develop a CNN's based model that supports prediction of stock volume changes in the capital markets. Moreover, the authors intend to compare the quality measures of the generated predictive model for different types of stock charts. In relation to the aforementioned, a noteworthy contribution to the existing body of knowledge involves comparing the utility of two types of charts for analyzing volume values.

Keywords: Artificial Intelligence · Convolutional Neural Networks · Stock Exchange · Stock Charts · Transactions Volume Analysis

1 Introduction

AI and data science in financial technology (FinTech) pose many research opportunities. They relate to fundamental business systems, operations, processes, regulations and their planning, monitoring, decision-making and optimalization [4, 9, 17, 38]. They cover forecasting and prediction, sentiment and intention modelling, portfolio planning and optimalization, risk management, objective and operations optimalization [10, 11, 24, 40]. Among the aforementioned fields of application forecasting and prediction seem to be of key importance in the capital market [1, 2, 12, 15, 29, 32]. They include classification, estimation and prediction of trend, movement (scale and direction), value (price or volatility), and sequential or temporal change [11]. Modern analytics and learning methods for presenting and managing dynamics, uncertainty, scale and expectations

M. Hernes and J. Wątróbski (Eds.): ECAI 2023, LNNS 1079, pp. 97–108, 2024.
https://doi.org/10.1007/978-3-031-66761-9_9

connected with the capital market include optimization methods, advanced representations learning, deep learning systems, visualization and interpretation techniques and reinforcement learning systems [8, 9]. They play critical role in making FinTech intelligent by describing, modelling and optimizing the intrinsic patterns, trends, sequential movements, structures and relations in financial systems [14–16, 20, 22, 42].

Stock market forecasting has been recently drawing a lot of attention from researchers. It is a challenging task, mainly due to dynamic, high-noise, non-parametric and non-linear properties of stock data [13]. There are numerous methods that can be applied to deal with stock trend prediction problem, from conventional statistical methods such as regression [30] and classification [7] to machine learning techniques [46]. The last ones have been introduced as a consequence of the disruption of AI to handle comprehensive capital market data and predict the future trends of stock. Artificial neural networks are one of the most commonly used algorithms that prove to be effective in finding the non-additive and non-linear relations in data [25, 49]. They have been applied for automatic feature extraction from raw financial data and making predictions of prices movements and trading strategies. Among many deep learning (DL) algorithms (such as recurrent neural network, long short-term memory network, restricted Boltzmann machine, autoencoder or deep multilayer perceptron) utilized for this purpose, convolutional neural networks (CNN) are those that have a promising potential, demonstrating great performance in predicting stock trend [6, 43].

CNN is one of the most popular DL networks with a multilayer neural structure, solving the problem that the traditional deep network parameters are hard to train [18]. It is based on concepts of shared weights and local receptive fields, therefore greatly reduces the number of network parameters [50]. CNN has a powerful pattern recognition ability, therefore it is widely used to perform image classification tasks and image recognition problems. These patterns are extracted by CNN's directly form image datasets, which makes CNN very useful [3]. Relying on the above-mentioned characteristics of CNN only several researchers began to transform stock market data into images and then applying CNN for future trends prediction [18, 21]. Therefore, with this work authors seek to apply CNN for stock market prediction domain. We develop our framework based on CNN model that supports prediction of stock volume changes in the value of stock exchange transaction volumes. Moreover, the authors intend to compare the quality measures of the generated predictive model for different types of stock charts.

2 Related Works

Among previously indicated DL methods, CNN has been recently introduced to handle the stock trend prediction problem. As stated by many researchers [28, 44] CNN produces better learning capabilities and has powerful patter recognition ability. Although CNN is getting popularity in capital market forecasting domain, its application is still rare. So far only a few studies have examined CNN potential for stock market forecasting [13, 18, 19, 21, 25, 45, 50]. They explored the use of CNNs to predict stock prices and volumes. CNNs have been applied to various financial data, including stock prices, trading volumes, and news articles. Based on these experiments, CNNs prove to dramatically improve the prediction accuracy over the expected performance.

A comprehensive and detailed analysis of 32 research works that applied artificial neural networks for stock price/volume prediction, including also CNN has been presented by Htun et al. [26]. Their work provides an overview of various CNN-based models that have been proposed for stock price prediction. The authors review several studies that use CNNs to analyze stock chart images, including experiments that predict trading volume. Thakkar and Chaudhari presented another intriguing study that examines the importance and necessity of deep neural networks in the realm of stock price and trend forecasting [47].

Applying CNNs to predict trading volume changes based on stock charts is recently very attractive field of research. For example, one study proposed a CNN-based model that takes historical stock chart images as input and predicts the trading volume for the next day [50]. The model achieved a prediction accuracy of 70% for the next day's trading volume. Overall, the study suggests that CNNs can effectively predict changes in trading volume based on stock charts, but further research is needed to improve the accuracy and generalizability of these models.

Some authors proposed an interesting approach, combining CNN and LSTM (long short-term memory) neural networks. The study by Lu et al. [37] proposes a hybrid model that links CNNs and LSTM to predict stock prices. The model uses historical stock price data and technical indicators as input features, achieving a prediction accuracy of around 70%. Another study by Orsel and Yamada [39] compares the performance of CNNs and LSTM models for stock price prediction. The authors find that CNNs are better at capturing spatial patterns in stock chart images, while LSTM models are better at modelling temporal dependencies in stock price time series data. On the other hand, the study by Liang et al. [36] proposes a CNN-based model that uses wavelet transforms to extract features from stock price data. The model achieves a prediction accuracy of around 75% for the next day's closing price. These studies demonstrate the diverse ways in which CNNs can be applied to stock price/volume prediction and suggest that CNN-based models have the potential to outperform traditional machine learning techniques for these tasks.

Another interesting field of research has been dedicated to exploration of the quality of different types of stock charts for prediction accuracy, using machine learning models, including CNNs. Usmani et al. [48] note that the quality of the input data (including stock charts) is critical to achieving accurate predictions. Moreover, the study by Kim and Kim [33] compared the performance of different types of stock chart images (line charts, bar charts, and candlestick charts) for predicting stock prices using CNNs. The study found that candlestick charts outperformed line and bar charts, achieving a prediction accuracy of 78.36% compared to 66.76% and 68.48%, respectively. Another work [31] covered the issue of the impact of different charting parameters (such as chart type, timeframe, and technical indicators) on the prediction accuracy of a machine learning model. The study revealed that chart type (candlestick charts outperforming line and bar charts) and technical indicators (moving averages and relative strength index) had the most significant impact on prediction accuracy. Moreover, according to some reported experiments, various technical indicators (such as moving averages and stochastic oscillators) have the impact on the accuracy of stock price prediction models. The study by Lee et al. [35] proved that some technical indicators (such as Bollinger Bands and the

Moving Average Convergence/ Divergence) were more effective when combined with candlestick charts as input.

In addition to these studies, some researchers have also explored the effect of chart quality on human traders' decision-making. For example, a study [27] found that traders were more likely to make profitable trades when using candlestick charts than line charts. Overall, these studies suggest that the quality of stock charts can significantly impact the accuracy of machine learning models and the decision-making of human traders. Candlestick charts are often considered the most informative chart type due to their ability to capture detailed price and volume information, but the impact of other charting parameters should also be considered. The reviewed studies demonstrate the importance of considering the quality of stock charts when using machine learning models for stock price prediction or technical analysis. Different chart types and charting parameters can significantly impact the accuracy of these models; therefore researchers continue to explore the most effective ways to use stock chart images for predictive purposes.

Different machine learning techniques, including CNNs, Support vector regression (SVR), and DNNs, have been used to predict trading volume levels with varying degrees of success. The issue that has been covered in the relevant literature is the performance of various machine learning models for predicting stock trading volume levels using both numerical data and stock chart images as input. The work by [23] presented the use of candlestick charts as input for SVR model to predict stock trading volume levels. It has been revealed that using candlestick charts improved the accuracy of the SVR model compared to using only numerical data. Another study [34] proposed a deep neural network-based model for predicting trading volume levels based on both numerical data and candlestick chart images. The model achieved a prediction accuracy of around 92%, outperforming other machine learning models that used only numerical data. Similar conclusions have been drawn by Qian and Gao [41] and Bansal et al. [5]. The first mentioned work introduced a DL model for predicting stock trading volume levels based on time-series data and technical indicators (such as moving averages and relative strength index) as well as stock chart images. Another experiment by Bansal et al. [5] benchmarked the performance of various machine learning models, including decision trees, k-nearest neighbors, and SVMs for predicting trading volume levels. Both authors reached the conclusion that the proposed models outperform other machine learning models that solely relied on numerical data as input features. Therefore, DNN-based models seem to have great potentialities to model the noisy and complex financial data.

Most discussed above experiments highlight the potential benefits of using stock chart images (particularly candlestick charts) as input for machine learning models that predict trading volume levels. Researchers have developed models that can achieve high prediction accuracy by combining different types of input data (such as time-series data, technical indicators, and stock chart images). Previous studies have proven the potential of various DL algorithms, including CNN to forecast and study stock market trends. Therefore, with this study we are going to introduce a new framework based on CNN, that supports prediction of stock volume changes in the capital markets.

3 Methodology of the Research Experiment

The first step of the research procedure was to download stock prices for the year 2022 for the 20 most traded companies on NASDAQ. This particular year was characterized by high volatility and, unlike previous years, the absence of a dominant trend, such as a sustained upward trend typically observed (excluding the Covid episode) over a span of several years in the USA. The companies considered for analysis were: AAPL, AMZN, MSFT, GOOGL, META, TSLA, NVDA, INTC, CMCSA, CSCO, ADBE, NFLX, PYPL, PEP, ZM, MRNA, ASML, BIDU, TMUS, and AMGN. In total, the dataset subject to analysis comprised 5,020 quotations. The labels for individual charts were generated based on the change in the average turnover over the past 5 days in relation to the turnover over a consecutive 5-day period, as standard stock market trading week. As illustrated in Fig. 1, the distribution of turnover changes can be compared to a normal distribution.

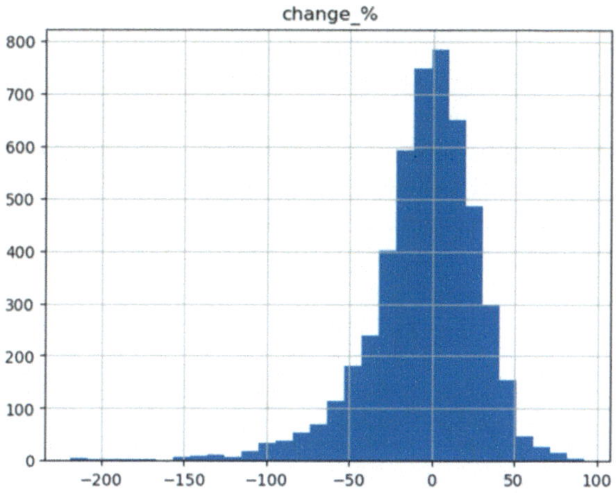

Fig. 1. Data distribution of percentage change in turnover

In order to divide the data into classes, it was decided to label individual quotations with labels A, B, C, D, E (a part of dataset is shown on Table 1) according to the following ranges: A: up to –23% (1140 quotations), from B: –23% to –6% (997), C: from –6% to 6% (955), D: from 6% to 23% (1043), E: over 23% (885). The purpose of such a division was to obtain the frequency for each of the labels in a similar proportion not differing by more than 30%.

Table 1. A set of data on the basis of which the analysis was carried out

Ticker	Date	Volume	last_5_days_avg	next_5_days_avg	change_%	change_class
AAPL	2022-01-03	104487900	73963280.0	96845340.0	23.63	E
	2022-01-04	99310400	77996500.0	92210920.0	15.42	D
	2022-01-05	94537600	84434240.0	88264440.0	4.34	C
	2022-01-06	96904000	91860440.0	85784800.0	-7.08	B
	2022-01-07	86709100	96389800.0	84531140.0	-14.03	B
...	...	...	...	...	...	...

In the subsequent step, two sets of graphics consisting of 5,020 images were generated: the first set comprised candlestick charts, while the second set consisted of Point-and-Figure charts. Each set of graphics was divided into groups based on classes. These two sets of data were used as input for an algorithm utilizing a convolutional neural network. The objective of this algorithm was to create two classifiers: one for the analysis of candlestick charts and the other for the analysis of Point-and-Figure charts (refer to Fig. 2). The charts encompassed a period of 92 trading days.

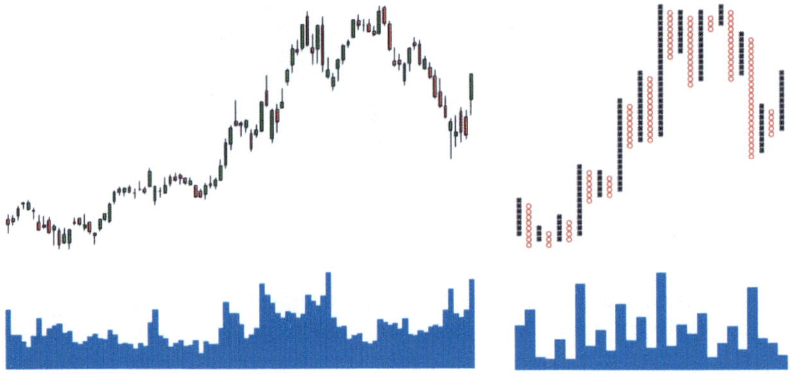

Fig. 2. An example of a candlestick chart and Point-and-Figure chart for the AAPL share price as of 2022-01-03.

The data was divided into training and validation in a ratio of 0.2 to 0.8.

The model itself was configured by code:

```
model = Sequential([
  layers.Rescaling(1./255, input_shape=(img_height, img_width, 3)),
  layers.Conv2D(32, 3, padding=padding, activation='relu'),
  layers.MaxPooling2D((2,2)),
  layers.Conv2D(32, 3, padding=padding, activation='relu'),
  layers.MaxPooling2D((4,4)),
  layers.Conv2D(16, 5, padding=padding, activation='relu'),
  layers.MaxPooling2D((4,4)),
  layers.Conv2D(4, 4, padding=padding, activation='sigmoid'),
  layers.MaxPooling2D((3,3)),
  layers.Flatten(),
  layers.Dense(2, activation='relu'),
  layers.Dense(num_classes)])
```

4 Results

The training of each model was run for 200 epochs. It turned out that the model that analyzed graphics with candlestick charts obtained accuracy at the level of 0.7388 (Fig. 3), while the model for Point-and-Figure charts at the level of 0.8020 (Fig. 4).

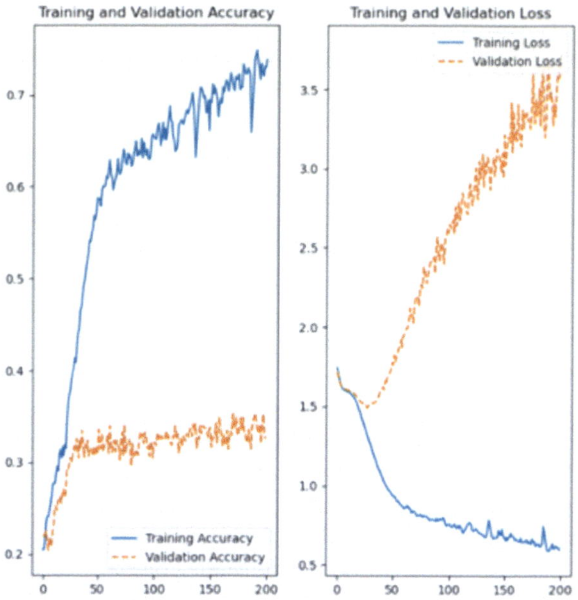

Fig. 3. Training and validation accuracy and loss for bar chart

As it can be observed from the charts depicting the learning performance in relation to the training data, it is evident that the model trained using data derived from Point-and-Figure charts exhibited an improvement of over 6% compared to the classifier trained

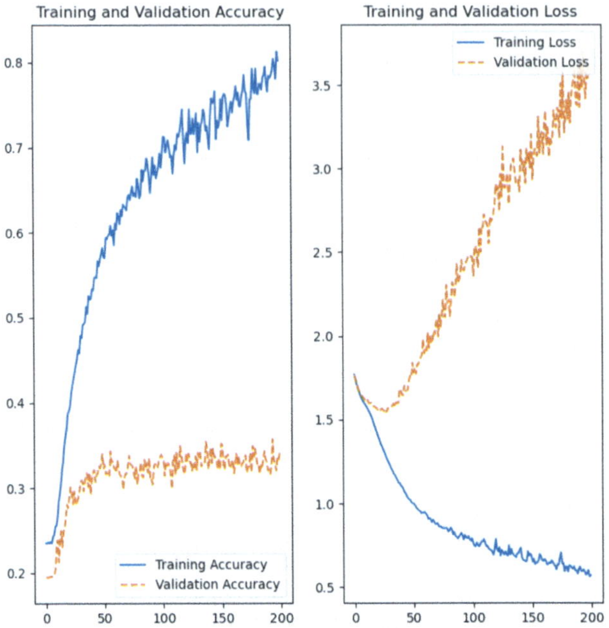

Fig. 4. Training and validation accuracy and loss for Point-and-Figure chart

using candlestick charts. This outcome may be somewhat unexpected, considering the widespread popularity of candlestick charts for analyzing stock market data.

In the authors' perspective, the obtained results for two types of charts, namely candlestick charts and Point-and-Figure charts, which were used and generated from real stock market data for the purpose of the experiment, provide a solid foundation for constructing a model that operates with convolutional neural networks to analyze stock charts. The experiment's findings demonstrated that, in accordance with the adopted assumptions, convolutional neural networks can be employed to develop solutions for processing stock market data and, specifically, for highlighting differences between various types of charts. While the authors' approach to comparing the usefulness of different stock charts has not been previously presented in other studies, it is important to acknowledge that utilizing an image in the form of a stock chart for analyzing stock price behavior or other financial instruments, adhering to the notion that "a picture is worth a thousand words," holds promise for creating targeted predictive models based on convolutional neural networks. Such models could facilitate more accurate predictions concerning stock price behavior, particularly in the context of volume analysis.

Considering that stock charts often predominantly consist of curves depicting stock price fluctuations, the experiment revealed that convolutional neural networks, as highlighted in the literature review section, undoubtedly possess the capability to capture crucial nuances depicted in the graphs, which are relevant from a problem-solving perspective. Naturally, the question arises as to whether there are artifacts (groups of pixels) that can indicate subsequent volume behavior. The study, however, provided a positive response to this question, as it demonstrated differences between two distinct chart types that present the same underlying data.

5 Conclusion

The conducted research experiment revealed differences depending on the type of chart and the accuracy of the trained convolutional neural network in analyzing charts that symbolically reflect stock market data. Therefore, it can be inferred that studying different types of stock charts is advisable in terms of their importance for analyzing trading volumes. Hence, the conducted experiment should be considered as the initial stage of extended research on the utilization of convolutional neural networks for analyzing various stock charts across various aspects of stock market data analysis. Undoubtedly, the model training as observed through the validation data could be improved. According to the authors, this aspect provides an opportunity for further research concerning the fine-tuning of the presented model. Further research should aim to address questions such as the specific types of charts to analyze and the specific stock market data to consider.

It is important to note that this study provides a base to delve deeper into the selection of chart types and the information to be included on the charts. That would enable the convolutional neural network to be effectively trained and add value to the investment decision making process, including predictions and volume analysis. Naturally, further research will be necessary to refine the data selection, chart types, and the intricate design of the convolutional neural network structure. Additionally, exploring the potential use of fundamental data available for selected stocks or financial instruments and their depiction on charts should be considered, as it could enhance the training process of the convolutional neural network and yield even better results.

In subsequent research, the computational capacity of the system used to train the neural network may present limitations. However, in the era of accessible and highly scalable cloud solutions, this challenge can be overcome. Although this preliminary study represents one of the first attempts to apply such an unconventional approach, there is undoubtedly room for improvement in the obtained results. Overall, the promising outcomes of this study warrant further exploration and more extensive research aimed at analyzing the potential use of stock charts as a foundation for constructing a system or methodology that leverages convolutional neural networks to support more accurate investment decisions and the analysis of stock market data, with a particular emphasis on volume analysis.

References

1. Abe, M., Nakayama, H.: deep learning for forecasting stock returns in the cross-section. In: Phung, D., Tseng, V.S., Webb, G.I, Ho, B., Ganji, M., Rashidi, L. (eds.) Advances in Knowledge Discovery and Data Mining. 22nd Pacific-Asia Conference, PAKDD 2018 Melbourne, VIC, Australia, June 3–6, 2018 Proceedings, Part I. pp. 273–284. Springer, Cham (2021). https://doi.org/10.48550/arXiv.1801.01777
2. Ahn, J.J., Lee, S.J., Oh, K.J., Kim, T.Y.: Intelligent forecasting for financial time series subject to structural changes. Intell. Data Anal. **13**(1), 151–163 (2009). https://doi.org/10.3233/IDA-2009-0360

3. Alamri, N.M.H., Packianather, M., Bigot, S.: Deep learning: parameter optimization using proposed novel hybrid bees Bayesian convolutional neural network. Appl. Artif. Intell. **36**(1), 1720–1745 (2022). https://doi.org/10.1080/08839514.2022.2031815
4. Arslanian, H., Fischer, F.: The Future of Finance: The Impact of FinTech, AI, and Crypto on Financial Services. Palgrave Macmillan (2019). https://doi.org/10.1007/978-3-030-14533-0
5. Bansal, M., Goyal, A., Choudhary, A.: A comparative analysis of k-nearest neighbour, genetic, support vector machine, decision tree, and long short-term memory algorithms in machine learning. Decis. Anal. J. **3**, 100071 (2022). https://doi.org/10.1016/j.dajour.2022.100071
6. Barra, S., Carta, S.M., Corriga, A., Podda, A.S., Recupero, D.R.: Deep learning and time series-to-image encoding for financial forecasting. IEEE/CAA J. Autom. Sin. **7**(3), 683–692 (2020). https://doi.org/10.1109/JAS.2020.1003132
7. Cagliero, L., Garza, P., Attanasio, G., Baralis, E.: Training ensembles of faceted classification models for quantitative stock trading. Computing **102**(5), 1213–1225 (2020). https://doi.org/10.1007/s00607-019-00776-7
8. Cao, L., Data Science Thinking: The Next Scientific, Technological and Economic Revolution. Springer (2018). https://doi.org/10.1007/978-3-319-95092-1
9. Cao, L., Yang, Q., Yu, P.S.: Data science and AI in Fintech: an overview. Int. J. Data Sci. Anal. **12**(2), 81–99 (2021). https://doi.org/10.1007/s41060-021-00278-w
10. Cao, L.: AI in finance: a review. SSRN 1–36 (2020). https://doi.org/10.2139/ssrn.3647625
11. Cao, L.: AI in finance: challenges, techniques, and opportunities. ACM Comput. Surv. **55**(3), 1–38 (2022). https://doi.org/10.1145/3502289
12. Cao, W., Zhu, W., Demazeau, Y.: Multi-layer coupled hidden Markov model for cross-market behavior analysis and trend forecasting. IEEE Access. **7**, 158563–158574 (2019). https://doi.org/10.1109/ACCESS.2019.2950437
13. Chen, W., Jiang, M., Zhang, W.G., Chen, Z.: A novel graph convolutional feature based convolutional neural network for stock trend prediction. Inf. Sci. **556**, 67–94 (2021). https://doi.org/10.1016/j.ins.2020.12.068
14. Chen, Y., Hao, Y.: A novel framework for stock trading signals forecasting. Soft. Comput. **24**, 12111–12130 (2020). https://doi.org/10.1007/s00500-019-04650-8
15. Chong, E., Han, C., Park, F.C.: Deep learning networks for stock market analysis and prediction: methodology, data representations, and case studies. Expert Syst. Appl. **83**, 187–205 (2017). https://doi.org/10.1016/j.eswa.2017.04.030
16. Dami, S., Esterabi, M.: Predicting stock returns of Tehran exchange using LSTM neural network and feature engineering technique. Multimed. Tools Appl. **80**, 19947–19970 (2021). https://doi.org/10.1007/s11042-021-10778-3
17. Dhar, V., Stein, R.M.: FinTech platforms and strategy. Commun. ACM **60**(10), 32–35 (2017). https://doi.org/10.1145/3132726
18. Di Persio, L., Honchar, O.: Artificial neural networks architectures for stock price prediction: comparisons and applications. Int. J. Circ. Syst. Signal Process. **10**, 403–413 (2016)
19. Dinh, T.A., Kwon, Y.K.: An empirical study on importance of modeling parameters and trading volume-based features in daily stock trading using neural networks. Informatics **5**(3), 36 (2018). https://doi.org/10.3390/informatics5030036
20. Farahani, M.S., Hajiagha, S.H.R.: Forecasting stock price using integrated artificial neural network and metaheuristic algorithms compared to time series models. Soft. Comput. **25**(13), 8483–8513 (2021). https://doi.org/10.1007/s00500-021-05775-5
21. Gunduz, H., Yaslan, Y., Cataltepe, Z.: Intraday prediction of borsa istanbul using convolutional neural networks and feature correlations. Knowl.-Based Syst. **137**, 138–148 (2017). https://doi.org/10.1016/j.knosys.2017.09.023
22. Haq, A.U., Zeb, A., Lei, Z., Zhang, D.: Forecasting daily stock trend using multi-filter feature selection and deep learning. Expert Syst. Appl. **168**, 114444 (2021). https://doi.org/10.1016/j.eswa.2020.114444

23. Henrique, B.M., Sobreiro, V.A., Kimura, H.: Stock price prediction using support vector regression on daily and up to the minute prices. J. Finance Data Sci. **4**(3), 183–201 (2018). https://doi.org/10.1016/j.jfds.2018.04.003
24. Hilpisch, Y.: Artificial Intelligence in Finance. O'Reilly Media (2020)
25. Hoseinzade, E., Haratizadeh, S.: CNNpred: CNN-based stock market prediction using a diverse set of variables. Expert Syst. Appl. **129**, 273–285 (2019). https://doi.org/10.1016/j.eswa.2019.03.029
26. Htun, H.H., Biehl, M., Petkov, N.: Survey of Feature selection and extraction techniques for stock market prediction. Finance Innov. **9**, 26 (2023). https://doi.org/10.1186/s40854-022-00441-7
27. Hu, G., Hu, Y., Yang, K., Yu, Z., Sung, F., Zhang, Z., et al.: Deep stock representation learning: from candlestick charts to investment decisions. In: 2018 IEEE International Conference on Acoustics, Speech, and Signal Processing (ICASSP), pp. 2706–2710. IEEE, Calgary (2018). https://doi.org/10.1109/ICASSP.2018.8462215
28. Hu, Z., Zhao, Y., Khushi, M.: A survey of forex and stock price prediction using deep learning. Appl. Syst. Innov. **4**, 1 (2021). https://doi.org/10.3390/asi4010009
29. Huang, W., Lai, K.K., Nakamori, Y., Wang, S.: Forecasting foreign exchange rates with artificial neural networks: a review. Int. J. Inf. Technol. Decis. Mak. **3**, 145–165 (2004). https://doi.org/10.1142/S0219622004000969
30. Jiang, M., Jia, L., Chen, Z., Chen, W.: The Two-stage machine learning ensemble models for stock price prediction by combining mode decomposition extreme learning machine and improved harmony search algorithm. Ann. Oper. Res. **309**, 553–585 (2022). https://doi.org/10.1007/s10479-020-03690-w
31. Jin, G., Kwon, O.: Impact of chart image characteristics on stock price prediction with a convolutional neural network. PLOS ONE, **16**(6), e0253121 (2021). https://doi.org/10.1371/journal.pone.0253121
32. Juszczuk, P., Kozak, J., Kania. K.: Using similarity measures in prediction of changes in financial market stream data – experimental approach. Data Knowl. Eng. **125,** 101782 (2020). https://doi.org/10.1016/j.datak.2019.101782
33. Kim, T., Kim, H.Y.: Forecasting stock prices with a feature fusion LSTM-CNN model using different representations of the same data. PLOS ONE. **14(2)**, e0212320 (2019). https://doi.org/10.1371/journal.pone.0212320
34. Kusuma, R.M.I., Ho, T.T., Kao, W.C., Ou, Y.Y., Hua, K.L.: Using deep learning neural networks and candlestick chart representation to predict stock market. arXiv:1903.12258 [q-fin.GN], pp. 1–13 (2019). https://doi.org/10.48550/arXiv.1903.12258
35. Lee, M.C., Chang, J.W., Hung, J.C., Chen, B.L.: Exploring the effectiveness of deep neural networks with technical analysis applied to stock market prediction. Comput. Sci. Inf. Syst. **18**(2), 401–418 (2021). https://doi.org/10.2298/CSIS200301002L
36. Liang, X., Ge, Z., Sun, L., He, M., Chen, H.: LSTM with wavelet transform based data preprocessing for stock price prediction. Math. Probl. Eng. 1340174, 1–8 (2019). https://doi.org/10.1155/2019/1340174
37. Lu, W., Li, J., Li, Y., Sun, A., Wang, J.: A CNN-LSTM-based model to forecast stock prices. Complexity, 6622927, 1–10 (2020). https://doi.org/10.1155/2020/6622927
38. Lynn, T., Mooney, J.G., Rosati, R., Cummins, M.: Disrupting finance: fintech and strategy in the 21st century. Palgrave Pivot (2019). https://doi.org/10.1007/978-3-030-02330-0
39. Orsel, O.E., Yamada, S.S.: Comparative study of machine learning models for stock price prediction. arXiv:2202.03156 [q-fin.ST], pp. 1–6 (2022). https://doi.org/10.48550/arXiv.2202.03156
40. Qi, Y., Xiao, J.: Fintech: AI powers financial services to improve people's lives. Commun. ACM **61**(11), 65–69 (2018). https://doi.org/10.1145/3239550

41. Qian, X.Y., Gao, S.: Financial Series Prediction: Comparison Between Precision of Time Series Models and Machine Learning Methods. arXiv:1706.00948, pp. 1–9 (2017). https://doi.org/10.48550/arXiv.1706.00948
42. Selvin, S., Vinayakumar, R., Gopalakrishnan, E.A., Menon, V.K., Soman, K.P.: Stock price prediction using LSTM, RNN and CNN-sliding window model. In: 2017 International Conference on Advances in Computing, Communications, and Informatics (ICACCI), pp. 1643–1647. IEEE Press (2018). https://doi.org/10.1109/ICACCI.2017.8126078
43. Sezer, O.B., Ozbayoglu, A.M.: Algorithmic financial trading with deep convolutional neural networks: time series to image conversion approach. Appl. Soft Comput. **70**, 525–538 (2018). https://doi.org/10.1016/j.asoc.2018.04.024
44. Singh, A.K., Ganapathysubramanian, B., Sarkar, S., Singh., A.: Deep learning for plant stress phenotyping: trends and future perspectives. Trends Plant Sci. **23**(10), 883–898 (2018). https://doi.org/10.1016/j.tplants.2018.07.004
45. Siripurapu, A.: Convolutional networks for stock trading. Comput. Sci. 1–6 (2015). http://vision.stanford.edu/teaching/cs231n/reports/2015/pdfs/ashwin_final_paper.pdf
46. Tang, H., Dong, P., Shi, Y.: A new approach of integrating piecewise linear representation and weighted support vector machine for forecasting stock turning points. Appl. Soft Comput. **78**, 685–696 (2019). https://doi.org/10.1016/j.asoc.2019.02.039
47. Thakkar, A., Chaudhari, K.: A Comprehensive survey on deep neural networks for stock market: the need, challenges, and future directions. Expert Syst. Appl. **177**, 114800 (2021). https://doi.org/10.1016/j.eswa.2021.114800
48. Usmani, M., Adil, S.H., Raza, K., Ali, S.S.A.: Stock market prediction using machine learning techniques. In: 2016 3rd International Conference on Computer and Information Sciences (ICCOINS), pp. 322–327. IEEE, Kuala Lumpur (2016). https://doi.org/10.1109/ICCOINS.2016.7783235
49. Wang, J., Wang, J.: Forecasting stock market indexes using principal component analysis and stochastic time effective neural networks. Neurocomputing **156**, 68–78 (2015). https://doi.org/10.1016/j.neucom.2014.12.084
50. Wu, J.M.T., Li, Z., Herencsar, N., Vo, B., Lin, J.C.W.: A graph-based CNN-LSTM stock price prediction algorithm with leading indicators. Multimed. Syst. **29**, 1751–1770 (2023). https://doi.org/10.1007/s00530-021-00758-w

Forecasting Fuel Retail Sales Volume Using Machine Learning for Sustainable Decision-Making

Tomasz Zema(✉), Filip Wójcik, Adam Sulich, and Marcin Hernes

Wroclaw University of Economics and Business, Ul. Komandorska 118/120, 53-345 Wroclaw, Poland

tomas2.zema@ue.wroc.pl

Abstract. *Purpose*: Forecasting fuel retail sales volume is important for sustainable decision-making and process management. Prediction of the petrol products sales volume influences the planning of the effective distribution of fuels. Therefore, it impacts all other supply chains and each country's economy. *Method*: This research uses machine learning to forecast fuel retail sales volume. The choice of this method and the research subject are based on the complexity of the decision-making process in energy carriers' sales. Machine learning model (LightGBM) and two other state-of-the-art models (ARIMA, Holt-Winters) were compared. Performed calculations are based on real-world data from single fuel company distributors. *Results*: The three machine learning models are proposed and examined to address this research gap. There is a baseline model proposed along with state-of-the-art models and the results are based on comparative analysis of MSE and RMSE. Finally, a real-world sales data set from a fuel dispenser is analyzed to determine the veracity of the model argument. *Conclusion*: The best results were obtained by the LightGBM model, utilizing both future covariates and forecasted fuel prices. This model demonstrated the best performance according to all metrics. Performed experiments proved that the best model can support strategic reasoning in organizations interested in forecasting fuel sales. The use of machine learning to forecast retail fuel sales volume for sustainable decision-making is a good example of leveraging modern technology to optimize business processes and make more informed decisions.

Keywords: Sales · Prediction · LightGBM model · Randomforest · Fuel

1 Introduction

Balanced and sustainable decision-making is often opposed to temporary or day-to-day management processes in the unstable business environment to present its characteristics [1]. Unlike ongoing and day-to-day decision-making, sustainable decision-making requires a reliable plan, strategy and models to forecast its outcomes [2]. Sustainable decision-making aims to overcome business instability and offers businesses a sustainable competitive advantage relying on the multi-perspective models [1]. Although

M. Hernes and J. Wątróbski (Eds.): ECAI 2023, LNNS 1079, pp. 109–120, 2024.
https://doi.org/10.1007/978-3-031-66761-9_10

sustainable decision-making does not exclude the environmental component or sustainability concept, it focuses more on the longitudinal approach, strategic approach and mechanism design for multi-perspective models. In this context, the energy and fuel sector are an example of the most challenging sector of the economy. There are environmental issues related to the energy sector transformation and political tensions revolving around energy security, altogether with social expectations related to the fuels' accessibility [3]. Therefore, sustainable decision-making regarding the retail sale of fuels requires processing large amounts of data, covering different variables influencing the fuels' market. One of the most significant problems in this area is fuel retail sales volume forecasting [4], and usually, this is addressed in the scientific literature by combination of classic statistical and machine learning methods. However, the outcomes of those forecasting procedure types depend on the subject of analysis and its accuracy. The forecasting procedures accuracy is measured by MSE (Mean Square Error) and MAPE (Mean Absolute Percentage Error) to check fuel retail sales volume prediction.

Accurate forecasting allows for more sustainable supply chain management and sustainable fuel stocks decision-making. Thanks to the fuels retail sales forecasting, it is possible to reduce the Forrester effect, which is unexpected or unpredicted demand distortion that travels upstream in the supply chain [5]. Scientific and business literature combining sustainable decision-making in fuel retail sales volume are also scarce. However, the existing researches are primarily concerned with collectively forecasting volumes and prices in the fuel market. The classical relation between fuel price and retail sales volume has rarely been addressed in scientific publications. However, among those publications, classic statistical and machine learning methods are used. Even though the fuel market plays a pivotal role in the modern economy [6], only a few scientific studies forecast retail fuel sales volume.

This research aims to develop approaches for forecasting the volume of retail sales of fuels. The main contributions of this research are as follows: (I) developing classical statistical approaches for forecasting retail sales of fuels, (II) developing machine learning approaches for forecasting retail sales of fuels, (III) comparing and analyzing developed approaches, and (IV) extending the existing literature on the subject. These scientific research results can inspire other researchers and business practitioners looking for the experiments' results and want to improve their fuel retail sales volume forecasting models.

This paper is organized into five sections. Besides this introduction, the scientific background is discussed in the second section. Next the pre-processing of the input data and methods are presented. The next part of the paper presents the results of the research experiments. The last sections are discussion and conclusions, addressing possible future research directions.

2 Literature Background

Sustainable decision-making is a consequence of the adopted organization management style [1], organizational routines [7] and culture [8, 9]. Energy sector organizations understand that sustainability is more comprehensive and can be supported by sustainable decision-making as environmental challenges become more urgent. This creates an incentive for implementing comprehensive intelligent, data-driven decision models that support the interactive evaluation of decision alternatives required by the business environment [10].

Sales forecasting volume can use statistical methods, historical data, and market analysis. This involves identifying trends and patterns in historical sales data and utilizing these insights to make accurate projections regarding future sales volumes, trends, and other relevant metrics [11]. Sales forecasting is a crucial element of business planning because it enables businesses to set realistic goals and make informed decisions regarding resource allocation, marketing strategies, and other crucial aspects of their operations. Incorporating external factors such as economic indicators, competitor analysis, and customer behavior can improve sales forecast accuracy. This is a canonical problem with a fairly classic and well-established spectrum of methods, which seems to have stagnated prior to the introduction of deep learning-based methods. Recently, a new machine learning technique has been implemented, which supports the explanation of cutting-edge black-box prediction models [12]. However, non-transparent black-box machine learning (ML) models such as random forests, boosting, support vector machines (SVM), and neural networks achieve significantly better predictive performance than simple, interpretative models such as decision trees, naive Bayes, or decision rules [13]. This is one explanation for the low utilization and acceptability of predictive ML models in sectors requiring transparency and clearness of choices. Dietterich [14] evaluates a number of well-established ML techniques for sequential data prediction on a variety of time series. The idea of ML techniques on sequential data expand to Corporate Bankruptcy Prediction [15] and use of Recurrent Neural Networks (RNN) [16]. Gamboa offers a recent overview of deep learning for time-series forecasting [17]. Due to their effectiveness with nonlinear and high-frequency time series, neural networks have gained popularity in regression models.

The petroleum retail industry is a classic example of price and demand fluctuations caused by a variety of factors, such as shifts in global oil prices due to political unrest in oil-producing nations and currency devaluation. Retail gasoline (gasoline) stations alter gasoline prices to remain competitive [18]. When one gas station lowers its price, another gas station responds by lowering its price even further, resulting in a cycle of price reductions that continues until the profit breaks even. At this point, a gas station may "relent," after which all other gas stations follow suit, and the cycle begins again [19]. Therefore, accurate fuel station sales forecasting is essential for station owners to make well-informed business decisions.

A fuel consumption forecasting system was developed in the early 2000s to predict fuel consumption at Petronas petrol stations in Malaysia [20]. This study utilized a rule-based expert system consisting of two logical components: a knowledge base and an inference engine. The results were satisfactory, but results from the testing showed the flaws of this method. Fuzzy linear regression tree [21] was introduced in problem of forecasting sale volume of petroleum products. In the study from 2012 The sales forecasting methodology presented in this study utilizes an adaptive fuzzy modeling technique (fuzzy linear regression trees). With a provided training dataset, the algorithm progressively assesses whether a fuzzy linear regression tree, which is a weighted combination of several local linear regression models, can outperform the initial global linear model [22]. Another research from 2018 focused on forecasting petrol products primarily investigates various approaches, using clustering and classical decision tree. However, these methods tend to identify patterns in forecasting schemes through clustering, rather than accurately predicting specific sales volumes [23].

3 Method

This method section, organized into several subchapters, comprehensively describes the different forecasting methods utilized for time-series analysis. These methods include ARIMA, Holt-Winters, and LightGBM. Each subchapter provides in-depth detail about each method's characteristics and presents specific examples of their applications in various fields. First subchapter describes the data collected from fuel distributors, the process of transforming raw data into a usable format, and the creation of a data pipeline for a machine learning model. Fuel price data as additional covariates subchapter explains the use of historical fuel prices as additional covariates in the model and the method of using percentage price differences for more effective forecasting. Fuel price differences prediction section focuses on the separate forecasting experiment conducted to predict price changes and prevent data leakage, comparing the performance of LightGBM, ARIMA, and NBeats algorithms. The final section discusses the training of the LightGBM model on historical refueling data, outlining the model's use of future covariates and forecasted fuel prices.

ARIMA is a versatile and expressive model that combines autoregressive (AR), differencing (I), and moving average (MA) components to represent time series. The AR component captures temporal dependence on past values, the MA component captures dependence on past white noise error terms, and the I component represents differencing to transform non-stationary time series into stationary ones [24]. ARIMA has been widely used in diverse fields, such as forecasting electricity demand [25] and stock prices [25–27], and can be combined with advanced methods like Deep Learning for tasks such as high-frequency financial time-series forecasting [28].

Holt-Winters models are a class of exponential smoothing methods that are widely used in time series forecasting. The Holt-Winters models are named after Charles C. Holt and Peter G. Winters, who introduced the method in their seminal paper [29]. The Holt-Winters models are a type of exponential smoothing method that uses three smoothing parameters to capture the level, trend, and seasonality of a time series. The basic idea behind exponential smoothing is to give more weight to recent observations and less weight to older observations. This method has been widely used in forecasting applications in various fields, for example, to forecast demand for products, sales of consumer goods [30], and energy consumption [31, 32].

Machine learning method utilizing Light Gradient Boosting (LightGBM), as it was proven to be effective and robust in tabular data modeling with the possibility to handle time series forecasting after proper data preprocessing. LightGBM is a gradient-boosting framework that uses a tree-based learning algorithm for classification and regression tasks, similar to XGBoost [33]. It was developed by Microsoft Research and was introduced in 2017 [34]. The main principle of LightGBM is to build decision tree leaf-wise rather than level-wise. This approach can lead to faster training times and better accuracy than other gradient-boosting frameworks. LightGBM also uses a technique called gradient-based one-sided sampling (GOSS) to reduce the number of samples used in the training process while maintaining [35] accuracy [34]. These principles are effective in various applications, including stock price predictions [35], cryptocurrency price trend forecasting [36], anomaly (malware) detection [37], or machine-learning-assisted medical diagnosis [38]. The experiment aimed to accurately predict refueling volume in liters on a predefined period, according to mean squared error (MSE) and mean absolute percentage error (MAPE) metrics.

3.1 The Data

This section describes the general meta-data knowledge discovery of company-collected data. In subsequent sections of this paper were discussed the methodologies and potential applications of data mining on time series collected from fuel distributors.

The data to be described is from one of the company's stations. Taking advantage of recent developments in information technology, this fuel distributor has modernized its business operations at the store level by connecting to PC-based servers via a specific interface, and at the corporate level by implementing highly scalable parallel processing architectures. Figure 1 describes generally the data pipeline for a machine learning model using gas distributor data involves several steps: raw data collection, data warehousing, data export, data transformation, and the machine learning model itself. These steps work together to convert raw data into meaningful insights and predictions, thus enabling better decision-making and process management.

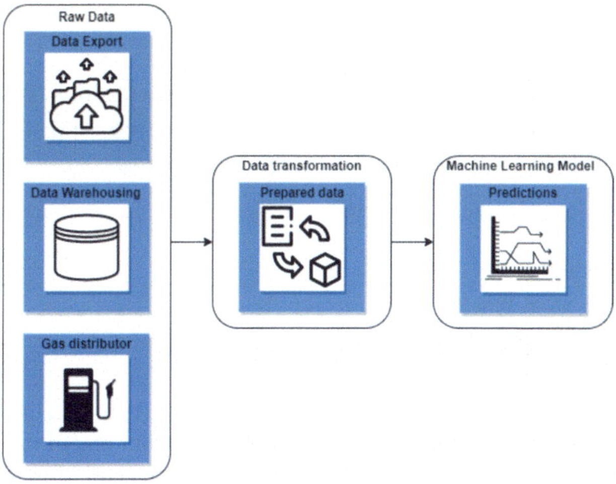

Fig. 1. The data pipeline. Source: Author's elaboration

The data obtained from the company's fuel volume sales have not been processed and it was mandatory to curate it. Raw data consists of 492178 data points from all distributors and type of gas. Table 1 show differences between obtained raw data and processed data to further experiment. After aggregation it was 1177 data points.

Table 1. Comparison between raw and processed data.

	Raw data	Transformed data
Data types	20 int64, 9 categorical, 4 float64	2 float 64,4 int64
Missing data	Yes* (categorical)	No
Aggregation	No	Daily sales data by gas distributor

Source: Author's elaboration

Refueling volume data has been used as the main time series during the modeling. The data spanned across the following periods with has x data points:

- Whole volume consisted of daily refueling volume from one distributor.
- Training data spanned from 2017-04-13 to 2020-03-05
- Testing data spanned from 2020-03-06 to 2020-07-01

Formally:

$$Y = \left[y_{t_1}, y_{t_2}, \ldots, y_{t_T}\right] \tag{1}$$

where, y_t is the refueling volume as the time t, and T is the maximal date in a time series. Therefore, the future-covariates matrix $X^{\{Cal\}}$ can be defined as follows:

$$X^{Cal} := \begin{bmatrix} Yr_{t_1} & M_{t_1} & D_{t_1} & W_{t_1} & DY_{t_1} & Q_{t_1} \\ Yr_{t_2} & M_{t_2} & D_{t_2} & W_{t_2} & DY_{t_2} & Q_{t_2} \\ \vdots & \vdots & \vdots & \vdots & \vdots & \vdots \\ Yr_T & M_T & D_T & W_T & DY_T & Q_T \end{bmatrix}, \forall Yr_t, M_t, D_t, W_t, DY_t, Q_t \in N \tag{2}$$

LightGBM algorithm requires additional predictors for handling time-series data. Therefore, the matrix of future covariates was constructed, consisting of the following indicators (for each date in a training and testing data), where:

- Year –*Yr*
- Month number - *M*
- Day of week number - *D*
- Number of the week in a year - *W*
- Number of the day in a year - *DY*
- Quarter number – *Q*

Fuel price can strongly influence the refueling volume. Therefore, the historical fuel prices per cubic meter were used. The prices originated from one of the fuel products suppliers and included the period from 2017-01-01 until the end of training data. The data originates from other source than fuel retail volume data Raw fuel prices can be challenging to forecast. Therefore, a time series of percentage price differences were used instead. Formally:

$$\begin{gathered} \Delta p_t = \frac{t - p_{t-1}}{p_{t-1}}, \Delta p_t \in R \\ \Delta p = \left[\Delta p_1, \Delta p_2, \ldots, \Delta p_T\right] \end{gathered} \tag{3}$$

3.2 Fuel Price Differences Prediction

To avoid the data leakage problem during forecasting - using fuel price differences Δp that should not be known at the prediction time - an additional meta-model was trained to predict price change, used then as covariates $X^{\Delta p}$.

The figures presented below illustrate the relationship between the refueling volume prediction y and fuel price differences Δp treated as additional covariates. Figure 1 depicts a situation when values of both time series are known; the model uses the following elements for prediction:

- Past values of the primary (predicted) time series - refueling.
- Past values of covariate time series - fuel prices.

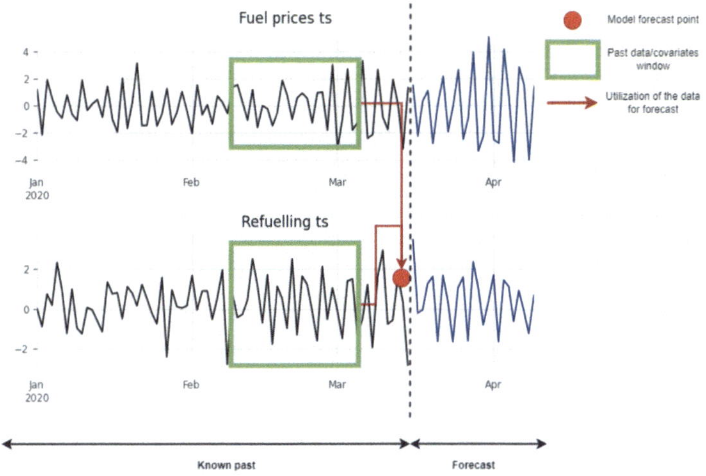

Fig. 2. Refueling time series prediction using known data. Source: Author's elaboration

Figure 2 depicts a situation when only some of the two-time series values are known. This scenario happens when the forecast is done out-of-sample. Then the model uses its own past predictions (rolling forecasts, sometimes called 'out-of-sample') and known historical data if it fits inside the prediction window (Fig. 3).

This modeling was a separate forecasting experiment, where the following algorithms were compared in terms of MSE and MAPE metrics: LightGBM, ARIMA, andNBeats. The fuel price difference prediction experiment included a separate training-validation loop independent from the refueling prediction. Based on the experimental results (presented in the Table 1) LightGBM was selected for forecasting (Table 2).

The large error values, in summary, are the result of very high price variability, although the inspection of LightGBM residuals indicates normality, and forecasts capture the overall price change trends, as shown in the figure below. Figure 4 on the left displays the difference between predicted and actual fuel price differences over time, offering insights into the model's prediction accuracy, where y is real data and yhat LightGBM is the LightGBM model. Figure 4 on right describes the distribution of LightGBM residuals

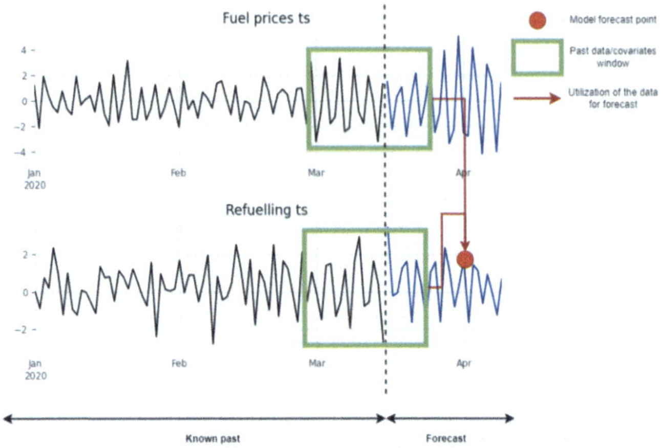

Fig. 3. Refueling time series prediction using rolling forecasts. Source: Author's elaboration

Table 2. Fuel price differences forecasting results.

Model	MSE	MAPE
LightGBM	0.0001	1059.8537
NBeats	0.0001	1460.3736
ARIMA	0.0001	inf

Source: Author's elaboration

could provide evidence of normality, indicating a good fit of the model to the data. The residuals mostly hold in range -100 to 100.

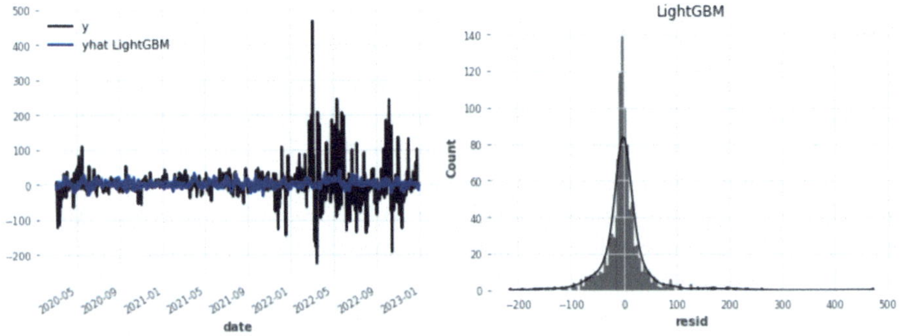

Fig. 4. LightGBM price difference prediction (left) and residuals (right). Source: Author's elaboration

4 Results

The results of the experiment are summarized in the Table 3. The best results were obtained by the LightGBM model, utilizing both future covariates and forecasted fuel prices. This model demonstrated the best performance according to all metrics (MSE and MAPE).

Table 3. Fuel price differences forecasting results

Model	MSE	MAPE
LightGBM	**88694,4968**	**17,9563**
Naive seasonal model	299658,2014	34,2128
Holt-Winters	384454,1543	41,9970
ARIMA	428295,8068	43,8615

Source: Author's elaboration

5 Conclusion

The LightGBM model developed in this research allows for achieving the lower values of MSE and MAPE than "classic" forecasting method. Therefore, the results of the research contribute to sales forecasting area by extending existing approaches about machine learning model for sales volume forecasting in fuel retail. The elaborated model can be used for improving volumes sales prediction in decision support systems performing in fuel retail. In consequence developed model allows for making more sustainable decisions by, among others, reducing the Forrester effect in the supply chain. Accurate petrol product sales forecasting enables businesses to reduce fuel inventory and improve companies' sustainable decision-making over uncertain data and models. These sales processes are affected by numerous interrelated factors, but most often considered is the classical influence of fuel price. Therefore, multi-perspective process models are rare. The main limitation of research is using only LightGBM machine learning model. Another limitation is lack of statistical analysis between models. Another limitation is the quality (lack of data continuity) and quantity of data (small number of refueling stations) analyzed in this study. Therefore, the future research works should be related to developing also other machine learning approaches for sales volume forecasting in fuel retail, including deep neural networks. Also, deeper statistical analysis of results should be performed.

Acknowledgements. (Tomasz Zema and Adam Sulich) This article has been written within the framework of the PROM Programme for International Exchange of Doctoral Students and Academic Staff Contract No. PPI/PRO/2019/1/00049/U/00001.

(Adam Sulich) This project is financed by the National Science Centre in Poland under the program "Business Ecosystem of the Environmental Goods and Services Sector in Poland", implemented in 2020–2022; project number 2019/33/N/HS4/02957; total funding amount PLN 120,900.00.

References

1. Sołoducho-Pelc, L., Sulich, A.: Between sustainable and temporary competitive advantages in the unstable business environment. Sustainability **12**, 8832 (2020). https://doi.org/10.3390/su12218832
2. Dalton, B.J., et al.: Forecasting of fuel oil supply using the transfer function approach (Case study: PT. Agrabudi Karyamarga Gas Station Division 64.706.07). J. Phys. Conf. Ser. **2106**, 012007 (2021). https://doi.org/10.1088/1742-6596/2106/1/012007
3. Galankashi, M.R., Fallahiarezoudar, E., Moazzami, A., Helmi, S.A., Rohani, J.M., Yusof, N.M.: An efficient integrated simulation–Taguchi approach for sales rate evaluation of a petrol station. Neural Comput. Appl. **29**, 1073–1085 (2018). https://doi.org/10.1007/S00521-016-2491-5/TABLES/12
4. Zema, T., Sulich, A., Kulhanek, L.: Energy sales forecasting in a sustainable development context: bibliometric review. In: Nedelko, Z., Korez-Vide, R. (eds.) 7th FEB International Scientific Conference: Strengthening Resilience by Sustainable Economy and Business – Towards the SDGs, pp. 99–108. University of Maribor, Maribor, Slovenia (2023). https://doi.org/10.18690/um.epf.3.2023
5. Rafati, E.: The bullwhip effect in supply chains: Review of recent development. J. Futu. Sustain. **2**, 81–84 (2022). https://doi.org/10.5267/J.JFS.2022.9.007
6. Ayvazyan, H.L., Aksyonova, O.P.: Comparative Analysis of Multi-agent Approaches for Planning Fuel Delivery of Petrol Stations
7. Stańczyk-Hugiet, E., Kozyra, C., Piórkowska, K., Stańczyk, S.: Organisational routines and interfirm collaboration: measurement dilemmas and recommendations for further research steps. European Res. Stud. J. **23**, 1086–1117 (2020). https://doi.org/10.35808/ersj/1927
8. Ge, Q., Sun, R.: Gas station customer relationship management research under market price of refined petroleum products conditions. In: Proceedings - 2011 4th International Conference on Information Management, Innovation Management and Industrial Engineering, ICIII 2011 **1**, 547–550 (2011). https://doi.org/10.1109/ICIII.2011.139
9. Zema, T., Sulich, A.: Models of Electricity Price Forecasting: Bibliometric Research. Energies (Basel) **15**, 5642 (2022). https://doi.org/10.3390/en15155642
10. Giri, C., Johansson, U., Lofstrom, T.: Predictive modeling of campaigns to quantify performance in fashion retail industry. In: Proceedings - 2019 IEEE International Conference on Big Data, Big Data **2019**, 2267–2273 (2019). https://doi.org/10.1109/BIGDATA47090.2019.9005492
11. Papers, M., Armstrong, J.S.: Part of the Marketing Commons, and the Sales and Merchandising Commons Recommended Citation Recommended Citation Armstrong. J. S. 278–290 (1999)
12. Bohanec, M., Robnik-Šikonja, M., Borštnar, M.K.: Decision-making framework with double-loop learning through interpretable black-box machine learning models. Ind. Manag. Data Syst. **117**, 1389–1406 (2017). https://doi.org/10.1108/IMDS-09-2016-0409
13. Caruana, R.: An Empirical Comparison of Supervised Learning Algorithms
14. Dietterich, T.G.: Machine learning for sequential data: A review. Lecture Notes in Computer Science (including subseries Lecture Notes in Artificial Intelligence and Lecture Notes in Bioinformatics) **2396**, 15–30 (2002). https://doi.org/10.1007/3-540-70659-3_2/COVER
15. Kim, H., Cho, H., Ryu, D.: Corporate bankruptcy prediction using machine learning methodologies with a focus on sequential data. Comput. Econ. **59**, 1231–1249 (2022). https://doi.org/10.1007/S10614-021-10126-5
16. Ostmeyer, J., Cowell, L.: Machine learning on sequential data using a recurrent weighted average. Neurocomputing **331**, 281–288 (2019). https://doi.org/10.1016/J.NEUCOM.2018.11.066

17. Gamboa, J.C.B.: Deep Learning for Time-Series Analysis (2017). https://doi.org/10.48550/arxiv.1701.01887
18. Younan, C.: Morrisons slashes the price on petrol and diesel at all UK petrol stations - Daily Star, https://www.dailystar.co.uk/real-life/morrisons-supermarket-fuel-diesel-offer-17215884. Last accessed 30 January 2023
19. Anderson, E.: A new model for cycles in retail petrol prices. Eur. J. Oper. Res. **210**, 436–447 (2011). https://doi.org/10.1016/J.EJOR.2010.10.009
20. Shafee Kalid, K., Ahmad, J., Suet Peng, Y., Kwang Hooi, Y.: Petronas Petrol Station Fuel Consumption Forecast System
21. Lemos, A., Caminhas, W., Gomide, F.: Fuzzy evolving linear regression trees. Evol. Syst. **2**, 1–14 (2011). https://doi.org/10.1007/S12530-011-9028-Z/TABLES/10
22. Lemos, A., Leite, D., MacIel, L., Ballini, R., Caminhas, W., Gomide, F.: Evolving fuzzy linear regression tree approach for forecasting sales volume of petroleum products. IEEE Int. Conf. Fuzzy Sys. (2012). https://doi.org/10.1109/FUZZ-IEEE.2012.6250809
23. Sun, L., Xing, X., Zhou, Y., Hu, X.: Demand forecasting for petrol products in gas stations using clustering and decision tree. J. Adv. Computat. Intelligence and Intelligent Informat. **22**, 387–393 (2018). https://doi.org/10.20965/JACIII.2018.P0387
24. Hyndman, R.J., Athanasopoulos, G.: Forecasting: principles and practice. OTexts, Melbourne, Australia (2018)
25. Kumar, G., Jain, S., Singh, U.P.: Stock market forecasting using computational intelligence: a survey. Archi. Computat. Methods in Eng. **28**, (2021). https://doi.org/10.1007/s11831-020-09413-5
26. Khan, S., Alghulaiakh, H.: ARIMA model for accurate time series stocks forecasting. Int, J. Adv. Comp. Sci. Appli. **11**, (2020). https://doi.org/10.14569/IJACSA.2020.0110765
27. Mondal, P., Shit, L., Goswami, S.: Study of effectiveness of time series modeling (Arima) in forecasting stock prices. Int. J. Comp. Sci. Eng. Appl. **4** (2014). https://doi.org/10.5121/ijcsea.2014.4202
28. Li, Z., Han, J., Song, Y.: On the forecasting of high-frequency financial time series based on ARIMA model improved by deep learning. J Forecast. **39**, (2020). https://doi.org/10.1002/for.2677
29. Holt, C.C.: Forecasting trends and seasonals by exponentially weighted moving averages. ONR Memorandum **52**, (1957)
30. Aohan, L.: An empirical analysis of total retail sales of consumer goods based on holt-winters. In: ACM International Conference Proceeding Series (2020). https://doi.org/10.1145/3436286.3436298
31. Bissing, D., Klein, M.T., Chinnathambi, R.A., Selvaraj, D.F., Ranganathan, P.: A hybrid regression model for day-ahead energy price forecasting. IEEE Access **7**, (2019). https://doi.org/10.1109/ACCESS.2019.2904432
32. Aurna, N.F., et al.: Time series analysis of electric energy consumption using autoregressive integrated moving average model and Holt Winters model. Telkomnika (Telecommunication Computing Electronics and Control) **19**, (2021). https://doi.org/10.12928/TELKOMNIKA.v19i3.15303
33. Bentéjac, C., Csörgő, A., Martínez-Muñoz, G.: A comparative analysis of gradient boosting algorithms. Artif Intell Rev **54**, (2021). https://doi.org/10.1007/s10462-020-09896-5
34. Ke, G., et al.: Lightgbm: a highly efficient gradient boosting decision tree. In: Advances in Neural Information Processing Systems. pp. 3146–3154 (2017)
35. Tian, L., Feng, L., Yang, L., Guo, Y.: Stock price prediction based on LSTM and LightGBM hybrid model. Journal of Supercomputing **78**, (2022). https://doi.org/10.1007/s11227-022-04326-5
36. Sun, X., Liu, M., Sima, Z.: A novel cryptocurrency price trend forecasting model based on LightGBM. Financ. Res. Lett. **32**, (2020). https://doi.org/10.1016/j.frl.2018.12.032

37. Al-kasassbeh, M., Abbadi, M.A., Al-Bustanji, A.M.: LightGBM Algorithm for Malware Detection. In: Advances in Intelligent Systems and Computing (2020)
38. Zhang, D., Gong, Y.: The comparison of LightGBM and XGBoost coupling factor analysis and prediagnosis of acute liver failure. IEEE Access (2020). https://doi.org/10.1109/ACCESS.2020.3042848

The Effectiveness of AI Methods Compared to Statistical Analysis and Data Modeling in the Construction of Scoring Models

Boleslaw Borkowski[1](✉), Marek Karwanski[2], Wieslaw Szczesny[2], and Artur Wilinski[2]

[1] Department of Quantitative Methods, Warsaw University, Warsaw, Poland
bborkowski@wz.uw.edu.pl

[2] Institute of Information Technology, Warsaw University of Life Sciences, Warsaw, Poland

Abstract. In the work, we made a comparative analysis of the effectiveness of two analytical cultures, statistical analysis and data modeling (DMC) and algorithmic AI analysis (AMC) in scoring models for calculating LGD (Losses Given Default) – the bank's percentage share of credit losses (in case of non-payment or delayed repayment of the loan). Analysis based on simulation data generated on the basis of probability distributions, which were obtained from empirical data in a credit risk study at a Polish bank. In the article, we proposed variable selection methods in the DMC and AMC models by analyzing the importance of variables (drop-out and permutation). The proposed methods of variable importance (used for the variable selection) in both cultures showed discrepancies. Each considered model, regardless of the number of observations, indicated different information about the importance of the variables.

Keywords: AI methods · machine learning · methods of statistical data analysis and modeling · importance of variables

1 Introduction

An important innovation in the consolidation of prudential regulations on financial markets was the Basel Capital Accord published in July 1988 (Basel I [1]). The next step was the publication of the new International Financial Reporting Standard 9 (IFRS 9), which has been in force since January 1, 2018 [2]. The IFRS 9 standard assumed the possibility of using artificial intelligence algorithmic models in parallel with statistical models [3].

One of the parameters subject to modeling credit risk is LGD (Loss Given Default) [4]. We discretized the LGD variable by transforming it into a binary. This allowed to directly compare traditional statistical classification models and machine learning models using simple statistics.

M. Hernes and J. Wątróbski (Eds.): ECAI 2023, LNNS 1079, pp. 121–130, 2024.
https://doi.org/10.1007/978-3-031-66761-9_11

The aim of the work is to compare the effectiveness of two analytical cultures in building scoring models to determine LGD. Two methodologies were considered: statistical analysis and data modeling (DMC) and algorithmic modeling (AMC) [6], using Machine Learning methods (Neural Networks, Random Forests, Decision Trees). The comparative analysis concerns above all:

1. comparison of the results of estimating the bank's losses when granting a loan in the corporate segment (regression models were replaced by classification models),
2. determining the importance of individual features (variables) in risk models,
3. selection of explanatory variables for the analyzed models using proposed uniform methods based on the drop-out and permutation strategies.

The proposed methods of evaluating the importance and order the potential explanatory variables make it possible to select variables [5]. The research material consists of simulation data generated on the basis of probability distributions of actual bank credit data.

Two samples were generated $n = 2000$ and $n = 20000$ observations, which were randomly split into two parts t - (training) and v - (validation) in a *50/50* proportion. Each sample contained 92 explanatory variables, of which 10 variables were generated as independent variables (confounding effect). The binary explanatory variable LGD *(0/1)* was obtained as follows: above and below *the median.* To solve the problem of overfitting, we usually split the data sample D into two subsets [7]: $D_{TRAINING}$ - training data (t) and $D_{VALIDATE}$ - validation data (v). In both statistical and machine learning approaches, classifiers minimize the loss function $l(x, y|h())$, where y is the actual value and $h()$ the result obtained from the model on the element x. Estimation of model parameters $h^*()$ was performed on training data by the loss function optimization.

Models from two analytical cultures were used to analyze the empirical material [6,8,9], namely: (1) statistical methods of data analysis (DMC - the data modeling culture) - logistic regression model, (2) algorithmic methods (AMC - the algorithmic modeling culture), i.e. Random Forests (RF), Gradient Boosting Machine (GBM), two classes Neural Networks (ANN1 and ANN2 - differing in regularization techniques: ANN1 regularization network, through the penalty function, ANN2 drop out - random zeroing of weights). Therefore, we used the most popular methods in the field of machine learning (ML) to analyze the empirical material. We performed a comparative analysis of the usefulness of the models included in these two analytical cultures for forecasting the size of the parameter used in credit risk management in banks.

We determined the importance of variables using computer simulation techniques, evaluating the change in model fit, after removing individual explanatory variables in two variants, i.e. logical and physical. In the logical variant, the model remained the same, and the fit values were calculated after applying a random permutation of the value of the selected explanatory variable (more precisely: we used 100 random permutations and calculated the average fit). In the variant of physical elimination of the explanatory variables, the model was estimated after removing the selected variable, so-called drop-out method.

We selected variables both in the logistic model and in the AI models only on the training set. The evaluation of fit the models to empirical data was done using the proposed criteria, i.e. the statistical significance of the odds ratio in the logistic models, i.e. AUC and R2 Snell on validation sets.

2 Methods of Analysis

For the comparative analysis of the effectiveness of two analytical cultures in the construction of scoring models, we used the logit model and the most popular machine learning methods.

The logit model was used as a statistical model. This model can be described by the function $f : X \to Y$, which estimates the distribution $P(Y|X)$ when Y has discrete values and $P(Y|X)$ is a vector of variables containing discrete or continuous values [11].

Among the machine learning methods, we used RF, ANN1, ANN2 and GB models. Random Forest (RF) is an ensemble classifier - based on decision trees. The basic idea is to average multiple noisy, approximately unbiased decision tree models, thereby achieving error reduction.

Neural networks are collections of affine transformations - neurons, consisting of multiplication by weight matrices and addition of the bias vector, alternating with a non-linear transformation [12]. We used two types of networks $ANN1$, $ANN2$. In the $ANN1$ model the loss function was: Euclidian distance and in $ANN2$ we determined the loss function using the Kullback-Leibler distance. Regularizations are used to reduce overfitting. We used two metrics to compare the effectiveness of the methods: (1) the Area Under ROC curve (AUC), and (2) Cox-Snell coeffcient of determination (R^2).

3 Discussion

In the study of the logit model (DMC method), we used the statistical significance of the odds ratio in the selection of explanatory variables. To compare all the considered models, we proposed two strategies for assessing the importance of variables, by permuting and drop-out individual variables.

The permutation strategy consists in perturbing the value of the selected variable. We performed 100 random permutations of each variable (calculating AUC and R^2 for each permutation) [9]. We determined the importance of the variable based on the difference between the AUC R^2 for models estimated on the prior simulated data and their mean value (calculated on premutation). The drop-out strategy was to remove the variable from the model and re-estimate the model. The procedure is then repeated for each variable. We determined the meaning of the variables, as before, based on changes in AUC and R^2. First, we generated 2000 observations and 92 explanatory variables for analysis. This set was randomly split into two equal parts of 1000 observations each. The first of them (1000 observations) was the training set (t), while the second seperable part was the validation set (v). Based on the set t, 28 statistically significant variables were selected in the logit model (odds ratio) for a predefined significance level of $alpha = 0.025$. We performed a comparative analysis of the effectiveness of the models on a two sets of numbers of observations (2000 and 20000 and a different number of explanatory variables-see Fig. 1, 2).

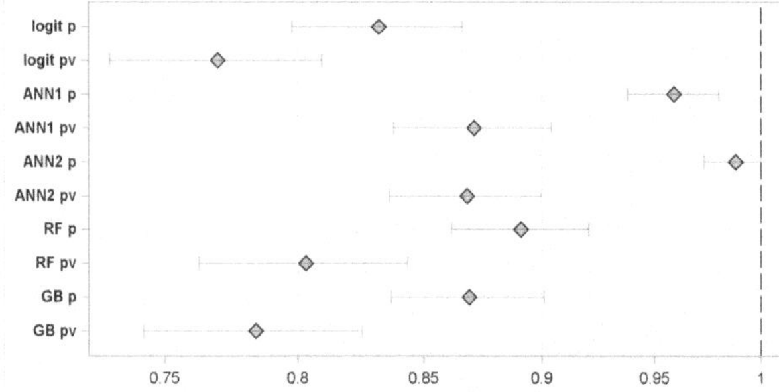

	AUC	LCL	UCL	R-square
logit p	0.832	0.798	0.866	0.302
logit pv	0.770	0.731	0.809	0.218
ANN1 p	0.959	0.938	0.980	0.597
ANN1 pv	0.871	0.838	0.904	0.354
ANN2 p	0.988	0.973	1.000	0.660
ANN2 pv	0.868	0.836	0.900	0.333
RF p	0.891	0.862	0.920	0.435
RF pv	0.803	0.763	0.843	0.287
GB p	0.869	0.837	0.901	0.387
GB pv	0.784	0.743	0.825	0.260

Fig. 1. Model stability on a set of 1000 observations with 28 statistically significant explanatory variables, where: p - test set, pv - validation set. Source: own study.

Our calculations showed that for models built on the number of observations of 1000 and 28 variables, the best-fit models (compared on the validation set) turned out to be neural network models (ANN1 and ANN2 - see in Fig. 1). In the case of models: logit, RF and GB, the results were worse than for neural networks. It should be noted, however, that the neural networks for $n = 1000$ showed high instability (a large difference between the values on the test and validation sets shown by separable confidence intervals). We observed stability only in the logit model. In the next step, we estimated the models on a large number of observations 10000 and with the same set of 28 variables (see Fig. 2). The results confirm a better fit of neural network models over other models on the similar level of fit. With the increased number of observations, all models are already stable. Secondly, analogically, on the sets of 1000 units and 10000, we estimated the parameters of the models with all variables (92 explanatory variables, without variable selection procedure, see Fig. 3, 4). The results for n = 1000 for all models were unstable (significantly different confidence intervals). In turn, for models estimated on 10000 observations, we obtained similar stability results as for models with 28 variables.

In conclusion, our research showed a better fit of neural network models than logit models, especially on a set with a large number of observations. On the other hand, in a set with a small number of observations, neural networks showed instability compared to the high stability of logit models.

The second important methodological issue analyzed in the article was our proposal for a common method of selecting variables for both DMC and AMC models. For the selection of variables in the DMC and AMC models, we proposed

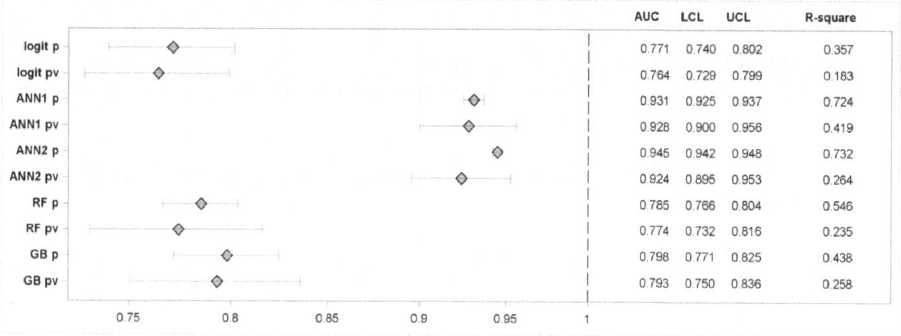

Fig. 2. Model stability on a set of 10000 observations with 28 statistically significant explanatory variables, where: p - test set, pv - validation set. Source: own study.

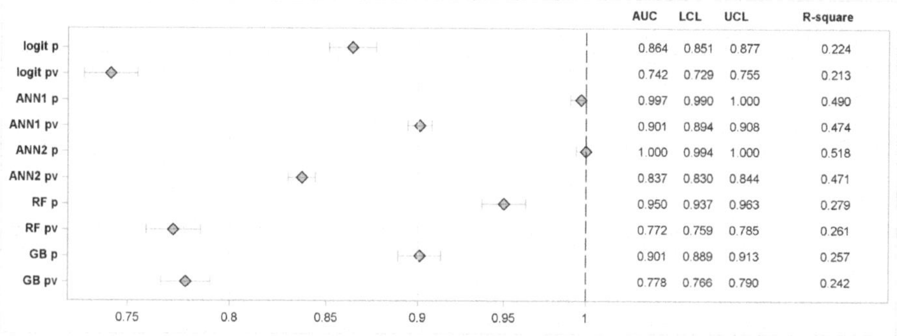

Fig. 3. Model stability on a set of 1000 observations with 92 explanatory variables, where: p - test set, pv - validation set. Source: own study.

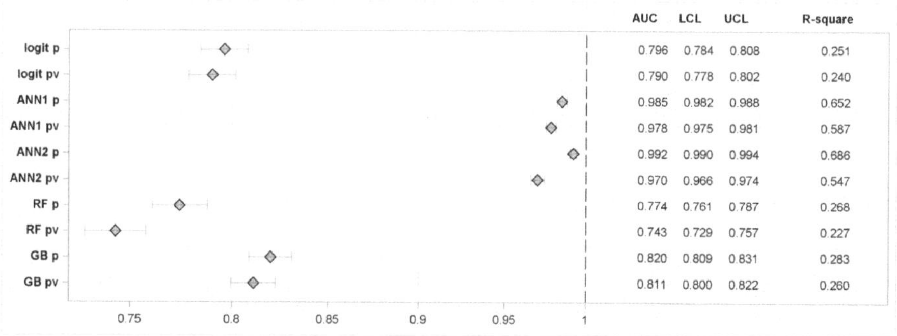

Fig. 4. Model stability on a set of 10000 observations with 92 explanatory variables, where: p - test set, pv - validation set. Source: own study.

two strategies for evaluating the importance of variables, by permuting and drop-out individual variables. First, based on the statistical logit model, we verify the results of variable selection using two proposed strategies in comparison to the commonly known stepwise regression method (see Table 1).

Table 1. Selection of explanatory variables in the logit model using the proposed methods. Source: own study.

->	Logistic n = 1000 k = 28	Decrease of Statistics drop-out method		Decrease of Statistics permutations method	
no	**Var.** vs. odds ratio	**Var.** vs. R^2	**Var.** vs. AUC	**Var.** vs. R^2	**Var.** vs. AUC
1	**31** | 3.56(2.13–5.95)	**31** | 0.03	**31** | 0.03	**31** | 0.03	**31** | 0.06
2	**49** | 3.07(1.16–8.11)	**34** | 0.02	**51** | 0.02	**34** | 0.02	**34** | 0.05
3	**34** | 2.76(1.85–4.12)	**44** | 0.02	**34** | 0.02	**44** | 0.02	**44** | 0.04
4	**36** | 2.29(1.57–2.86)	**42** | 0.02	**36** | 0.01	**42** | 0.02	**51** | 0.03
5	**44** | 2.09(1.52–2.86)	**36** | 0.01	**42** | 0.01	**36** | 0.01	**55** | 0.02
6	**28** | 1.79(1.15–2.79)	**51** | 0.01	**55** | 0.01	**51** | 0.01	**36** | 0.03
7	**50** | 1.67(1.17–2.37)	**20** | 0.01	**44** | 0.01	**20** | 0.01	**74** | 0.02
8	**20** | 1.59(1.18–2.14)	**33** | 0.01	**74** | 0.01	**33** | 0.01	**42** | 0.02
9	**42** | 1.54(1.27–2.19)	**50** | 0.01	**50** | 0.01	**50** | 0.01	**20** | 0.02
10	**51** | 1.48(1.25–1.75)	**40** | 0.01	**28** | 0.01	**40** | 0.01	**28** | 0.01

Comparative analysis of results between a set of variables selected by the stepwise method regression based on the logit model (statistically significant odds ratio) and the methods proposed by the authors showed a high convergence. Differences in the set of the first 10 variables selected by the stepwise method regression and the proposed methods of selecting variables were small and in the set of the first 10 statistically significant variables, the differences concerned only two variables. Comparison of the set of explanatory variables between the two proposed methods (drop-out and permutation) in different models (logit, ANN1, ANN2, RF and GB) did not show large differences (in ordering the importance of variables). The permutation method gave results more similar between the considered models than the drop-out method (see in Table 2). As can be seen in the case of the logit model, there are 9 identical variables among 10 variables selected on the training data set, both by one and the other methods. In the case of neural networks, it is only 6 variables. When using the drop-out technique, variables 36, 44, 51 and 55 (numbering of variables in the data generator) appear in all models among the "most important" 10 variables. In the permutation technique, these are only 3 variables: 34, 36, and 55. Behind the top ten there are two variables 34 and 44, which are common to only three models (2). Therefore, if we wanted to choose the 10 most important variables, we would get a different sets depending on the model used. Therefore, in the further part of the analysis, we controlled whether the process of evaluating importance of individual variables is stable.

The analysis of the stability of variable importance evaluation in the models in the same way as in the case of fitting to empirical data. We measured the similarity of vectors containing the importance of variables on the training

Table 2. Ranking of the importance of variables in models using the drop-out (d) and permutation (p) method on a training set with 1000 records and 28 variables. Source: own study.

Methods										
no	logit p	ANN1 p	ANN2 p	RF p	GB p	logit d	ANN1 d	ANN2 d	RF d	GB d
1	31	55	55	55	55	31	55	36	55	55
2	34	44	44	34	34	51	44	55	74	51
3	44	58	34	74	74	34	36	7	62	39
4	51	34	65	31	31	36	58	74	51	28
5	55	36	51	36	28	42	49	56	44	61
6	36	42	56	62	36	55	28	34	36	44
7	74	65	42	51	51	44	51	39	28	74
8	42	28	36	28	65	74	56	32	61	36
9	20	31	61	44	42	50	62	44	42	45
10	50	61	58	65	62	28	34	51	65	32

and validation data sets in both techniques. For this purpose, in both the drop-out and permutation techniques (to study the similarity of variable importance scoring), we used the cosine measure of similarity of vectors whose coordinates were the importance of individual variables. Before calculating the similarity, all vectors were transformed by standardizing the coordinate values of individual vectors (see in [13,14]). For identical vectors, the value of this measure is 1. The further away from 1, the less and less the similarity. The measure assumes values in the range $[-1, 1]$. In this situation (after standardizing the coordinates of individual vectors), the cosine measure is equal to the linear correlation coefficient. Thus, we can use the Guillford scale to evaluate the similarity value (see in [15]). This means that values in the range $(0.7, 0.9]$ mean high similarity, and in the range $(0.9, 1]$ very high. Before calculating the similarity, all vectors were transformed by standardizing the coordinate values of individual vectors (see in [13,14]). For identical vectors, the value of this measure is 1.0. The further away from 1.0, the similarity decreases. The measure accepts values in the range from -1.0 to 1.0. After standardizing the coordinates of the individual vectors, the cosine measure is equal to the linear correlation coefficient. Then, we can use the Guilford scale to evaluate the similarity value. This means that the values in the range $(0.7, 0.9]$ mean high similarity, and in the range $(0.9, 1.0]$ very high.

The results of the stability of the process of evaluating importance of individual variables (i.e. the similarity of the importance of variables on sets t and v) for the permutation and drop-out technique with a size of 1000 and 10000 and 28 variables are presented in Table 3) (see col. $p(t, v)$ and $d(t, v)$). The measures indicate the lack of stability of the studied process (only a moderate similarity according to the Guillford scale, the values of the cosine measure belong to the range $[0.4, 0.7]$ - see. col $d(t, v)$ for n = 1000). We observed greater stability for sets with more observations (10000). However, the stability results for the drop-out technique remained much lower than for the permutation technique (i.e. for sets of 10000 observations). It should be emphasized that the increase in the number of observations (up to 10000) resulted in an increase in stability (similarity of vectors) in all the considered models. However, the ANN1 and ANN2 models for the drop-out technique still show weaker stability than for the permutation technique. Therefore, in the following part of the article we mainly refer to the permutation technique.

Table 3. Compatibility of the importance of variables in the models (stability and comparison of validation processes). Source: own study.

	n = 1000, k = 28)			n = 10000, k = 28)		
	p(t, v)	d(t, v)	*p&d(t)*	p(t, v)	d(t, v)	*p&d(t)*
logit	0.891	0.942	0.750	0.988	0.962	0.951
ANN1	0.902	0.406	0.657	0.992	0.835	0.826
AR2	0.874	0.224	0.465	0.996	0.776	0.873
RF	0.756	0.530	0.654	0.979	0.976	0.967
GB	0.972	0.664	0.514	0.999	0.988	0.709

Table 4. Consistency of the importance of variables between models on the set t according to the permutation strategy (upper triangle for n = 1000, k = 28, and lower triangle for n = 10000 and k = 28). Source: own study.

	methods				
	logit	ANN1	AR2	RF	GB
logit		0.408	0.446	0.604	0.590
ANN1	0.402		0.880	0.753	0.743
AR2	0.384	0.986		0.703	0.686
RF	0.650	0.534	0.561		0.913
GB	0.682	0.747	0.781	0.757	

In Table 3 we have additionally included information on the consistency of the importance of variables determined with both techniques (permutation and drop-out - col. *p&d(t)* on the training data set. Comparative analysis (vector similarities) between the considered methods showed consistency in the case of logit models, regardless of the number of observations, and less consistency in the case of AI models. The increase in the number of observations to 10000 resulted in an increase in the similarity of the importance of the variables also for the AI models, but it is still lower than in the case of the logit and RF models. In the case of 92 variables according to the permutation strategy, we obtained similar results as in the case of 28 variables: for n = 1000 similarity values $p(t, v)$, were respectively (in the order of models as in Table 3): 0.922, 0.780, 0.695, 0.616, 0.971 and for n = 10000 respectively 0.984, 0.994, 0.991, 0.950, 0.998. Thus, with a greater number of variables and n = 1000, in the case of neural networks, we have significantly worse stability of the process of validating variables (similarity values are more than 0.1 lower than in the case of the logit model). We observed an even greater difference in the comparison the selectios techniques (permutation and drop-out) of the evaluating importances of variables on the training data set (the $p\&d(t)$ column), we obtained the following values: (1) for n = 1000 respectively: 0.952, 0.178, 0.027, 0.297, 0.048 and (2) for n = 10000 respectively: 0.999, 0.845, 0.530, 0.895, 0.523.

Table 4 shows the similarity values between the vectors containing the importance of variables in the models (determined on the training set) in the case of n = 1000 observations and k = 28 variables (upper triangle) and n = 10000 and k = 28 (lower triangle). The results indicate that the consistency of the importance of variables in the logit model compared to AI models is not high, both for the set with the number of observations n = 1000 and the set with a large number of observations n = 10000. On the other hand, the consistency of the importance of variables between the models of neural networks (ANN1 and (ANN2) was high (n = 1000) and very high (n = 10000). In addition, we compared how significantly the importance of variables changes in the case of the permutation strategy when we change the number of observations in the considered models for the entire set of 92 variables. For n = 1000 and n = 10000, we have a logit value of 0.748 for the model, and 0.472, 0.224 for the ANN1 and ANN2 models, respectively. This experiment suggests that the number of 1000 observations considered large by the DMC culture is not large enough by the AMC culture. This is probably due to the much larger number of estimated parameters in the ANN models than in the logit model. Then, we compared how the importance of variables changes in the case of the permutation strategy when we change the number of observations in the considered models for the entire set of 92 variables. For n = 1000 and n = 10000, we have value of 0.748 for the logit model, and 0.472, 0.224 for the ANN1 and ANN2 models, respectively. This experiment suggests that the number of 1000 observations considered as large by the DMC culture is not large enough in the AMC culture. This is probably due to a much larger number of estimated parameters in ANN models than in the logit model.

4 Summary

Studies have shown that the use of AMC models compared to DMC models in the bank's credit risk analysis improves LGD prediction regardless of the sample size. In particular, studies have shown a better fit of AI models to empirical data than logit models on datasets with a large number of observations. On the other hand, in sets with fewer observations, AI models showed high instability compared to the high stability of logit models (see Fig. 1).

Comparative analysis of the results of variable selection using the stepwise method regression based on the logit model (statistically significant odds ratio) and the strategies examined by the authors (drop-out and permutation) showed a high similarity of the results. High and very high agreement of the importance of variables (calculated by the permutation method) on the training and validation data sets with different sizes (1000 and 10000) indicated the stability of the strategy used. Thus, it can be used to select a set of core variables, regardless of the analytic culture used. The selection of a set of a few or a dozen of the most important variables describing the phenomenon described by the model is very important from a business point of view. It allows to interpret the results in business terms. We performed the experiment on classification models, so there remains a problem to be solved in the field of parameter interpretation and generalization of the results to regression models and more complex data models (e.g. survival models).

References

1. Engelmann, B., Rauhmeier, R.: The Basel II Risk Parameters: Estimation, Validation, Stress Testing - with Applications to Loan Risk Management. Springer, Heidelberg (2011)
2. Bellini, T.: IFRS 9 and CECL Credit Risk Modelling and Validation. AP (2019)
3. EBA, Discussion Paper on Machine Learning for IRB Models. EBA/DP (2021)
4. Wytyczne dotyczace ograniczania ryzyka kredytowego dla instytucji stosujacych metodę IRB z wykorzystaniem własnych oszacowan LGD. EBA/GL (2020)
5. Heinze, G., Wallisch, C., Dunkler, D.: Variable selection - a review and recommendations for the practicing statistician. Biom. J. **60**(3), 431–449 (2018). https://doi.org/10.1002/bimj.201700067
6. Breiman, L.: Statistical modeling: the two cultures. Stat. Sci. **16**(3), 199–231 (2001)
7. http://www.cs.cornell.edu/courses/cs4780/2018fa/lectures/ . Accessed 01 June 2023
8. Alonso, A., Carbó, J.: Accuracy of explanations of machine learnings models for credit decision. Banco de Espana, Documentos de Trabajo (2022)
9. Cox, D.R., Snell, E.J.: The Analysis of Binary Data. Chapman and Hall, London (1989)
10. Draper, N.R., Smith, H.: Applied Regression Analysis. Wiley, New York (1973)
11. Hastie, T., Tibshirani, R., Friedman, J.: The Elements of Statistical Learning: Data Mining, Inference, and Prediction. Springer, New York (2009)
12. Calin, O.: Deep Learning Architectures: A Mathematical Approach. Springer, Cham (2020)
13. Tan, P., Steinbach, M., KumarCalin, V.: Wprowadzenie do eksploracji danych. Addison-Wesley (2005)
14. Kryszkiewicz, M.: Determining cosine similarity neighborhoods by means of the euclidean distance. In: Skowron, A., Suraj, Z. (eds.) Rough Sets and Intelligent Systems-Professor Zdzisław Pawlak in Memoriam. ISRL, vol. 43, pp. 323–345. Springer, Heidelberg (2013). https://doi.org/10.1007/978-3-642-30341-8_17
15. Guilford, J.: Fundamentalna statystyka w psychologii i edukacji (wyd. 3). McGraw-Hill (1956)

Supporting Algorithmic Trading with Machine Learning: Progress in Backend Technology

Justyna Szydłowska-Samsel(✉)

Department of IT in Management, University of Szczecin, Szczecin, Poland
justyna.szydlowska-samsel@usz.edu.pl

Abstract. Over the last decades, the development of artificial intelligence revolutionized the finance sector. Algorithmic trading can be more beneficial not only because of the speed and frequency but also by ruling out the influence of human emotions. Over the years many programming languages were used for machine learning research and system development. Python being easy to learn but still powerful has become popular within the scientific computing community, with most recent machine learning libraries being Python-based. In this paper, Python optimization is tested on the machine learning models frequently used in algorithmic trading. The results are compared between four different Python versions: Python 3.7, Python 3.8, Python 3.9, and Python 3.11.

Keywords: Machine learning · Algorithmic trading · Python · Performance measurement

1 Introduction

Over the last decades, the development of computer technology and artificial intelligence revolutionized the finance sector by transforming it from a human-dominated industry to one highly influenced by technology. Key modern financial players are part human part machine, which came both from advances in technology as well as regulatory reforms [6].

Algorithmic trading can be defined as a computer program that follows a defined set of instructions to place a trade. In theory, such trade can generate much higher profits at a speed and frequency impossible for human traders. The defined set of instructions can be based on timing, quantity, price, or any mathematical model - not only generating opportunities for the trader but also by ruling out the influence of human emotions rendering markets more liquid and trading more systematic [7].

Financial institutions across the world have employed algorithms to trade shares since the 1970s resulting in algorithmic trading becoming a dominant force in global financial markets. Using complex artificial intelligence and machine learning systems, computers execute extremely fast trading decisions benefiting

M. Hernes and J. Wątróbski (Eds.): ECAI 2023, LNNS 1079, pp. 131–141, 2024.
https://doi.org/10.1007/978-3-031-66761-9_12

the execution of trades at the best possible prices, increased accuracy and a likelihood of reducing mistakes, the ability to automatically and simultaneously check multiple market conditions, and the potential reduction of human errors caused by psychological or emotional conditions. With all these benefits, it's no surprise that in 2015 algorithmic trading contributed to approximately one-half of UK hedge fund Man Group's profits [3].

Over the years, many different programming languages and environments were used for machine learning research and system development, one of which is general-purpose Python language. It comes as no surprise that in recent decades there has been tremendous growth in popularity within the scientific computing community, most recent machine learning and deep learning libraries are now Python-based. Python is a high-level interpreted programming language, which guiding idea is readability. It is widely recognized as easy to learn yet powerful when necessary. Not only that, the community surrounding the Python language and its many available tools and libraries makes it very attractive for many use cases, including workloads in data science, machine learning, and scientific computing.

One such library is "Scikit-learn" which has become the industry standard Python library for feature engineering and classical ML modeling on small to medium-sized data sets. Another library worth mentioning is "Pandas", which extends the "NumPy" library by providing a data frame-like object supporting heterogeneous column types and row and column metadata. In recent years, the "Pandas" format has become the norm for representing tabular data in Python in data analysis among others [9].

With the recent optimization of Python language, it is of high interest to see how much it will impact implementations of machine learning solutions, which are known to be computationally intensive.

In this publication, the optimization of the newer Python version and its impact on solutions used in algorithmic trading is evaluated. To accomplish this, four Python versions are tested on eight combinations consisting of two different data sets, two different machine-learning models, and two different cross-validations. In the second section of this publication a short literature review is conducted to familiarize the reader with the current field of interest in machine learning within an algorithmic trading environment. Next, the tested Python versions used data sets and libraries are described. The results are then presented in graphical form and analyzed. In the last section, everything is summarised, and conclusions are drawn with suggestions for further research.

2 Short Literature Review

A literature assessment was conducted by searching available scientific publications in the Web Of Science database. The phrases "Machine learning" and "Algorithmic trading" were used in a query limiting the search to publication title, abstract, and authors keywords, which resulted in finding 83 publications that matched the set criteria.

In Fig. 1 the publication year distribution among the scientific papers are presented, beginning in 2010 with only one study and ending in 2023 with 11 studies already printed. It indicates a clear increase in the number of studies, with the highest growth rate between the years 2018 (5 publications) and 2021 (16 publications).

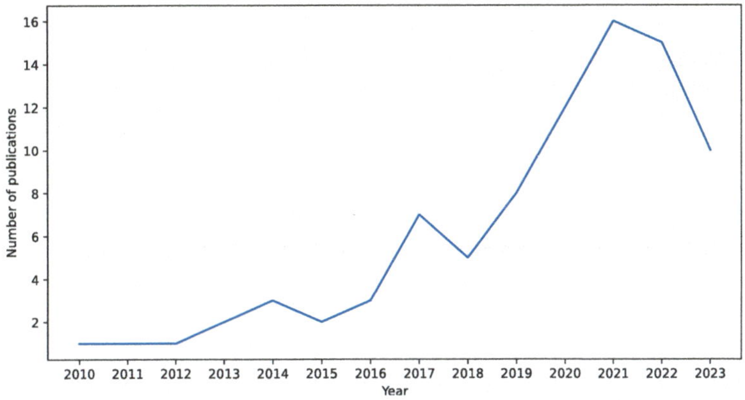

Fig. 1. Number of publications on the topic of Machine Learning and Algorithmic Trading by year

To better depict places of higher interest in the topic of Machine Learning and Algorithmic Trading the five most cited publications are presented in Table 1.

Table 1. Five the most cited publications.

Author	Title	Year	Times cited
Booth A., Gerding E., Mcgroarty F	Automated trading with performance weighted random forests and seasonality	2014	81
Jiang Zy., Liang Jj.	Cryptocurrency Portfolio Management with Deep Reinforcement Learning	2017	77
Shynkevich Y. et al.	Forecasting price movements using technical indicators: Investigating the impact of varying input window length	2017	72
Ozbayoglu Am., Gudelek Mu., Sezer Ob.	Deep Learning for financial applications: A Survey	2020	57
Huang Dj. et al.	Robust Median Reversion Strategy for Online Portfolio Selection	2016	54

The most cited paper in the analyzed set is "Automated trading with performance weighted random forests and seasonality" by A. Booth, E. Gerding, and F. Mcgroarty [2]. It proposes a system that uses machine learning techniques to predict the price return over seasonal events which is then able to develop a profitable trading strategy. Analyzing various regression techniques as well as exploring the merits of various techniques for expert weighting showed that recency-weighted ensembles of random forests produce superior results in terms of both profitability and prediction accuracy compared with other ensemble techniques.

The second most cited publication is "Cryptocurrency Portfolio Management with Deep Reinforcement Learning" by Zy. Jiang and Jj. Liang [5], in which authors present a model-less convolutional neural network with historic prices of a set of financial assets as its input, outputting portfolio weights of the set. The network is not limited to cryptocurrency and can be applied to any other financial market.

In third place is "Forecasting price movements using technical indicators: Investigating the impact of varying input window length" authored by Y. Shynkevich et al. [10]. As a creation of a predictive system that correctly forecasts future changes in a stock price is crucial for investment management and algorithmic trading, this study explores how the performance of the-predictive system depends on a combination of a forecast horizon and an input window length for forecasting variable horizons.

Fourth in the ranking is "Deep Learning for financial applications: A Survey" by Am. Ozbayoglu, Mu. Gudelek and Ob. Sezer [8]. In this paper, the authors try to provide a state-of-the-art snapshot of the developed deep learning models for financial applications and identify possible future implementations and research within the field.

The last spot in the top five most cited publications is taken by "Robust Median Reversion Strategy for Online Portfolio Selection" by Dj. Huang et al. [4] in which the revision phenomenon was exploited using robust L1-median estimators to design novel online portfolio selection strategy named "Robust Median Reversion" (RMR). Empirical results showed that RMR can overcome the drawbacks of existing mean reversion algorithms and achieve significantly better results, additionally being suitable for large-scale real-time algorithmic trading applications.

3 Examination

3.1 Python Version

To properly examine the impact of Python version on the speed of a script execution, four different interpreter versions were used: Python 3.7.4 (Jul 8, 2019), Python 3.8.2 (Feb 24, 2020), Python 3.9.17 (Jun 15, 2023) and Python 3.11.4 (Jun 15 2023). The oldest version - Python 3.7 is no longer supported with Python 3.7.17 being the last security update on Jun 6, 2023.

The chosen Python 3.7.4 version is a middle bugfix release. Python 3.8 will reach its end of life in October of 2024, with the 3.8.2 version being one of many bugfix releases. Whereas Python 3.9 is expected to receive no further updates before October 2023, with 3.9.17 being the latest available security update. Python 3.11 is the latest released interpreter version with 3.11.4 being the newest bugfix. It is worth noting that Python 3.11 is measured to be between 10–60% faster than the previous version [1].

3.2 Data Sets and Python Libraries

The tested script consists of two different machine learning models: Decision Tree and Random Forest which are combined both with Train Test Split method and K-fold cross-validation. Each model and method/cross-validation combination are used in both Regression and Classification problems each using two different data sets.

Both data sets are imported by Python library "yfinance" which allows data from Yahoo! Finance's API to be downloaded. For this research the retrieved data is historical market prices for Google and Amazon. It is then preprocessed by creating the needed predictors for trend following and range trading, including EMA (Exponential Moving Average), ART (Average True Range), ADX (Average Directional Index), RSI (Relative Strength Index) and MACD (Moving Average Convergence/Divergence), by using the "TA-Lin" Python library. The chosen Python libraries and stock indicators are commonly used in algorithmic trading, hence their use in this research.

Python libraries, "Pandas" and "Scikit-learn", updated to the latest possible version are used for machine learning models implementation with Python built-in "time" module used for time measurements. The newest 23-rd "NumPy" library version is used as it is the latest version compatible with other libraries.

Measurement of the execution time is done using Pythons "time" module and repeated ten times for each model, cross-validation/method and data set combination which also includes the data cleaning and prediction. The computer used for the tests have macOS 13.5 (22G74) operating system installed, a Quad-Core Intel Core i5 processor (2,4 GHz) and 16 GB RAM. The mean of the collected results is taken and presented on charts.

3.3 Results

The first two combinations are the Decision Tree model with Train Test Split method used on both classification (DTC + TTS) Fig. 2 and regression (DTR + TTS) Fig. 3 problems. In the first arrangement, Python 3.9 and 3.11 were the fastest with Python 3.9 being faster only by 0.6% with the execution time of 0.0318 s. Followed by Python 3.8 with 0.0331 s of execution time and Python 3.7 - 0.0900 s, which are accordingly 4% and 95% slower than Python 3.9. The situation is similar when using DecisionTreeRegressor class, where the disparity between the fastest - being Python 3.11, and the slowest -Python 3.7, is around 101%. This time the second place was taken by Python 3.8 (2% slower), followed by Python 3.9 (3% slower). Overall, this combination was not time-consuming for newer Python versions although still slight distinctions can be made, with the oldest tested version being visibly much slower then the rest.

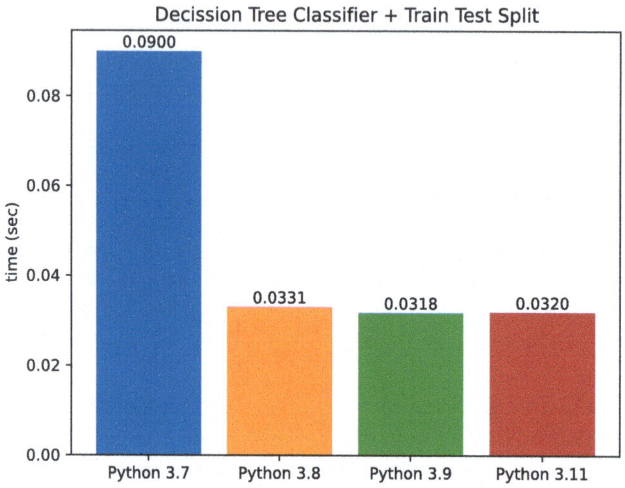

Fig. 2. Execution time for DTC + TTS

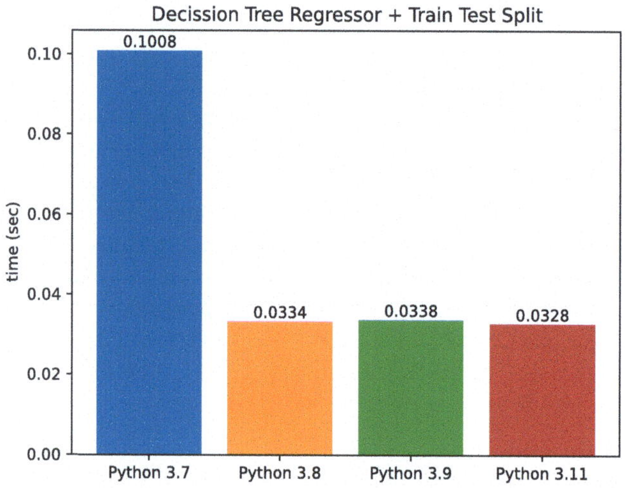

Fig. 3. Execution time for DTR + TTS

Next, the Decision Tree model was paired with K-fold cross-validation- DTC + Kf results are presented in Fig. 4, DTR + Kf on Fig. 5. Although the measured elapsed times are visibly much longer than the previously described, the differentiation between Python versions at first seem to still apply with Python 3.11 outperforming Python 3.8 by 0.6% and Python 3.9 by 3% dealing with classification solution. This example however sees Python 3.7 deal with the task 78% slower than the fastest Python 3.11 which is a significant decrease compared with previously described results. The regression model looks very similar

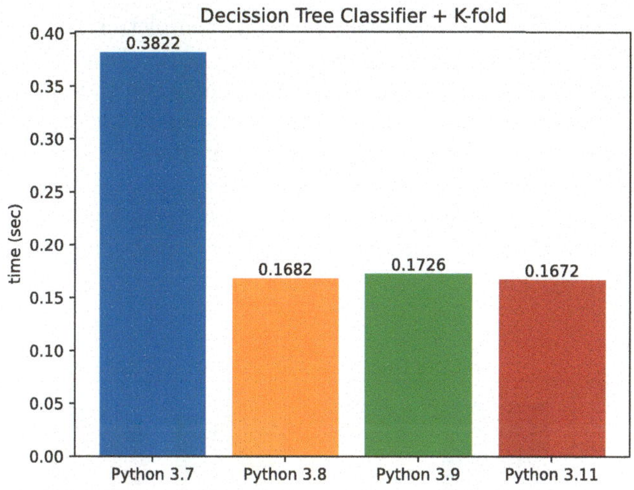

Fig. 4. Execution time for DTC + Kf

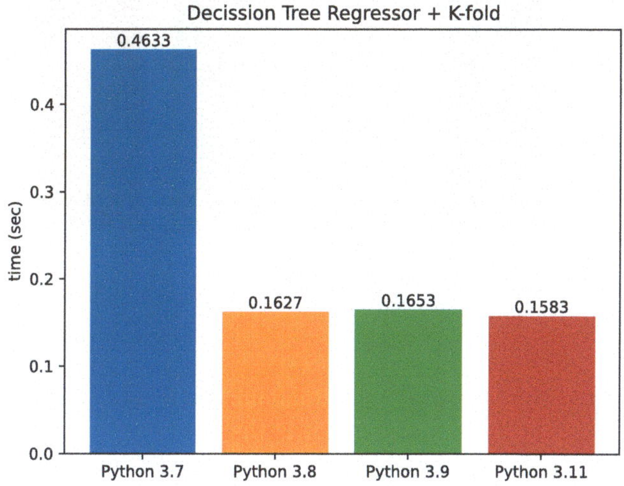

Fig. 5. Execution time for DTR + Kf

to the classification with the only noticeable difference being between the fastest Python 3.11 and slowest Python 3.7 being again around 98%.

The second examined machine learning model is Random Forest, which likewise was paired with Train Test Split method and used for classification (RFC + TTS, 6) as well as regression (RFR + TTS 7) solutions. From the start, it is immediately visible that the Random Forest model is much more time-consuming than the formerly described one, which is to be expected as it is constructing several Decision Trees during training time. Figure 6 shows again Python 3.11

outperforming older versions being the only one completing the task in a little over 0.5 s - 6% faster than the slowest in this test Python 3.7. Interestingly, Python 3.8 and 3.9 came ex aequo second running the script on average for 0.6552 s. The same model and method combination used for regression analysis turned out to be even more time-consuming with Python 3.8 surprisingly taking, on average, the least time to complete it - 1.9602 s. The rest of the Python versions maintaining the pattern of Python 3.11 overtaking Python 3.9, this time only by 0.2%, with Python 3.7 being the worst performing interpreter version - 11% slower than Python 3.8 (Fig. 7).

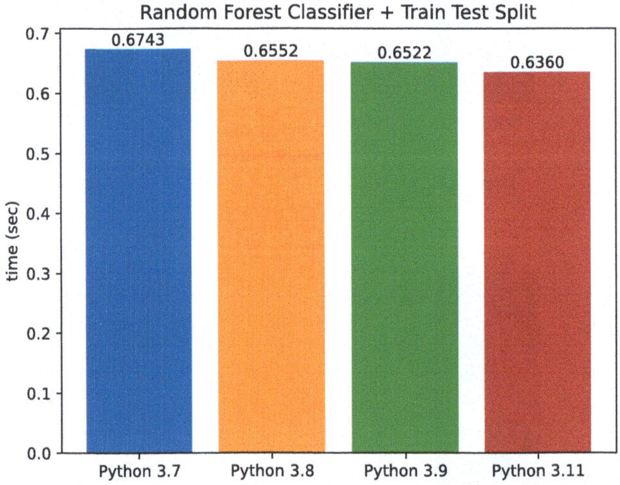

Fig. 6. Execution time for RFC + TTS

The last tested combination is the Random Forest model with K-fold cross-validation. As seen in Fig. 8 and Fig. 9 it is the most time-consuming variation of all described in this study with the Python 3.8 achieving the best results in regression analysis running on average for 9.2234 s, and surprisingly Python 3.7 suppressing other tested interpreter versions in classification by needing on average 3.2087 s complete its run. In the classification case, Python 3.9 required the most time - being 10% slower than the best-performing Python version, whereas in the regression case the slowest was Python 3.7 but only by 2% difference from the best best-performing version.

In Fig. 10 and Fig. 11 all measured average times are presented for classification and regression analysis respectively. It shows Python 3.7 being the least optimized for such tasks, within both classification and regression versions in almost all tested cases. For both classification and regression solutions, the achieved performance of Python 3.8, 3.9 and 3.11 are comparable.

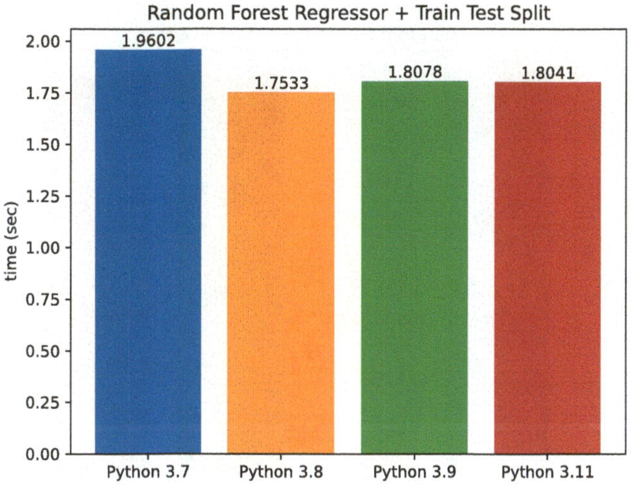

Fig. 7. Execution time for RFR + TTS

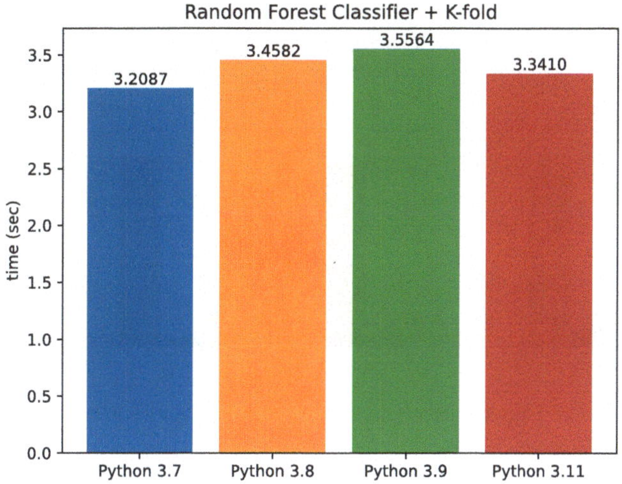

Fig. 8. Execution time for RFC + Kf

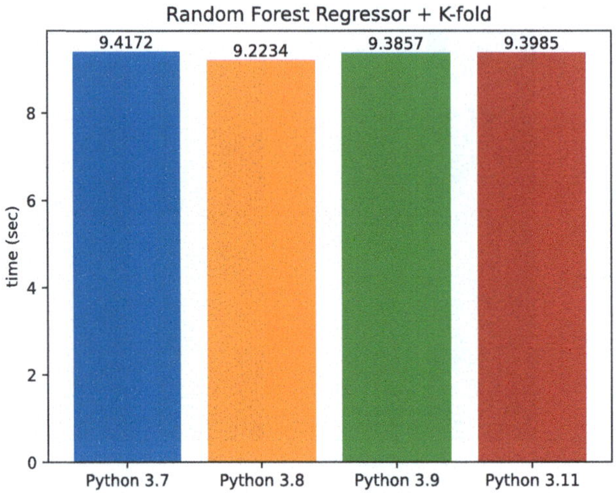

Fig. 9. Execution time for RFR + Kf

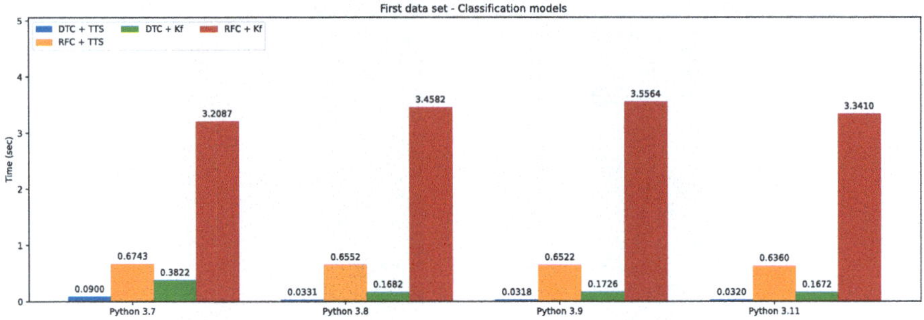

Fig. 10. Execution time for DTC and RFC

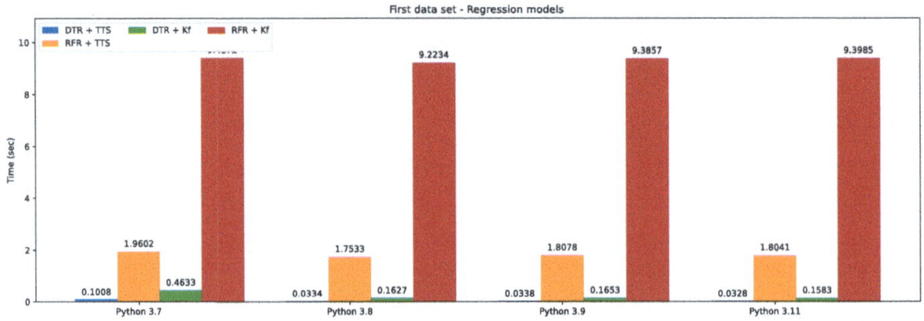

Fig. 11. Execution time for DTR + RFR

4 Conclusions

Over the last decades, artificial intelligence has revolutionized many aspects of our lives, one of which is the way we look at the finance sector and trading. Algorithmic trading, thanks to its many benefits, has become almost a standard practice with the use of machine learning solutions. With such fast paced change, Python language- being easy to learn and readable but also powerful enough to address machine learning computations- has become a popular choice within the scientific computing community. The changes in its latest version concentrated on optimisation makes it interesting to see how much it improves the run time of machine learning solutions for algorithmic trading. The results show improvement of Python 3.11 over Python 3.9 and Python 3.8 performance, however, in some cases the oldest Python 3.7 achieved slightly better results. It still paints a bright future for Python use in machine learning and algorithmic trading as it confirms the positive directions of changes made in the newest versions. The optimizations not only shorten the execution time but also improve the computer resources management, which can also have an impact on carbon footprint within companies. With new Python 3.12 being in a prerelease stage of development it will be interesting to see if the optimization improvements will continue allowing more demanding tasks to be done in Python, which may be explored in future research.

References

1. PEP 664 – Python 3.11 Release Schedule — peps.python.org. https://peps.python.org/pep-0664/#features-for-3-11
2. Booth, A., Gerding, E., McGroarty, F.: Automated trading with performance weighted random forests and seasonality. Expert Syst. Appl. **41**(8), 3651–3661 (2014)
3. Buchanan, B., Wright, D.: The impact of machine learning on UK financial services. Oxf. Rev. Econ. Policy **37**(3), 537–563 (2021)
4. Huang, D., Zhou, J., Li, B., Hoi, S., Zhou, S.: Robust median reversion strategy for online portfolio selection. IEEE Trans. Knowl. Data Eng. **28**(9), 2480–2493 (2016)
5. Jiang, Z., Liang, J.: Cryptocurrency portfolio management with deep reinforcement learning, pp. 905–913. IEEE (2017). https://doi.org/10.1109/IntelliSys.2017.8324237
6. Lin, T.C.W.: The new investor (2013). https://papers.ssrn.com/abstract=2227498
7. Napate, S., Thakur, M., B-School, D.: Algorithmic trading and strategies (2020)
8. Ozbayoglu, A., Gudelek, M., Sezer, O.: Deep learning for financial applications: a survey. Appl. Soft Comput. **93**, 106384 (2020)
9. Raschka, S., Patterson, J., Nolet, C.: Machine learning in python: main developments and technology trends in data science, machine learning, and artificial intelligence. Information **11**(4), 193 (2020)
10. Shynkevich, Y., McGinnity, T., Coleman, S., Belatreche, A., Li, Y.: Forecasting price movements using technical indicators: investigating the impact of varying input window length. Neurocomputing **264**, 71–88 (2017)

Forecasting e-learning Course Purchases Using Deep Learning Based on Customer Retention

Paweł Golec[1], Marcin Hernes[1], Tomasz Sajewski[2], Ewa Walaszczyk[1](✉), Artur Rot[1], Marcin Fojcik[3], Tomasz Turek[4], and Damian Dziembek[4]

[1] Wroclaw University of Economics and Business, Wrocław, Poland
{pawel.golec,marcin.hernes,ewa.walaszczyk,artur.rot}@ue.wroc.pl
[2] CRP We Wrocławiu Sp. Z o.o, Wrocław, Poland
t.sajewski@crpwewroclawiu.pl
[3] Western Norway University of Applied Sciences, Bergen, Norway
marcin.fojcik@hvl.no
[4] Czestochowa University of Technology, Częstochowa, Poland
{tomasz.turek,damian.dziembek}@wz.pcz.pl

Abstract. Retention analysis is essential from the point of view of customer behavior, including purchases of products. Existing purchasing forecasting approaches are based on analyzing many user profile characteristics, which are only sometimes available to e-learning course providers. The aim of this research is to develop a method for forecasting e-learning course purchases using deep learning based on customer retention. The multi-layer perceptron and long short-term memory are used. Analyzing the research results, it can be seen that the long short-term memory network achieved significantly better results than the multi-layer perceptron. Using a developed approach, there is no need to acquire additional data characterizing the user, such as demographic data. From the business point of view, it is a crucial benefit for new customers who are not registered in the customer relationship management system, but use anonymously this system via the website.

Keywords: purchases forecasting · e-learning courses · customer retention · deep learning

1 Introduction

Nowadays, online shopping has transformed the traditional shopping trend enormously. This shift has taken an altogether enlarged view after the COVID-19 pandemic, which provided a world full of opportunities for customers to purchase online at home. Understanding the needs of the customer is the key to success in online shopping [3]. To promote business, sales prediction and a sense of future consumer behavior can help business decision-makers in marketing campaigns, budgets, and resource planning [4]. The main objective of consumer behavior analysis is to identify valuable products consumers are likelier to buy. Consumer behavior analysis is a vital part of accurate recommendations, and researchers in different fields have provided different ideas to solve the

M. Hernes and J. Wątróbski (Eds.): ECAI 2023, LNNS 1079, pp. 142–155, 2024.
https://doi.org/10.1007/978-3-031-66761-9_13

problem from the perspective of consumer behavior processes [5]. Purchase prediction has gained significant attention in recent years, especially in e-commerce. Deep learning methods like neural networks have become increasingly popular in this field.

One of the critical parameters of users' behavior is retention, defined as returning to the application or performing one of the assumed actions by the application user at least once within a selected period [18]. From the service provider's point of view, retention is a very important issue in the life cycle of the enterprise-customer relationship, which includes:

- Identification,
- Acquisition,
- Retention (including retention analysis),
- Development.

Thus, retention analysis is important from the point of view of customer behavior, which is the primary goal of customer relationship management, since retaining existing customers is 5 to 20 times more profitable than acquiring new ones (depending on business fields) [1].

In addition, the paper [2] found that statistical methods and traditional data mining methods, such as regression analysis and decision trees, are mainly used to study customer retention in the mobile app market. The limitations of these methods were also demonstrated, such as:

1. Data mining methods have a relatively short shelf life. The application market, including mobile applications, faces new technologies on a daily basis. As a result, historical data becomes less valuable for forecasting.
2. The level of analysis in data mining methods reduces the ability to capture the heterogeneity of customer behavior.

New methods are therefore needed to support customer retention analysis to circumvent the drawbacks of traditional data mining techniques.

Based on an analysis of the literature [e.g., 17–19] on retention research in other areas of the market, it can be assumed that the limitations presented can be largely eliminated through the use of machine learning (ML) methods, including deep learning (DL). The authors of this work conclude that ML techniques such as artificial neural network (ANN), AdaBoost, and support vector machine (SVM) are most promising for studying customer retention in various business areas.

The aim of this research is to develop a method for forecasting e-learning course purchases using deep learning based on customer retention.

The rest of the paper is structured as follows: the next part presents the background, including the indication of the research contribution. Next, the research methodology is presented. The last part of the paper concerns the analysis of research results and conclusions.

2 Background

Analyzing the literature on the application of deep learning in predicting customer purchases reveals several key findings and insights. Researchers have explored various approaches, methodologies, and challenges associated with using deep learning models for customer purchase prediction [6]. Forecasting future purchases on a customer level offers businesses a great opportunity to personalize marketing efforts and focus their resources efficiently. In many cases, historical customer data is readily available, and Machine Learning and Deep Learning methods can be used to make valuable predictions, for example, the time of a customer's next order. Analyzing the literature, it can be assessed that most retail stores currently rely on manual order size estimation when it comes to sales forecasting models [7].

In the article [8], the authors have compared their deep learning approach's predictive efficacy to that of some of the most popular supervised learning methods for forecasting, such as Decision Trees, Support Vector Machines, Random Forests, and ANNs. The deep learning approach performed better on the same dataset compared to other machine learning methods. The Authors of the work [9] compared the predictive powers of different techniques to predict actual purchases on an online shopping platform. To this end, they analyzed the actual purchases of online customers across more than 50,000 sessions on an online shopping platform and predicted them accurately with the help of DL and ML techniques. In the [4] study, deep learning based on LSTM has been developed for the NASDAQ stock exchange prediction. The proposed model has been extensively tested and compared with ARIMA. In the work [10], the authors propose the Deep Interest Network (DIN) model based on Recurrent Neural Networks (RNNs) and Convolutional Neural Networks (CNNs). This model considers user interests and product information, leading to improved effectiveness in predicting click-through rates and conversions in recommender systems. In the paper [11], the authors present a model based on Recurrent Neural Networks (RNNs) for predicting purchases based on user session data. RNNs are used to model user activity sequences within sessions, enabling personalized recommendations. The paper [12] addresses the problem of personalized recommendation in e-commerce, focusing on predicting the next item a customer is likely to purchase. The authors propose a deep learning model called Caser, which combines convolutional neural networks (CNNs) and RNNs to capture both the sequential and hierarchical features of customer behavior.

Predicting customers' decisions to purchase has been considered by many researchers in different studies, including management, marketing, information systems, neuroscience, and so on [13, 14]. Various models were developed to determine purchase decisions, and neuroscience scholars have used different techniques to predict purchases. While each of these different approaches has found some success, there still exists an immense opportunity to further extend our knowledge in this area, in particular, the prediction of purchases of online courses, which the literature treats very briefly and to a limited extent.

Considering customer retention, the papers [15, 16] state that mainly statistical methods and data mining methods, such as regression analysis and decision trees, are used to study customer retention in the IT application market. For example, the works [17,

18] presented various methods for analyzing customer retention from the telecommunications sector. Machine learning is also being used to study retention in the retail sector [19]. Also, retention study approaches using machine learning are being applied to the higher education sector; for example [20], Work [21] deals with retention studies in the banking sector. There is also a need for approaches in the literature regarding the use of machine learning methods (including deep learning) to study customer retention of e-learning courses. It should be noted that IT applications, especially e-learning courses, have characteristics that distinguish them from products and services in other market sectors. These features include, among others, availability regardless of place and time, ease of personalization, a certain level of information coverage, and a specific user interface. These features mean that in the computer application market, particularly e-learning courses, more than the application of a general approach to retention analysis, or approaches used for other market sectors, may be required. At the same time, it should be noted that as machine learning methods, including deep learning, are used for other market segments, there is a high probability that they will allow to improve the retention analysis process and obtain features of innovative technologies, also for the segment of computer applications, including e-learning courses. Their advantage over other methods used in retention analysis is, among other things, their ability to classify, recognize, detect, describe data, and automatically analyze the causal relationships between data. These properties will, therefore, be possessed by methods for forecasting e-learning course purchases using deep learning based on customer retention developed within this research.

Summing up, previous work on purchase forecasting often uses user profile data. For example, in the work [16], a number of user profile characteristics are considered. In the work [22], on the other hand, many attributes of customer satisfaction, customer behavior, customer turnover costs, customer characteristics, marketing activities, and social relations are analyzed. Thus, the presented approaches are based on the analysis of many user profile characteristics, which are not always available to e-learning course providers. Often, users do not have the need (or do not want) to fill out data forms. Hence, this research contributes to analyzing customer behavior to shape pricing policy through machine learning methods using only retention levels.

3 Research Methodology

3.1 Data Processing

The research used data provided by a company selling e-learning courses in Poland. Customer data was collected from the existing Customer Relationship Management system and from the www.synerise.com system, which the company uses. The obtained data was saved in the form of a set of user data and markers describing their retention.

The "raw" data contained 50,327 records described by 107 fields (variables). The fields sale_id, email, date, and price were highlighted in the result set. The email field has been anonymized - and replaced by numeric tokens (IDs). An RFM (Recency, Frequency, Monetization) analysis was performed for these data. The result set was connected with the processed set via the email field. The fields selected for the final analysis are presented in Table 1.

Table 1. Fields selected for analysis

Field (variable) name	Type	Meaning
nazwa_kategoria	Nominal	Course category name
name_channel	Nominal	Sales channel name
cena	Metric	Course price
czas_trwania	Nominal	Course duration (as a text string)
type_course	Nominal (binary)	Online or on-site course
discount_used	Nominal (binary)	Whether a discount was applied
Recency	Metric	R - Relevance - time elapsed since the last purchase (destimulant[1])
Frequency	Metric	F - Purchase frequency (stimulant[2])
Monetization	Metric	M - Monetization (stimulant[3])

Null values (NA) were removed from the dataset. In the end, after initial preparation, the dataset had 31,667 records and ten fields defined.

The next step was data transformation. In several places, the data contained gaps or notes left by the administrator. It gave the impression that the dumps from the database were combined into one CSV file. Fortunately, this was not a recurring problem, so correcting it manually and then saving the file was easy. Thus, the data set was brought to the appropriate table.

3.2 Methods

We use the following methods in this research:

1. Multi-layer perceptron (MLP] is a complement to feed-forward neural networks. MLP consists of three types of layers [23]:
   - input layer - receives the input signal for processing,
   - hidden layer - the MLP computation engine,
   - output layer - performs a task such as prediction or classification.

   MLP can approximate any continuous function and solve linearly inseparable problems. MLP is used to classify, recognize, predict, and approximate patterns.
2. Long short-term memory (LSTM) is an artificial recurrent neural network architecture used in deep learning. LSTM networks compensate for the problem of fading gradients and short-term memory of the traditional recurrent neural network. The architecture of LSTM is the state of the cell (the memory unit of the network) and its controllers. The cell state carries information that can be stored, written, or read from the previous cell state by gates. LSTM is used in deep learning tasks such as prediction in the stock market or handwriting and speech recognition [24].

[1] The nature of the variable was not taken into account when constructing the tree.

[2] The nature of the variable was not taken into account when constructing the tree.

[3] The nature of the variable was not taken into account when constructing the tree.

4 Results

For retention analysis, a pivot table was developed, where the columns were dates set every week, while the values were the sum of transactions made in a given week. With data prepared this way, as many consecutive weeks as possible were extracted in sets of six and one predicted. The method of data preparation is shown in Fig. 1.

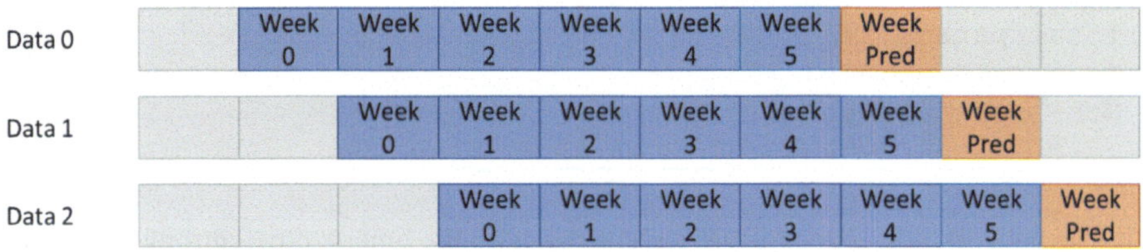

Fig. 1. Data selection method.

The last modifications of the dataset were the transformation of the PCA into input data (marked in blue in Fig. 1) and assigning the value 1 to those output weeks (red fields in Fig. 1) that had a value greater than 0. As a result, a new dataset was obtained, the histogram of which is shown in Fig. 2.

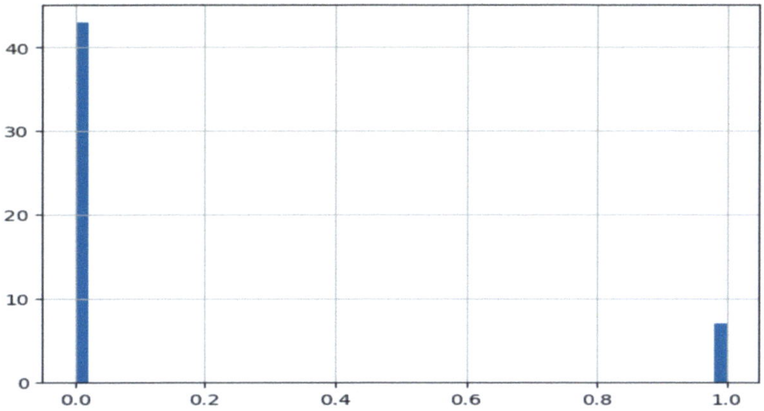

Fig. 2. Real data histogram.

4.1 Development of the Basic MPL

A basic MPL model was developed in the first stage of data analysis. Several hyper-parameters of MLP were tested. The best MLP network hyperparameters are as follows:

```
Model: "sequential_79"
_________________________________________________________________
 Layer (type)                Output Shape              Param #
=================================================================
 dense_92 (Dense)            (None, 6)                 42
 dense_93 (Dense)            (None, 3)                 21
 dense_94 (Dense)            (None, 1)                 4
=================================================================
Total params: 67
Trainable params: 67
Non-trainable params: 0
_________________________________________________________________
```

The network was trained using ten epochs, with the binary_crossentropy loss function, the Adam optimizer, and accuracy was the metrics for assessing network quality. The efficiency achieved by the model equaled 85.7%. The learning process is shown in Figs. 3 and 4.

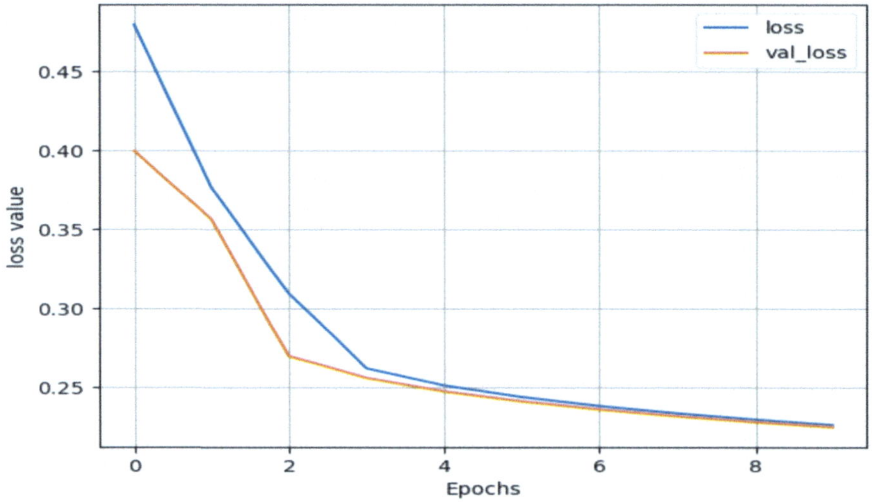

Fig. 3. MPL network learning plot - loss value.

Measures in Fig. 3 suggest model has not encounter any difficulties during training. Both training and validation losses achieved similar values at the end of training. Another important thing to mention is that as fitting was closing to an end, loss values didn't change as much as at the beginning. This provides a conclusion that the model's fitting has come to an end. However, as can be seen, further results reveal some serious mistakes.

By looking at Fig. 4, all conclusions could be confirmed. The accuracy that the model provides is not getting any better in time. Training and validation performance are similar to each other. Figure 5 shows the histogram of the forecasted data.

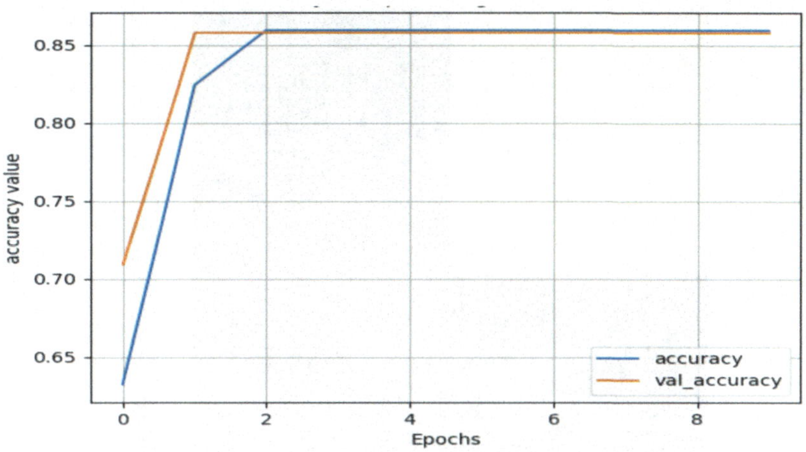

Fig. 4. MPL network learning plot - efficiency.

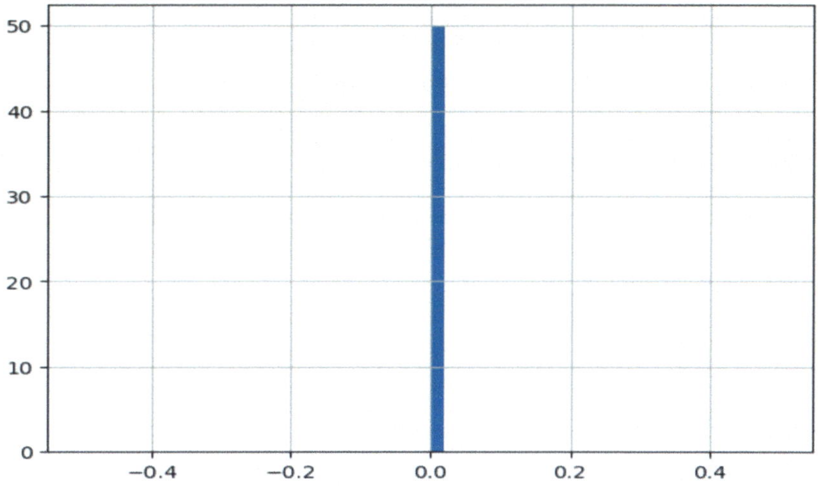

Fig. 5. Histogram of values predicted by the MPL model.

As mentioned before, the prediction values shown in Fig. 5 exposed that the MLP model has found only one class. This is an example of overfitting since the dataset used for training was dominated by class 0 (which means that customers won't purchase an item).

The last analysis of the results for the MLP network was the confusion matrix presented in Fig. 6.

The confusion matrix displayed in Fig. 6 dissolves any other doubts that the model couldn't predict any purchase example. It still had 85% accuracy, but only because of data used to train neural network.

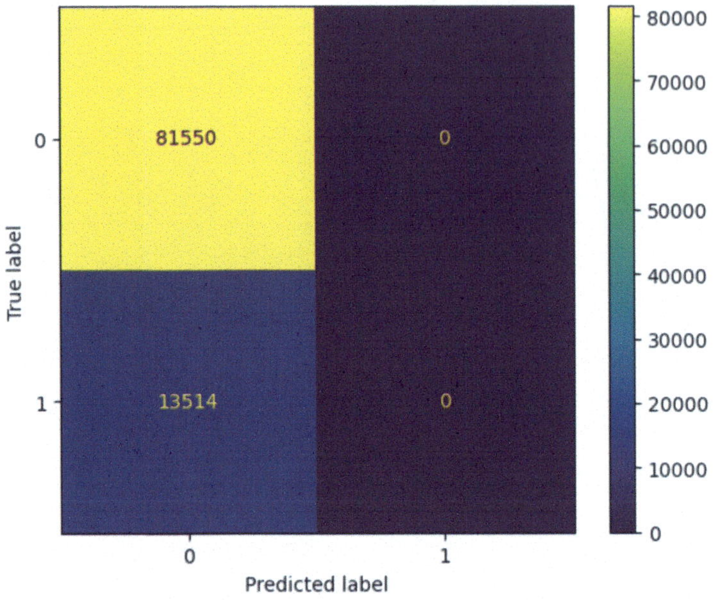

Fig. 6. Confusion matrix for the MLP model.

4.2 Development of a Classifier Based on LSTM

A classifier based on long-short term memory (LSTM) was developed in the next research stage. Several hyperparameters of LSTM were tested. The best LSTM network hyperparameters are as follows:

```
Model: "sequential_78"
_________________________________________________________________
 Layer (type)                Output Shape              Param #
=================================================================
 lstm_127 (LSTM)              (None, 1, 6)                312
lstm_128 (LSTM)             (None, 1, 3)                120
dropout_40 (Dropout)        (None, 1, 3)                0
flatten_66 (Flatten)        (None, 3)                   0
 dense_91 (Dense)             (None, 1)                    4
=================================================================
Total params: 436
Trainable params: 436
Non-trainable params: 0
_________________________________________________________________
```

Finally, the LSTM network achieved an efficiency of 99.2%. The training process was performed using the same parameters as in the case of the MLP network. Thus, the learning graphs shown in Figs. 7 and 8 look alike in Figs. 3 and 4.

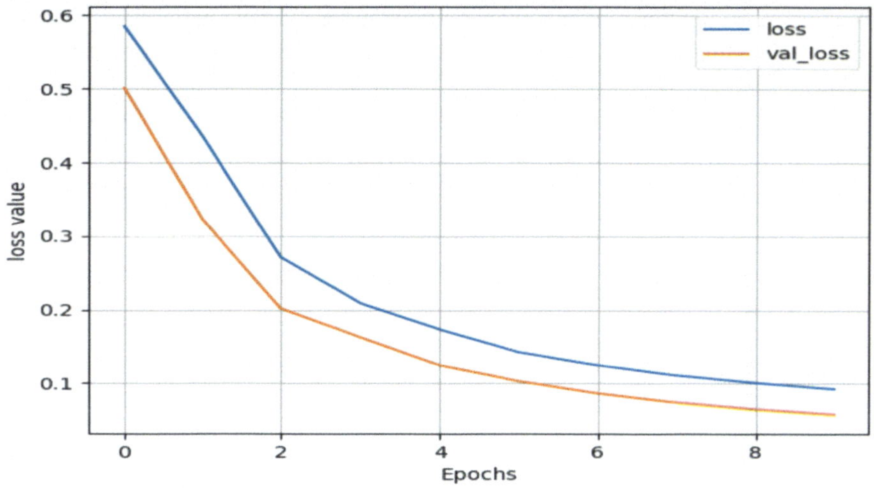

Fig. 7. LSTM network learning plot - loss value

In this case, the training process (as shown in Fig. 7) was similar to the MLP model. One can see the difference between training and validation measures. What is interesting is the fact that the loss is slightly lower than the training one. It can mean that the training dataset was more difficult for a model to predict, while on the other hand, validation cases were simple and easy to manage.

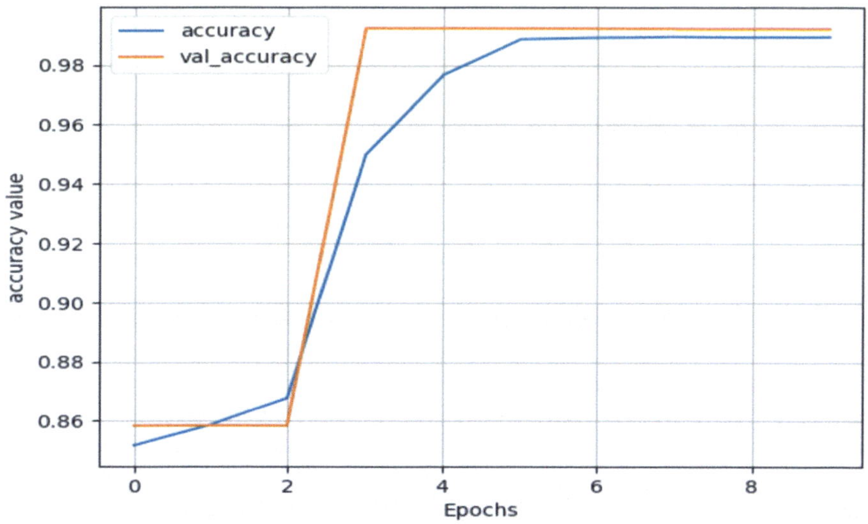

Fig. 8. LSTM network learning plot – efficiency

The accuracy of prediction results is displayed in Fig. 8. It can be seen that validation accuracy went way up after three epochs. This suggests that the test dataset contains, indeed, more straightforward data than the training one. Again, performance on validation and training datasets stands near each other because their accuracy values are similar.

The values predicted by the LSTM model are presented in Fig. 9 in the form of a histogram.

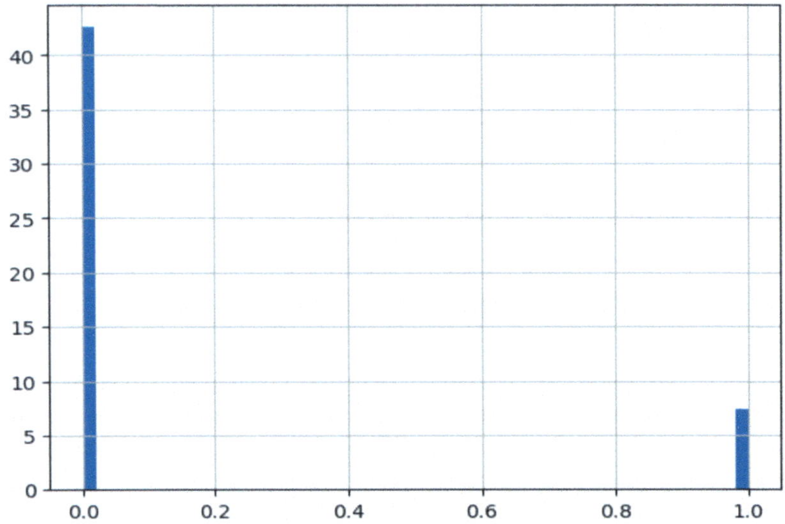

Fig. 9. Histogram of values predicted by the LSTM model.

As Fig. 9 shows, the LSTM model recognized all classes. More importantly, the distribution of predicted classes looks similar to the real one. However, to be specific, these results should be compared with the confusion matrix.

The confusion matrix for the LSTM network is presented in Fig. 10.

Looking at the confusion matrix shown in Fig. 10, one can easily see that most cases were labeled correctly. Even more interesting is that the model makes no mistake when customers buy an item. This ensures that the model won't miss any opportunity to sell a product.

Analyzing the research results, it can be seen that the LSTM network achieved significantly better results than the MLP network. This compares the accuracy results of the MLP model (85%) and the LSTM network (99.2%). The LSTM model was more effective and correctly predicted the purchase (13 446 correctly detected purchases). There is less such data, so it is a more difficult task. Comparing the histograms of accurate data and data predicted using the LSTM network, a similarity is noticeable in both graphs. This is also confirmed by the network confusion matrix, which shows a difference at 0.8% of dataset length (in this case, it is 725 examples).

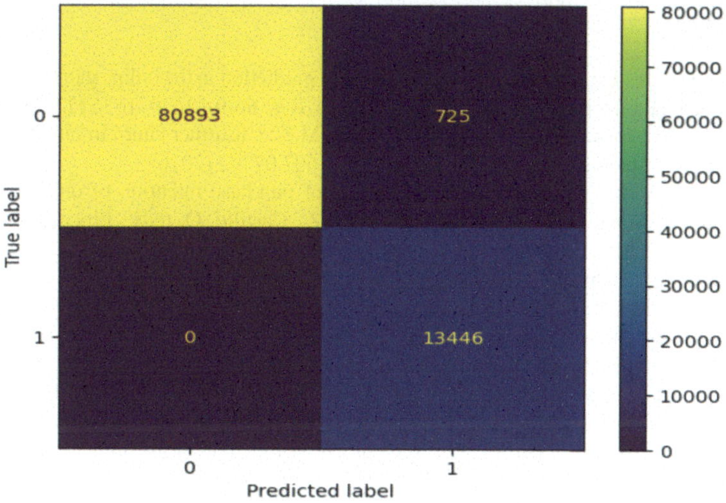

Fig. 10. Confusion matrix for the LSTM model.

5 Conclusion

The developed method for forecasting e-learning course purchases using deep learning performs solely based on retention data, contributing to the theory of the studied area. Therefore, there is no need to acquire additional data characterizing the user, such as demographic data (gender, age, etc.), the ability to operate in the lack of additional data availability. The best results have been achieved using LSTM deep neural network. The main implication of the theory is using only retention levels for analyzing customer behavior to shape pricing policy through machine learning methods. From the business point of view, it is a crucial benefit for new customers who need to be registered in the CRM system but use this system anonymously via the website.

The developed method allows for automatically forecasting customer purchases by companies selling e-learning courses. The main limitation of the research is the use of data from only one company. Therefore, future works should concern extending the research to other data sets, including data from companies operating in other areas of e-commerce. This will make it possible to increase the developed approach's generalization level.

Founding. The research was founded by a project: "Vouchers to support the innovation of Lower Silesian enterprises," Priority axis 1 Enterprises and innovations, Measure 1.2 Innovative Enterprises, Sub-measure 1.2.1 Innovative Enterprises – Horizontal Competition, Type 1.2.C.b Enterprise services - "Innovation voucher. Project title: Innovative technology to improve the offer personalization process, increasing the customer retention level of CRP in Wrocław.

Fundusze Europejskie Program Regionalny   Rzeczpospolita Polska   DOLNY ŚLĄSK   Unia Europejska Europejski Fundusz Rozwoju Regionalnego

References

1. Coussement, K., De Bock, K.W.: Customer churn prediction in the online gambling industry: The beneficial effect of ensemble learning. J. Bus. Res. **66**(9), 1629–1636 (2013)
2. Hassouna, M., Tarhini, A., Elyas, T., AbouTrab, M.S.: Customer churn in mobile markets a comparison of techniques. arXiv preprint arXiv:1607.07792 (2016)
3. Bansal, M., Vyas, V.: Analysis and prediction of purchase intention of online customers with deep learning. In: Khanna, A., Polkowski, Z., Castillo, O. (eds.) Proceedings of Data Analytics and Management. Lecture Notes in Networks and Systems, vol 572. Springer, Singapore (2023). https://doi.org/10.1007/978-981-19-7615-5_16
4. Kobiela, D., Krefta, D., Król, W., Weichbroth, P.: ARIMA vs LSTM on NASDAQ stock exchange data. Procedia Comp. Sci. **207**, 3836–3845 (2022)
5. Sundararaj, V., Rejeesh, M.R.: A detailed behavioral analysis on consumer and customer changing behavior concerning social networking sites. J. Retail. Cons. Ser. **58**, 102190 (2021). ISSN 0969-6989, https://doi.org/10.1016/j.jretconser.2020.102190
6. Guo, L., et al.: Buying or Browsing?: Predicting Real-time Purchasing Intent using Attention-based Deep Network with Multiple Behavior. In: Proceedings of the 25th ACM SIGKDD International Conference on Knowledge Discovery & Data Mining, pp. 1984–1992 (2019)
7. Kao, C.-Y., Chueh, H.-E.: Deep learning based purchase forecasting for food producer-retailer team merchandising. Scientific Programming 2022, Article ID 2857850, 9 (2022). https://doi.org/10.1155/2022/2857850
8. Nisha, Singh, A.S.: Customer behavior prediction using deep learning techniques for online purchasing. In: 2023 2nd International Conference for Innovation in Technology (INOCON), pp. 1–7. Bangalore, India (2023). https://doi.org/10.1109/INOCON57975.2023.10101102
9. Chaudhuri, N., Gupta, G., Vamsi, V., Bose, I.: On the platform but will they buy? Predicting customers' purchase behavior using deep learning. Decision Support Systems **149** (2021). https://doi.org/10.1016/j.dss.2021.113622
10. Zhou, G., et al.: Deep Interest Network for Click-Through Rate Prediction. In: Proceedings of the 24th ACM SIGKDD International Conference on Knowledge Discovery & Data Mining (KDD' 18), pp. 1059–1068. Association for Computing Machinery, New York, NY, USA (2018). https://doi.org/10.1145/3219819.3219823
11. Hidasi, B., Karatzoglou, A., Baltrunas, L., Tikk, D.: Session-based Recommendations with Recurrent Neural Networks. Computer Science, ICLR (2016). https://arxiv.org/abs/1511.06939
12. Tang, J., Wang, K.: Personalized Top-N Sequential Recommendation via Convolutional Sequence Embedding. WSDM '18: Proceedings of the Eleventh ACM International Conference on Web Search and Data Mining February 565–573 (2018). https://doi.org/10.1145/3159652.3159656
13. Singhal, R., et al.: Fast online 'next best offers' using deep learning. In: Proceedings of the ACM India Joint International Conference on Data Science and Management of Data. CoDS-COMAD '19, pp. 217–223. ACM, New York, NY, USA (2019)
14. Vieira, A.: Predicting online user behaviour using deep learning algorithms. arXiv preprint arXiv:1511.06247 (2015)
15. Huang, C., et al.: Online purchase prediction via multi-scale modeling of behavior dynamics. In: Proceedings of the 25th ACM SIGKDD International Conference on Knowledge Discovery & Data Mining, pp. 2613–2622 (2019)
16. Spanoudakis, N., Moriaitis, P.: Engineering an agent-based system for product pricing automation. Eng. Intell. Syst. **17**(2), 139 (2009)
17. Qureshi, S.A., Rehman, A.S., Qamar, A.M., Kamal, A., Rehman, A.: Telecommunication subscribers' churn prediction model using machine learning. In: Eighth International Conference on Digital Information Management (ICDIM 2013), pp. 131–136. IEEE (2013)

18. Premkumar, G., Rajan, J.: Customer retention in mobile telecom service market in india: opportunities and challenges. Ushus-Journal of Business Management **12**(2), 17–29 (2017)
19. Sakar, C.O., Polat, S.O., Katircioglu, M., Kastro, Y.: Real-time prediction of online shoppers' purchasing intention using multi-layer perceptron and LSTM recurrent neural networks. Neural Comput. Appl. **31**(10), 6893–6908 (2019)
20. DeLotell, P., Millam, L. yReinhardt, M.M.: The use of deep learning strategies in onlinebusiness courses to impact student retention. American J. Bus. Edu. **3**(12), 49–56 (2010)
21. Prasad, U.D., Madhavi, S.: Prediction of churn behaviour of bank customers using data mining tools. Indian J. Market. **42**(9), 25–30 (2011)
22. Misra, K., Schwartz, E.M., Abernethy, J.: Dynamic online pricing with incomplete information using multiarmed bandit experiments. Mark. Sci. **38**(2), 226–252 (2019)
23. Sivasankari, S.S., et al.: Classification of diabetes using multi-layer perceptron. In: 2022 IEEE International Conference on Distributed Computing and Electrical Circuits and Electronics (ICDCECE), pp. 1–5. IEEE (2022)
24. Ali, O.M.A., Kareem, S.W., Mohammed, A.S.: Evaluation of electrocardiogram signals classification using CNN, SVM, and LSTM algorithm: A review. In: 2022 8th International Engineering Conference on Sustainable Technology and Development (IEC), pp. 185–191. IEEE (2022)

Customer Segmentation and Anticipation of Consumer Behaviors Based on Machine Learning and CART

Dorota Jelonek[1](✉), Magdalena Graczyk-Kucharska[2], Magdalena Wyrwicka[2], and Robert Olszewski[3]

[1] Czestochowa University of Technology, Ul. Dabrowskiego 69, 42-201 Czestochowa, Poland
dorota.jelonek@pcz.pl

[2] Faculty of Engineering Management, Poznan University of Technology, Ul. Jacka Rychlewskiego 2, 60-965 Poznan, Poland

[3] Department of Geodesy and Cartography, Warsaw University of Technology, Pl. Politechniki 1, 00-661 Warsaw, Poland

Abstract. Companies should choose competitive markets for their products to maximize the efficiency of their resources. The younger generation has increasing demands for personalized products. The way their needs are met and understanding their consumer behaviors should allow companies to continuously analyze trends within the target group. In the current era of globalization, the Internet, and Big Data, using traditional methods alone may lead to companies selecting the wrong markets, resulting in significant financial and resource losses. Therefore, this article proposes the utilization of machine learning and CART for modeling customer needs to facilitate market segmentation among the younger generation, focusing on young individuals studying in the IT field. Implementing such modeling in practice can contribute to optimizing decision-making, minimizing financial losses, and resource efficiency. The study analyzed the needs of 1149 individuals aged 16–19 in the Wielkopolska region. The developed results provide an explanation of the "high salary" and "low salary" values in the form of a decision tree. Additionally, a spatial distribution map of expected salaries was visualized using the CART model. The extracted rules, which can be explicitly interpreted for each terminal node of the CART model, not only allow for spatial differentiation of the model but, most importantly, enable understanding of the motivations driving the survey respondents. The conducted research demonstrated that to comprehend the motivations of the surveyed individuals, it is crucial to consider several completely different independent variables in the process of modeling the spatial distribution of expected remuneration, including economic parameters. The conducted analyses have shown that machine learning and AI have broad applications in marketing, including customer and market segmentation.

Keywords: Customer Segmentation · Anticipation of Consumer Behaviors · Machine Learning · CART · AI

M. Hernes and J. Wątróbski (Eds.): ECAI 2023, LNNS 1079, pp. 156–165, 2024.
https://doi.org/10.1007/978-3-031-66761-9_14

1 Introduction

Artificial intelligence (AI) has rapidly evolved in recent years, with applications increasingly prevalent across various sectors of the economy, including business and management practice [1, 2], government and public sector [3], healthcare and medicine [4, 5], agriculture [6], education [7], and society at large [8].

We conceptualize AI as an ecosystem comprising three fundamental elements: data collection and storage, statistical and computational techniques, and output systems. These elements enable products and services to perform tasks typically associated with intelligence and autonomous decision-making on behalf of humans [9]. AI is a broad term encompassing various technologies such as Natural Language Processing (NLP), Machine Learning (ML), Deep Learning (DL), computer vision, and others. AI techniques empower machines to execute cognitive functions that require human intelligence, including learning, reasoning, and interaction with the environment [10]. Machine Learning (ML) and Deep Learning are two of the most well-known AI techniques.

In the conducted analyses, Machine Learning and Classification And Regression Tree (CART) were employed. CART is a predictive algorithm used in ML that explains how the values of the target variable can be predicted based on other factors.

AI technologies, supported by data analytics, are increasingly embraced by companies in response to growing customer demands in marketing [11, 12]. By integrating AI into marketing strategies, businesses can effectively utilize available data and reach potential customers with attractive advertisements at more convenient times [10].

AI has a significant impact on digital marketing, enabling marketers to understand consumer behavior, build client profiles, and comprehend the customer journey process. It allows brands to quickly and easily provide valuable personalized content tailored to various client profiles at different stages of the marketing funnel and across various channels [10]. Marketers can also leverage AI to gain insights into customer preferences and create tailored experiences based on individual preferences [13].

AI personalizes marketing, with firms utilizing AI to customize websites, emails, social media posts, videos, and other materials to align with the customer profile [14]. AI-driven tools facilitate data collection, the development of predictive models, and testing and validation with real customers. AI enables the delivery of personalized and customized emails to individual consumers [15] and assists marketers in identifying and forecasting trends [16].

The implementation of AI has the potential to revolutionize the way businesses interact with their customers [17]. AI enriches the consumer experience during the customer journey across multiple dimensions, including emotional, cognitive, behavioral, sensory, and social aspects [18–20].

This article explores the utilization of large customer datasets for customer segmentation and behavioral anticipation using ML and CART. In the context of the presented research directions on the use of AI in marketing, it is appropriate to present the results of the conducted analyses, which demonstrate the broad application of Machine Learning and CART in the field of marketing, including customer and market segmentation.

2 Literature Review

In the field of customer segmentation and behavior anticipation using AI, ML, CART, and data analysis methods, numerous studies have been conducted. In this section, we will explore some of these studies.

Customer segmentation, also known as market segmentation, plays a vital role in catering to the tastes and preferences of groups of individuals rather than individuals themselves. The primary goal of market segmentation is to identify a homogeneous customer group with similar consumption patterns. It consists of four components: identifiability, economic reachability, homogeneity of characteristics, and profitability [21].

In a study [22], customer segmentation was identified as one of the most important aspects in the competitive business world. The common approach for customer segmentation is clustering. To enhance the interpretability of cluster assignments, a decision tree-based explainability was implemented for customer segmentation in this research, considering both small and large datasets.

In the research [23], the authors examined advancements in E-commerce and created customer satisfaction brand value using AI. The analyzed results demonstrate that future online buying can be automated using AI based on customer expectations and preferences.

The study conducted by [24] focused on market segmentation using machine learning algorithms. The objective was to identify profitable products, gain insights into customer behavior, and forecast sales using the Auto Regressive Moving Average (ARIMA) model. The customers were grouped into clusters using the k-means approach based on their weekly sales. The ARIMA model was then applied to forecast the time series data, resulting in improved results.

In the research [25], an unsupervised deep learning model called Self-organizing Map with an Improved Social Spider Optimization approach was utilized for efficient customer segmentation. The experimental results confirmed the effectiveness of the proposed model in terms of clustering and segmentation, leading to improved business profits in the market.

The objective of the research [26] was to explore two data mining approaches, clustering and subgroup discovery, for customer segmentation. The study obtained six distinct market segments and identified 49 rules, providing valuable insights into customer preferences within a highly customized fashion manufacturing and e-tailoring context. These findings shed light on the scope and limitations of clustering data mining techniques, paving the way for future methodological refinements.

In the paper [27], the authors propose a general methodology called Market Segmentation Trees (MSTs) for learning market segmentations driven by identifying differences in user response patterns. Two new specialized MST algorithms were designed: Choice Model Trees (CMTs) for predicting user choice among multiple options and Isotonic Regression Trees (IRTs) for solving the bid landscape forecasting problem. The versatility of the methodology was demonstrated.

The benefits of segmentation are twofold: externally, it helps the company redefine its communication strategy, particularly sales promotions, and internally, it assists in product design, manufacturing process adjustments, and speedy delivery [26].

To achieve these benefits, it is crucial to characterize market segments based on preferred product attributes and customer profiles. A comprehensive analysis of research indicates that while statistical techniques can be employed for market segmentation, their effectiveness tends to diminish when dealing with large datasets. An alternative approach involves the utilization of decision trees, which are machine learning models capable of achieving high performance without relying on pre-existing hypotheses. Decision trees have the advantage of generating logical rules that may be less comprehensible to managers but offer valuable insights.

3 Materials and Methods

3.1 Data

In the conducted study, the expected value of the initial remuneration was chosen as the dependent variable in the developed model. Independent variables included data obtained from the survey (Group I), statistical yearbooks (Group II), and information derived from spatial analyses (Group III). Given the diverse spatial distribution of the subjects and the spatial differentiation of their financial expectations, a wide range of independent variables was utilized in the analysis.

For Group I, variables such as year, grade, place of residence, gender, maximum distance to a potential workplace, preference for individual or team work, significance of a positive work atmosphere, inclination towards working in an innovative or traditional company, and the importance of finding a job "with passion" were considered. Group II incorporated variables such as local unemployment rates (at the poviat level), migration balance in individual poviats, and the income-to-expenditure ratio in poviats. In Group III, geolocation data from the survey and associated information, such as affiliation with specific administrative divisions from Group II, distance from Poznań, and distance from the nearest city, were examined.

The analysis revealed that the expected value of the average salary for future IT specialists exceeds 3100.00 PLN, while for other professions, it is significantly lower, falling below 3000.00 PLN. Spatial distribution of the data indicated that individuals trained in IT professions either reside (represented by black dots) or attend school (represented by red dots) in 28 out of 36 poviats within the Wielkopolskie Voivodship.

To facilitate the analyses, geocoding and spatial data enrichment were performed. This involved obtaining additional attributes that characterized individual records in the database, such as calculating the distance from one's place of residence or study to Poznań and the nearest city. A custom computer program was developed using ESRI's ArcMap GIS software to calculate these distances along road networks, as opposed to straight-line distances. The analyses relied on the digital reference data from the BDOT 10 k Database of Topographic Objects, which corresponds to the accuracy of 1:10000 analog maps.

The process of data enrichment utilized tools within the ArcGIS environment, along with a custom application to determine road distances between individual objects. Subsequent analytical stages involved employing data mining algorithms. To accomplish this, the survey data enriched with the outcomes of spatial analyses were exported to the

Statistica software environment, specifically version 13.1 with the Data Miner module extension.

3.2 Methods

Due to significant variation in the dependent data observed during the study, attempts were made to explain all 1149 surveys and a dataset from which outliers were eliminated using a single model. This approach enables the analysis and development of a model that characterizes the preferences of the "dominant majority." However, removing atypical observations necessitates considering regional specificity and spatial differentiation of both the response variable and independent variables. The study employed four different approaches:

Analysis of all data (1149 subjects).

Data analysis with global omission of outlier surveys. Outliers are defined as data points deviating from the voivodeship average by more than ± two standard deviations calculated for the entire voivodeship (351 surveys).

Data analysis with local omission of outlier surveys. Outliers are defined as data points deviating from the poviat average by more than ± two standard deviations calculated for individual poviats (332 surveys).

Data analysis with local omission of outlier surveys. Outliers are defined as data points deviating from the average determined in the moving window by more than ± two standard deviations (calculated for the moving primary field with a 50-km radius scanning the voivodeship with a 10-km interval - 258 surveys).

For each of the four approaches, the study independently determined the five most important explanatory variables. The data selection and variable selection approach presented here have been previously employed in modeling HR challenges of Generation Z [28].

4 Results

4.1 CART Model

Considering the six most significant independent variables, a CART model was developed. The subsequent independent variables include:

- Year of graduation
- Unemployment rate
- Road distance to the main city
- Road distance to Poznan
- Level of passion
- Level of innovation

In previous analyses [28], the data were analyzed using methods such as linear polynomial regression, support vector method, neural network of a multi-layer perceptron type (MLP), and MARS method. In this analysis, CART was used to analyze the results. It turns out that this method exhibits the lowest error value in these analyses and additionally allows for easy data interpretation and decision-making with respect to the target group. Compared to the aforementioned methods, the model developed using the CART algorithm has the lowest error value (Table 1). This model not only allows for the development of the resulting spatial distribution map of expected salaries (Fig. 1) in the form of a regression model but also enables the creation of a classification-type CART tree explaining the influence of individual variables on the estimated value. Such a model allows, among other things, for the extraction of rules using SQL query syntax (Fig. 2).

Developing a classification CART model requires generalizing salary expectations, i.e., reducing the measurement scale in which the salary is expressed from a quantitative scale (data expressed in PLN) to an ordinal scale (expected salary "high" or "low"). This approach allows for the explanation of "high salary" and "low salary" values in the form of a decision tree (Fig. 2) and, consequently, SQL rules. An example SQL code for node 10, explaining relatively low financial expectations, is as follows:

```
SELECT * FROM <TABLE>
WHERE (("year of graduation" > 2019).
AND ("Road distance to main city" < = 50 "km").
AND ("unemployment rate" < = 7));
```

The interpretation of this formula is as follows: the expected salary is HIGH if the student graduated after 2019, lives within 50 km of a major city, and in a county with an unemployment rate less than 7%. Similarly, one can interpret relatively LOW salary expectations for node 12, characterizing individuals who graduated before 2020, live more than 23 km away from Poznan, and expect innovative work with "passion" from potential employers. This path is also the best from the point of view of the revenue available to the potential customer. If you choose the target group with the highest revenue, it is the one you should focus on first and target intensified promotional activities to. As a result, such a method of segmentation can increase the effectiveness of the marketing budget.

The extracted rules (which can be explicitly interpreted for each terminal node of the CART model) not only allow for spatial differentiation of the model but, above all, enable an understanding of the motivations driving the surveyed students.

Table 1. Error value for CART algorithm

Method name	Mean squared error [PLN]
Classification and regression trees (CART)	603

The conducted research has shown that understanding the motivation of surveyed technical school students requires the inclusion of several completely different independent variables in the modeling process of the spatial distribution of expected remuneration. In addition to the data collected directly in the survey, which characterizes Generation Z respondents, such as preferences for individual/team work, the importance of a positive work atmosphere, inclination towards innovative/traditional companies, and the significance of finding a job "with passion," it is also crucial to consider economic parameters that characterize individual counties (unemployment, migration balance, etc.), as

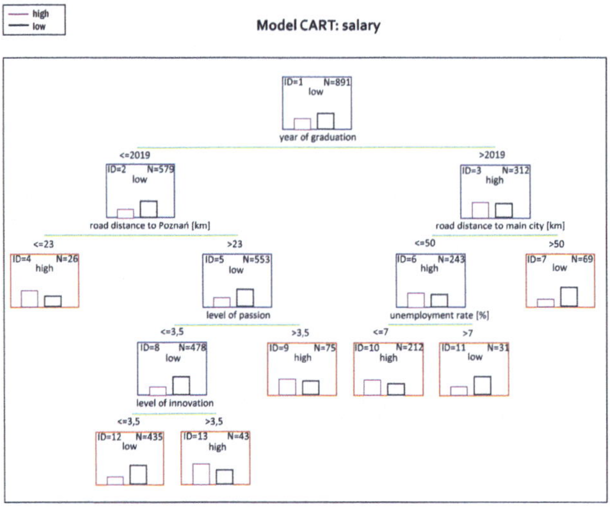

Fig. 1. CART decision tree explaining high/low salary expectations of respondents based on independent variables

well as attributes obtained through spatial data enrichment. Parameters like the geolocation of the survey and derived information, such as road distance to Poznan or the nearest major city, have proven to be significant for explaining the independent variable, which is the expected salary.

4.2 Spatial Distribution of a Dependent Variable

The spatial distribution map of the expected salary (Fig. 2), developed using the CART model, provides an overview of the collective influence of individual independent variables on the obtained outcomes. It is important to highlight that the colors on the resulting map depict the relative distribution of the expected salary. Red colors represent high expectations, while blue colors indicate relatively low expectations. The yellow color signifies an average level of expectations. In the case of analyzing all the data (Model I), the average value corresponds to the range of 3100–3300 PLN.

Interpreting the data, it can be said that those expecting the highest wages are those who graduated from high school after 2019, with a distance to a major city of less than 50 km, where unemployment is below 7%. Another possible case for those graduating from high school before 2019 is a distance from a major city of less than 23 km. When this distance exceeds 23 km additionally, one of the conditions must be met: “level of satisfaction” of more than 3.5, or in the case of a lower level of satisfaction “level of innovation” of more than 3.5.

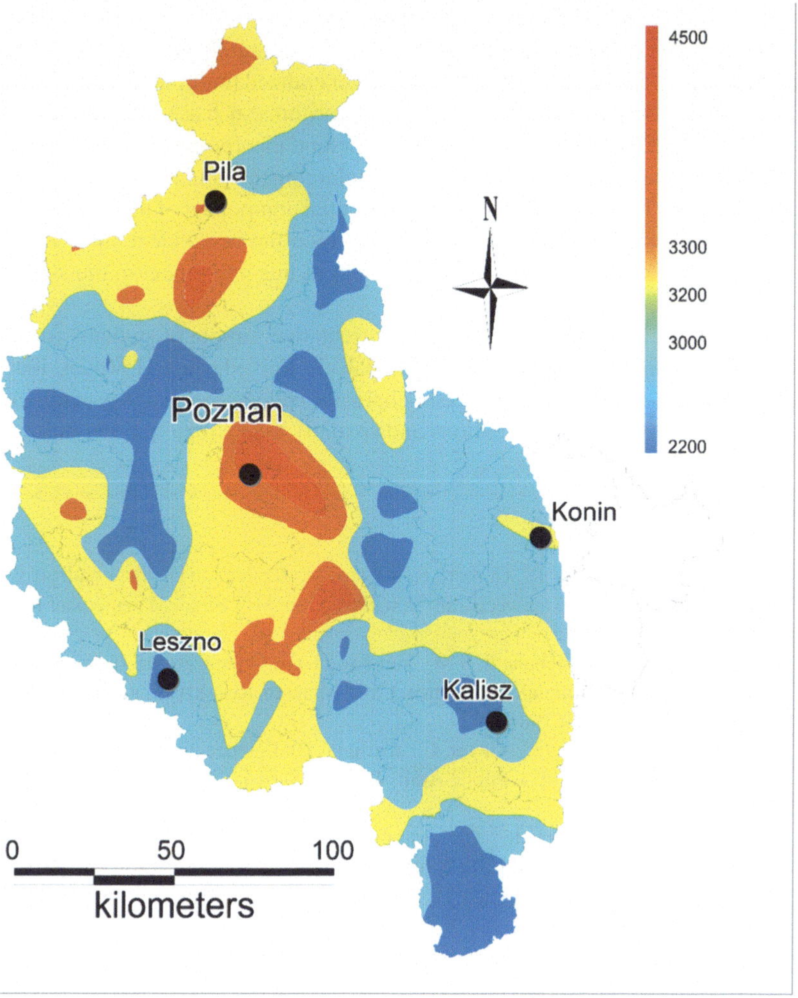

Fig. 2. Spatial distribution of the dependent variable estimated in the CART model

5 Conclusions and Discussion

The conducted analyses also revealed the significant importance of intangible goods for Generation Z individuals. Technical school graduates are willing to accept lower remuneration in exchange for working in an innovative company that provides a sense of satisfaction and allows them to pursue their passions. In modeling the multidimensional space of predictors' influence on the dependent variable, the age of the respondents (a temporal variable) also plays a significant role. The analysis results indicate that younger students (those who finish school later) have higher expectations regarding compensation and potential employers.

It is worth emphasizing that a substantial portion of exogenous predictors is nonlinearly correlated with the endogenous variable. To accurately and statistically model such relationships, it is necessary to use not only generalized multiple regression methods but also strongly nonlinear spatial data mining algorithms such as artificial neural networks, MARS, SVM [28], or the extended CART method incorporating spatial predictors. The use of classification trees also allowed for explaining the obtained model through explicitly defined IF-THEN decision rules. An essential component of these rules is the spatial dependencies, such as the road distance between the respondent's place of residence and the capital of the voivodeship or the level of unemployment in the student's home county.

The research was conducted on a representative sample of students from technical schools with an IT profile in the Wielkopolskie Voivodeship. However, the proposed research methodology has a universal character and can be easily generalized, both in terms of content to other professions and spatially to any region in Poland.

References

1. Ransbotham, S., Kiron, D., Gerbert, P., Reeves, M.: Reshaping business with artificial intelligence: Closing the gap between ambition and action. MIT Sloan Management Review **59**(1), (2017)
2. Shrestha, Y.R., Ben-Menahem, S.M., Von Krogh, G.: Organizational decision-making structures in the age of artificial intelligence. Calif. Manage. Rev. **61**(4), 66–83 (2019). https://doi.org/10.1177/0008125619862257
3. Wirtz, B.W., Weyerer, J.C., Geyer, C.: Artificial intelligence and the public sector - applications and challenges. Int. J. Public Adm. **42**(7), 596–615 (2019). https://doi.org/10.1080/01900692.2018.1498103
4. Davenport, T., Kalakota, R.: The potential for artificial intelligence in healthcare. Future Healthcare Journal **6**(2), 94 (2019). https://doi.org/10.7861/futurehosp.6-2-94
5. Rajpurkar, P., Chen, E., Banerjee, O., Topol, E.J.: AI in health and medicine. Nat. Med. **28**(1), 31–38 (2022). https://doi.org/10.1038/s41591-021-01614-0
6. Smith, M.J.: Getting value from artificial intelligence in agriculture. Animal Production Science **60**(1), 46–54 (2018). https://doi.org/10.1071/AN18522
7. Chen, L., Chen, P., Lin, Z.: Artificial intelligence in education: A review. Ieee Access **8**, 75264–75278 (2020). https://doi.org/10.1109/ACCESS.2020.2988510
8. Makridakis, S.: The forthcoming Artificial Intelligence (AI) revolution: Its impact on society and firms. Futures **90**, 46–60 (2017). https://doi.org/10.1016/j.futures.2017.03.006
9. Agrawal, A., Gans, J., Goldfarb, A.: Prediction machines: the simple economics of artificial intelligence. Harvard Business Press (2018). https://doi.org/10.1080/15228053.2019.1673511
10. Haleem, A., Javaid, M., Qadri, M.A., Singh, R.P., Suman, R.: Artificial intelligence (AI) applications for marketing: A literature-based study. Int. J. Intell. Netw. **3**, 118–132 (2022). https://doi.org/10.1016/j.ijin.2022.08.005
11. Kumar, V., Rajan, B., Venkatesan, R., Lecinski, J.: Understanding the role of artificial intelligence in personalized engagement marketing. Calif. Manage. Rev. **61**(4), 135–155 (2019). https://doi.org/10.1177/0008125619859317
12. Davenport, T., Guha, A., Grewal, D., Bressgott, T.: How artificial intelligence will change the future of marketing. J. Acad. Mark. Sci. **48**, 24–42 (2020). https://doi.org/10.1007/s11747-019-00696-0

13. Chatterjee, S., Chaudhuri, R., Vrontis, D., Thrassou, A., Ghosh, S.K.: Adoption of artificial intelligence-integrated CRM systems in agile organizations in India. Technol. Forecast. Soc. Chang. **168**, 120783 (2021). https://doi.org/10.1016/j.techfore.2021.120783
14. Murgai, A.: Transforming digital marketing with artificial intelligence, International Journal of Latest Technology in Engineering, Management & Applied. Science **7**(4), 259–262 (2018)
15. Sahai, S., Goel, R.: Impact of artificial intelligence in changing trends of marketing. Applications of artificial intelligence in business and finance, Modern Trends (2021). https://doi.org/10.1201/9781003129639-12
16. Forrest, E., Hoanca, B.: Artificial Intelligence: Marketing's Game Changer. In: Trends and Innovations in Marketing Information Systems, pp. 45–64, (2015). https://doi.org/10.4018/978-1-4666-8459-1.ch003
17. McLean, G., Osei-Frimpong, K.: Hey Alexa examine the variables influencing the use of artificial intelligent in-home voice assistants. Comput. Hum. Behav. **99**, 28–37 (2019). https://doi.org/10.1016/j.chb.2019.05.009
18. Dahl, D.W., Moreau, C.P.: Thinking inside the box: Why consumers enjoy constrained creative experiences. J. Mark. Res. **44**(3), 357–369 (2007). https://doi.org/10.1509/jmkr.44.3.357
19. Newman, D.: 5 ways AI is transforming the customer experience. https://www.forbes.com/sites/danielnewman/2019/04/16/5-ways-ai-is-transforming-the-customer-experience/#49e17a31465a. Last accessed 20 June 2023
20. Puntoni, S., Reczek, R.W., Giesler, M., Botti, S.: Consumers and artificial intelligence: an experiential perspective. J. Mark. **85**(1), 131–151 (2021). https://doi.org/10.1177/0022242920953847
21. Ghauri, P., Cateora, P.: International marketing, 2nd edn. Mc Graw Hill Education, New York, USA (2006)
22. Khan, R.H., Dofadar, D.F., Alam, M.G.R.: Explainable customer segmentation using k-means clustering. In: 2021 IEEE 12th Annual Ubiquitous Computing, Electronics & Mobile Communication Conference (UEMCON), pp. 0639–0643. IEEE (2021). https://doi.org/10.1109/UEMCON53757.2021.9666609
23. Nazim, S., Rajeswari, M.: Creating a brand value and consumer satisfaction in e-commerce business using artificial intelligence. In: Proceedings of International Conference on Sustainable Computing in Science. Technology and Management (SUSCOM) (2019). https://doi.org/10.2139/ssrn.3351618
24. Janardhanan, S., Muthalagu, R.: Market segmentation for profit maximization using machine learning algorithms. In: 1706. First International Conference on Advances in Physical Sciences and Materials. Journal of Physics (2020). https://doi.org/10.1088/1742-6596/1706/1/012160
25. Wang, C.: Efficient customer segmentation in digital marketing using deep learning with swarm intelligence approach. Info. Proces. Manage. **59**(6) (2022). https://doi.org/10.1016/j.ipm.2022.103085
26. Brito, P.Q., Soares, C., Almeida, S., Monte, A., Byvoet, M.: Customer segmentation in a large database of an online customized fashion business. Robot. Comp.-Integr. Manuf. **36**, 93–100 (2015). https://doi.org/10.1016/j.rcim.2014.12.014
27. Aouad, A., Elmachtoub, A.N., Ferreira, K.J., McNellis, R.: Market segmentation trees. Manuf. Serv. Oper. Manag. **25**(2), 648–667 (2023). https://doi.org/10.1287/msom.2023.1195
28. Graczyk-Kucharska, M., Olszewski, R., Weber, G.W.: The use of spatial data mining methods for modeling HR challenges of generation Z in greater Poland Region. CEJOR **31**(1), 205–237 (2023). https://doi.org/10.1007/s10100-022-00805-5

Prediction of Customer Purchases Using LSTM Deep Neural Network

Krzysztof Lutosławski[1(✉)], Marcin Hernes[1], Artur Rot[1], and Cezary Olejarczyk[2]

[1] Wroclaw University of Economics and Business, Komandorska 118/120, 53-345 Wrocław, Poland
{krzysztof.lutoslawski,marcin.hernes,artur.rot}@ue.wroc.pl
[2] Open Loyalty Ltd., Dmowskiego 17, 50-203 Wrocław, Poland
cezary.olejarczyk@openloyalty.io

Abstract. In the realm of business, accurately predicting customer purchases holds significant importance, as it enables the adept tailoring of marketing strategies, enhances promotional campaign effectiveness, optimizes inventory management, and reduces food loss and waste, ultimately leading to sales growth and profit maximization. The paper aimed to develop a deep learning model using the LSTM (long short-term memory) neural network to predict customer purchases in the next week, including transaction count and overall transaction gross value, based on their historical transaction data. In the research, to achieve the best predictions, various sizes of the training dataset, different model architectures, and its hyperparameters, as well as the hyperparameters of the learning algorithm, were tested. Prediction performance was assessed by RMSE and R^2. As a result, the optimal predictive model trained using 10% of customers' transaction sequences was characterized by very high performance at the levels of 1.4 (for transaction count) and 26.2 (for transaction gross value) in RMSE and at the levels of 99.9233% (for transaction count) and 99.4425% (for transaction gross value) in R^2. The obtained model can be implemented before validating its performance on another instance of production data.

Keywords: customer purchases prediction · deep learning · artificial neural network · long-short term memory

1 Introduction

Prediction of the consumer's purchasing behavior is essential. Having insights into the future, at least to a certain extent, gives businesses a competitive advantage in many areas. Starting from improving the sales process by providing personalized offers to customers, and predicting business performance, to better stock and food waste management.

As a business, you can use the knowledge about the likelihood of purchase in various ways. First and foremost, you can influence this likelihood through personalized communication and targeted offers. Hesitant customers can be transformed into actual buyers.

M. Hernes and J. Wątróbski (Eds.): ECAI 2023, LNNS 1079, pp. 166–181, 2024.
https://doi.org/10.1007/978-3-031-66761-9_15

Moreover, businesses often measure this by collecting data on whether and when the message was delivered and whether customers availed of the offer. This feedback loop can lead to improved future predictions, enhanced communication, and more targeted recommendations.

By predicting customers' purchases, you can also predict the business's overall performance. This helps businesses forecast future growth, optimize marketing budgets, and enhance cash flow and communication with the market.

Another way of using customers' purchase behavior is not apparent but is as essential as others. If, as a business, you can know who, when, and what will purchase, you can optimize your inventory and stock management in on-ground stores. Knowledge about customer purchase behaviors allows you to optimize and increase ROI (return on investment) by investing in the desired products and maintaining stock sizes according to the demand. Moreover, customers' behavior can vary geographically. People in large cities can have different purchase patterns than people living in the suburbs or rural areas. Businesses can optimize the routing of products to stores by adjusting product availability, whether through increasing or decreasing stock levels, and by adapting to purchasing habits.

Connecting various data, such as predictive purchase information along with product expiration dates, can help reduce food waste in FMCG (fast-moving consumer goods) businesses. Predicting patterns like increased sales of sandwiches can optimize its production, offer discounts and promotions for near-expired products to people likely to buy those products, and more.

This research aimed to develop a deep learning model using the LSTM (long short-term memory) neural network to predict customer purchases in the next week, including transaction count and overall transaction gross value, based on their historical transaction data.

The rest of the paper is structured as follows: the next part presents the background, including the indication of the research contribution. Next, the research methodology is presented. The last part of the paper concerns the analysis of research results and conclusions.

2 Background

The diversity and specificities of today's businesses have leveraged a wide range of prediction techniques. In particular, purchase prediction is a major economic concern for many companies [1]. A thorough understanding of online customers' purchase behavior will directly boost e-commerce business performance. While purchase prediction has received attention for a long time in consumer research, the rise of customer analytics has revived such issues in recent years. Many researchers from different fields like management, marketing, information systems, and neuroscience have been striving to determine how to predict customer behavior [2, 3]. They've made different models to understand these buying decisions. In most cases, those approaches are based on extracting customer's latent characteristics from their past purchase behavior with the mindset that observed behavior is the outcome of an underlying stochastic process [4, 5]. Existing studies have overtly focused on purchase intention and used sales rank as a natural proxy, which however has limited business application [6]. Analyzing the literature on the application of deep learning in predicting customer purchases reveals several key findings. Researchers have explored various approaches, methodologies, and challenges associated with using deep learning models for customer purchase prediction [7]. There are several works related to identifying customers with interest in a product [8]. The Authors of the work [9] presented a system for the prediction of user interest domains, which was based on features derived from Facebook profiles to predict from which category of products an eBay user is likely to make a purchase. The next work [10] used a rule-based approach for identifying sentences indicating "buy wishes" from forums with buy-sell sections. In the last years, various machine learning and deep learning methods for predicting customer retention and profitability have been analyzed in the academic field and some of them used by practitioners [4]. The application of machine learning techniques for predictive purposes on the customer base is often analyzed in the customer relationship management and expert systems domain, and customer churn prediction is the most popular objective in this field. Churn models are made to detect, as soon as possible, signals of potential churn and help to identify customers willing to abandon a given company voluntarily. It has been analyzed in many works [11–30]. In contemporary literature, we can observe two main trends concerning customer churn and purchase predictions [11]. According to [17], the first branch includes traditional classification methods such as decision tree (DT) and logistic regression (LR) [18–25]. The second mentioned line of thought is based on artificial intelligence methods such as neural networks [19, 25–27], evolutionary learning, genetic algorithms [19, 20], random forests [28, 29], improved balanced random forests, k-nearest neighbor, fuzzy logic systems, and support vector machines [30]. Our deep learning model using the LSTM network will allow for the accurate targeting of promotional campaigns depending on the purchasing activity of customers.

3 Research Methodology

3.1 Data

The data feeding the predictive model came from Open Loyalty Ltd. (Wrocław, Poland) and contained historical consumer transaction data (approximately 20,000 records) with the following attributes:

- Customer ID
- Customer purchase date – the date on which a transaction was executed by a customer
- Customer purchase gross value – the gross value of the transaction made by the customer.

The pre-processed data, derived from a.csv file, was imported into the database system, and then duplicates (records with the same values of all attributes) were removed from it. Data without duplicates were sub-analyzed to remove outliers in customer purchase gross value, i.e. above the value corresponding to the percentile equal to 99.95%.

After data selection, a data transformation was performed, in which data were initially aggregated within the same week. Then, cumulative values were determined for the aggregated data regarding the number of transactions and corresponding gross values, to increase the entropy of the data source for learning the developed model. For both attributes, maximum, minimum, average, and standard deviation values were determined. Based on the values of these parameters, the data were normalized and standardized to values in the interval [0 1] with mean value (m) and standard deviation (s = 0).

The processed data in this way was used in the learning of the LSTM deep neural network. The input of the model was the historical purchase weekly sequence data of a given consumer, while the predictive output of the model pertains to both the transaction count and the overall gross value of transactions for the next week. The predictions provided by the model were denormalized to reflect the real values.

3.2 Methods

A 4-layer neural network with an input layer defined as a sequence vector, an LSTM hidden layer, a layer for aggregating outputs from the hidden layer, and a regression layer was adopted to implement the prediction assumptions. The model was taught using the following assumptions:

- Learning algorithm - Adam Optimizer (adaptive moment estimation), do-tuned by the learning rate coefficient, the L2 regularization factor, and the number of shows in an epoch
- Stop condition - achievement of satisfactory evaluation rates during learning, i.e. min{RMSE, LOSS function} or achievement of the maximum number of learning epochs
- Criterion for evaluating the quality of the trained model - min{RMSE} and max{R^2} determined for the test dataset.

The first stage of the research investigated the influence on the quality of learning: the *size of the learning dataset* and the value of the *mini batch size* parameter, defined by the customer selection criterion, and hyperparameters strictly related to the learning algorithm used, i.e.: *initial learn rate*, *learn rate drop period*, *learn rate drop factor* and *gradient threshold*, as well as *max epochs*. The customer selection for the training dataset, thus determining the size of this dataset, was made according to the criterion relating to the width of customers' observation window (eq. to the length of the transaction sequence vectors, Δt). For the training dataset constituting 44% of customers, the criterion of the sequence length of not less than 3 weeks ($\Delta t \geq 3$ wk) was adopted, for 34% – $\Delta t \geq 8$ wk, for 24% – $\Delta t \geq 15$ wk, and for 10% – $\Delta t \geq 30$ wk. Taking the values of the aforementioned parameters obtained from the first stage, in the second stage of the research, the influence of the architecture of the LSTM neural network on the quality of the learning model was investigated. The variable number of neurons in the hidden layer (*num hidden units*) and the type of activation function defining the cell state (*state activation function*) were taken into account. Each experiment was repeated at least 10 times. As a final result of these studies, the set of parameter values that produced the most favorable model learning quality was established. The results of these studies are presented in the following section of the paper.

4 Results

4.1 Architecture and Training of the Deep Neural Network

To achieve the most effective LSTM predictive model, the influence of the learning algorithm hyperparameters were sequentially examined: *gradient threshold* (Table 1), *learn rate drop factor* (Table 2), *learn rate drop period* (Table 3), *initial learn rate* (Table 4) and *max epochs* (Table 5); as well as the influence of network architecture in terms of *state activation function* (Table 6) and *num hidden units* (Table 7). In addition, all tests were performed using different sizes of the training dataset, which constituted 44%, 34%, 24% and 10% of all customers' transaction sequences. Based on the obtained results, it can be concluded that in almost all examined cases, the quality of the trained model depended primarily on the chosen size of the training dataset, which also determined the size of the testing dataset, as well as the mini-batch size (sample shows per epoch). As a result, the greater the size of the training dataset, the lower the quality of the obtained model. For instance, if the training dataset size was 44%, the RMSE and R^2 values were worse, even by 4.4 and 3.0%, respectively (for the transaction count), and even by 13.8 and 4.3%, respectively (for the transaction gross value), than when the training dataset size was 10% (Tables 1–7). Indeed, it was likely due to the short transaction sequences of some consumers occurring mainly in the 44% dataset, with even only 3 weeks of observation ($\Delta t \geq 3$ wk). Such short observation windows, not containing much information for the network, clearly worsened the learning process in all conducted experiments (Tables 1–7).

In the research concerning the selection of hyperparameters of the learning algorithm, a noticeable influence of the values of *gradient threshold*, *learn rate drop factor*, and *initial learn rate* on the obtained model's quality was observed (Table 1, 2, and 4). The most advantageous results for the model evaluation metrics were achieved at the lowest tested value for *gradient threshold* (0.1) (Table 1), while at higher tested values for *learn rate drop factor* (1.0) (Table 2) and *initial learn rate* (0.01) (Table 4). With such settings, the model exhibited average minimum values of RMSE error and R^2 determination coefficient within the range of 1.9–2.25 and 99.4866–99.5770%, respectively, for the count feature, and 35.0–39.35 and 98.0026–98.4719%, respectively, for the gross feature. Too low values of *learn rate drop factor* and *initial learn rate* or too high values of *gradient threshold* slowed down the learning process, which negatively affected the values of model performance metrics (Tables 1, 2, and 4). However, there was no noticeable impact of the studied values of the *learn rate drop period* parameter on the LSTM network training (Table 3). Changes in the values of this learn algorithm coefficient in the range from 5 to 100 caused variations in the average values of RMSE and R^2 within a narrow range of variability, respectively 1.9–2.2 and 99.5250–99.6675% (for transaction count) and 34.2–36.6 and 98.2971–98.8479% (for transaction gross value) (Table 3). Nevertheless, the most favorable values of both metrics were observed at the highest value of *learn rate drop period* (100) (Table 3).

For optimal performance results, the model required conducting the deep learning process for 200 or 500 epochs (Table 5). This provided minimum RMSE and R^2 values, amounting to 1.4 and 99.9233% for count, and 24.1 and 99.5722% for gross, respectively (Table 5). However, it should be emphasized that the model trained for as many as 500

Table 1. The effect of the training dataset size and the ***gradient threshold*** on the model training efficiency expressed by RMSE and R^2.

Constant parameters: *num hidden units* = 100, *state activation function* = 'tanh', *max epochs* = 100, *initial learn rate* = 0.01, *learn rate drop period* = 25, *learn rate drop factor* = 0.5								
Training dataset size [%] (mini batch size)	*Gradient threshold*							
	0.1		0.5		1		2	
	RMSE	R^2	RMSE	R^2	RMSE	R^2	RMSE	R^2
44% (601)	2.6	99.2615%	2.8	99.1655%	3.1	99.0785%	3.3	99.0349%
	41.3	97.1118%	48.2	96.5676%	46.5	96.6743%	52.1	96.3369%
34% (275)	2.3	99.3715%	3.5	99.0468%	3.8	99.1016%	4.3	99.0032%
	43.8	97.4554%	51.5	96.6381%	53.2	96.6427%	55.4	96.6212%
24% (165)	2.5	99.3816%	3.0	99.1476%	2.4	99.4785%	2.6	99.3351%
	42.3	98.0770%	51.1	97.3478%	40.9	98.4073%	46.9	97.8371%
10% (105)	**1.6**	**99.9318%**	2.0	99.7727%	2.0	99.7946%	2.1	99.7493%
	30.0	**99.3661%**	46.3	98.3442%	47.4	98.2544%	50.9	97.9746%
Avg	**2.3**	**99.4866%**	2.8	99.2832%	2.8	99.3633%	3.1	99.2806%
	39.4	**98.0026%**	49.3	97.2244%	47.0	97.4947%	51.3	97.1925%

The first value of RMSE and R^2 refers to the transaction count and the second to the transaction gross.

Table 2. The effect of the training dataset size and the learn rate drop factor on the model training efficiency expressed by RMSE and R^2.

Constant parameters: *num hidden units* = 100, *state activation function* = 'tanh', *max epochs* = 100, *initial learn rate* = 0.01, *learn rate drop period* = 25, *gradient threshold* = 0.1								
Training dataset size [%] (mini batch size)	*Learn rate drop factor*							
	0.05		0.1		0.5		1	
	RMSE	R^2	RMSE	R^2	RMSE	R^2	RMSE	R^2
44% (601)	7.7	96.4715%	5.2	98.9149%	4.0	98.9832%	2.3	99.2054%
	84.8	92.2529%	68	95.6843%	53.8	95.6464%	41.2	97.2807%
34% (275)	3.3	99.2202%	4.0	99.0191%	2.7	99.1785%	1.9	99.4542%
	58.6	96.0198%	56.0	96.3425%	45.7	97.0738%	34.0	98.4468%
24% (165)	3.0	99.1547%	3.2	99.1541%	2.4	99.4291%	2.1	99.5127%
	52.7	97.2256%	56.1	96.9335%	43.4	98.0535%	34.5	98.7797%
10% (105)	3.3	99.5194%	3.7	99.5271%	2.1	99.7291%	**1.2**	**99.9243%**
	71.0	96.5594%	67.5	96.5983%	43.3	98.5457%	**30.4**	**99.3802%**

(*continued*)

Table 2. (*continued*)

Constant parameters: *num hidden units* = 100, *state activation function* = 'tanh', *max epochs* = 100, *initial learn rate* = 0.01, *learn rate drop period* = 25, *gradient threshold* = 0.1								
Training dataset size [%] (mini batch size)	*Learn rate drop factor*							
	0.05		0.1		0.5		1	
	RMSE	R^2	RMSE	R^2	RMSE	R^2	RMSE	R^2
Avg	4.3 66.8	98.5915% 95.5144%	4.0 61.9	99.1538% 96.3897%	2.8 46.6	99.3300% 97.3299%	**1.9** **35.0**	**99.5242%** **98.4719%**

The first value of RMSE and R^2 refers to the transaction count and the second to the transaction gross.

epochs did not surpass the model trained for 200 epochs in any way; it only extended the duration of the learning process.

Table 3. The effect of the training dataset size and the learn rate drop period on the model training efficiency expressed by RMSE and R^2.

Constant parameters: *num hidden units* = 100, *state activation function* = 'tanh', *max epochs* = 100, *initial learn rate* = 0.01, *learn rate drop factor* = 1, *gradient threshold* = 0.1								
Training dataset size [%] (mini batch size)	*Learn rate drop period*							
	5		20		50		100	
	RMSE	R^2	RMSE	R^2	RMSE	R^2	RMSE	R^2
44% (601)	2.5 39.9	99.1510% 97.3871%	2,5 41,6	99.3014% 97.0213%	2.5 42.8	99.1695% 96.7005%	2.5 37.9	99.2815% 97.6274%
34% (275)	2.3 34.9	99.4891% 98.3783%	1.9 35.6	99.4772% 98.2533%	2.0 36.4	99.4770% 98.2409%	1.9 35.0	99.4801% 98.4225%
24% (165)	2.3 34.4	99.5129% 98.7925%	2.1 39.1	99.5181% 98.9940%	2.0 35.0	99.5205% 98.9049%	2.1 34.3	99.5060% 98.7523%
10% (105)	1.5 33.0	99.9373% 99.3730%	1.3 35.0	99.9334% 99.2026%	1.4 30.4	99.9328% 99.3421%	**1.0** **29.4**	**99.9333%** **99.3630%**
Avg	2.2 35.6	99.6464% 98.8479%	1.8 36.6	99.6675% 98.3678%	2.0 36.2	99.5250% 98.2971%	**1.9** **34.2**	**99.5502%** **98.5413%**

The first value of RMSE and R^2 refers to the transaction count and the second to the transaction gross.

The research on the LSTM neural network architecture also revealed a noticeable influence of the activation function type in the hidden layer on the model's quality. Changing the *tanh* activation function to *softsign* resulted in an increase of over 50% in

Table 4. The effect of the training dataset size and the initial learn rate on the model training efficiency expressed by RMSE and R^2.

Constant parameters: *num hidden units* = 100, *state activation function* = 'tanh', *max epochs* = 100, *learn rate drop period* = 100, *learn rate drop factor* = 1, *gradient threshold* = 0.1								
Training dataset size [%] (mini batch size)	*Initial learn rate*							
	0.001		0.005		0.01		0.1	
	RMSE	R^2	RMSE	R^2	RMSE	R^2	RMSE	R^2
44% (601)	8.1 80.4	98.7042% 94.5937%	2.6 41.8	99.1141% 96.9864%	2.5 43.2	99.3713% 96.8372%	7.5 99.0	98.2717% 89.9435%
34% (275)	6.4 71.0	98.9095% 95.9687%	2.1 37.5	99.4706% 98.1770%	1.8 37.1	99.4978% 98.3016%	6.1 116.3	96.4318% 93.7022%
24% (165)	7.7 76.8	98.8320% 96.6048%	2.1 37.6	99.5227% 98.4812%	2.2 35.4	99.5131% 98.6772%	11.4 171.7	96.5501% 65.8458%
10% (105)	14.3 139.4	98.4407% 95.0539%	1.5 33.6	99.8887% 99.0776%	**1.3** **34.7**	**99.9258%** **99.2522%**	10.0 126.1	97.9764% 96.5364%
Avg	9.1 91.9	98.7216% 95.5553%	2.1 37.6	99.4990% 98.1806%	**2.0** **37.6**	**99.5770%** **98.2671%**	8.8 128.3	97.3075% 86.5070%

The first value of RMSE and R^2 refers to the transaction count and the second to the transaction gross.

the RMSE error and a decrease of even 0.9% in the R^2 determination coefficient (Table 6). However, no unequivocal discrepancies in the average values of model evaluation metrics were observed depending on the number of tested neurons in the hidden layer, within the range of 50 to 200 (Table 7).

Table 5. The effect of the training dataset size and the epochs number (max epochs) on the model training efficiency expressed by RMSE and R^2.

Constant parameters: *num hidden units* = 100, *state activation function* = 'tanh', *initial learn rate* = 0.01, *learn rate drop period* = 100, *learn rate drop factor* = 1, *gradient threshold* = 0.1								
Training dataset size [%] (mini batch size)	*Max epochs*							
	100		200		500		1000	
	RMSE	R^2	RMSE	R^2	RMSE	R^2	RMSE	R^2
44% (601)	2.5 43.2	99.3713% 96.8372%	2.3 35.9	99.3866% 98.1515%	2.0 35.3	99.3913% 98.2018%	2.4 33.1	99.1999% 98.4652%

(continued)

Table 5. *(continued)*

Constant parameters: *num hidden units* = 100, *state activation function* = 'tanh', *initial learn rate* = 0.01, *learn rate drop period* = 100, *learn rate drop factor* = 1, *gradient threshold* = 0.1

Training dataset size [%] (mini batch size)	*Max epochs*							
	100		200		500		1000	
	RMSE	R^2	RMSE	R^2	RMSE	R^2	RMSE	R^2
34% (275)	1.8 37.1	99.4978% 98.3016%	2.1 31.7	99.4770% 98.9006%	2.1 30.8	99.3883% 98.7976%	2.4 32.5	99.1627% 98.6302%
24% (165)	2.2 35.4	99.5131% 98.6772%	2.2 33.4	99.4948% 98.9176%	2.6 36.4	99.2857% 98.7448%	2.3 37.4	99.3994% 98.5110%
10% (105)	1.3 34.7	99.9258% 99.2522%	**1.4** **26.2**	**99.9233%** **99.4425%**	1.4 24.1	99.8595% 99.5722%	3.1 38.7	99.4703% 99.1959%
Avg	2.0 37.6	99.5770% 98.2671%	**2.0** **31.8**	**99.5704%** **98.8531%**	2.0 31.7	99.4812% 98.8291%	2.6 35.4	99.3081% 98.7006%

The first value of RMSE and R^2 refers to the transaction count and the second to the transaction gross.

Table 6. The effect of the training dataset size and the hidden layer activation function (state activation function) on the model training efficiency expressed by RMSE and R^2.

Constant parameters: *num hidden units* = 100, *max epochs* = 200, *initial learn rate* = 0.01, *learn rate drop period* = 100, *learn rate drop factor* = 1, *gradient threshold* = 0.1

Training dataset size [%] (mini batch size)	*State activation function*			
	tanh		softsign	
	RMSE	R^2	RMSE	R^2
44% (601)	2.3 35.9	99.3866% 98.1515%	3.0 47.6	98.7602% 96.7198%
34% (275)	2.1 31.7	99.4770% 98.9006%	3.2 45.6	98.8129% 97.6087%
24% (165)	2.2 33.4	99.4948% 98.9176%	3.4 49.7	99.4339% 98.6644%
10% (105)	**1.4** **26.2**	**99.9233%** **99.4425%**	3.4 52.8	99.4966% 98.6678%
Avg	**2.0** **31.8**	**99.5704%** **98.8531%**	3.3 48.9	99.1259% 97.9152%

The first value of RMSE and R^2 refers to the transaction count and the second to the transaction gross.

In conclusion, considering the results of the conducted research regarding the influence of both related to the LSTM network architecture and the learning process on the quality of the trained predictive model, it can be stated that, regardless of the chosen training dataset size, the most favorable results for RMSE and R^2 were achieved by employing the following hyperparameter values: *max epochs* = 200, *initial learn rate* = 0.01, *learn rate drop period* = 100, *learn rate drop factor* = 1, *gradient threshold* = 0.1, *num hidden units* = 100, *state activation function* = tanh. These optimal hyperparameter values were used to train the best LSTM network model, and then to assess its prediction efficiency for the transaction count and the gross value of customer purchases.

Table 7. The effect of the training dataset size and the number of neurons in the hidden layer (num hidden units) on the model training efficiency expressed by RMSE and R^2.

Parameters: *state activation function* = 'tanh', *max epochs* = 200, *initial learn rate* = 0.01, *learn rate drop period* = 100, *learn rate drop factor* = 1, *gradient threshold* = 0.1								
Training dataset size [%] (mini batch size)	*Num hidden units*							
	50		100		150		200	
	RMSE	R^2	RMSE	R^2	RMSE	R^2	RMSE	R^2
44% (601)	2.4	99.3723%	2.3	99.3866%	1.8	99.3918%	2.2	99.3841%
	33.7	98.1545%	35.9	98.1515%	34.2	98.1816%	33.4	98.3414%
34% (275)	2.0	99.4744%	2.1	99.4770%	2.1	99.4709%	2.0	99.4165%
	32.5	98.5961%	31.7	98.9006%	32.0	98.7287%	35.1	98.6138%
24% (165)	2.1	99.4918%	2.2	99.4948%	2.7	99.3785%	2.2	99.4942%
	32.3	98.9745%	33.4	98.9176%	39.5	98.6588%	32.4	98.8997%
10% (105)	1.9	99.9132%	**1.4**	**99.9233%**	1.6	99.9297%	2.0	99.9297%
	28.5	99.3243%	**26.2**	**99.4425%**	26.2	99.4653%	24.7	99.5238%
Avg	2.1	99.5629%	**2.0**	**99.5704%**	2.1	99.5427%	2.1	99.5561%
	31.8	98.7624%	**31.8**	**98.8531%**	33.0	98.7586%	31.4	98.8447%

The first value of RMSE and R^2 refers to the transaction count and the second to the transaction gross.

4.2 Prediction Performance

The best LSTM deep network architecture selected in the previous stage of the research (*num hidden units* = 100, *state activation function* = tanh) and the best learning hyper-parameters (*max epochs* = 200, *initial learn rate* = 0.01, *learn rate drop period* = 100, *learn rate drop factor* = 1, *gradient threshold* = 0.1) resulted in achieving an exceptional model performance. Sample courses of the transaction data for randomly tested customers, containing observations and predictions of count and gross value are presented in Fig. 1a-1b. The selected model trained using various sizes of training dataset was characterized by an R^2 effectiveness not lower than 99.3866% for count predictions and 98.1515% for gross value predictions (Table 7). The plots of observed values versus simulated deep LSTM neural network values for different training dataset sizes are shown in Figs. 2a-2d for the transaction count and in Figs. 3a-3d for the transaction gross value. Based on them, it should be noticed that the larger the size of the training dataset, the greater the number of outliers from the ideal realization (regression line). Undoubtedly, this was influenced by short sequences of observations, not carrying too much information for the network, occurring in particular in the training dataset with a

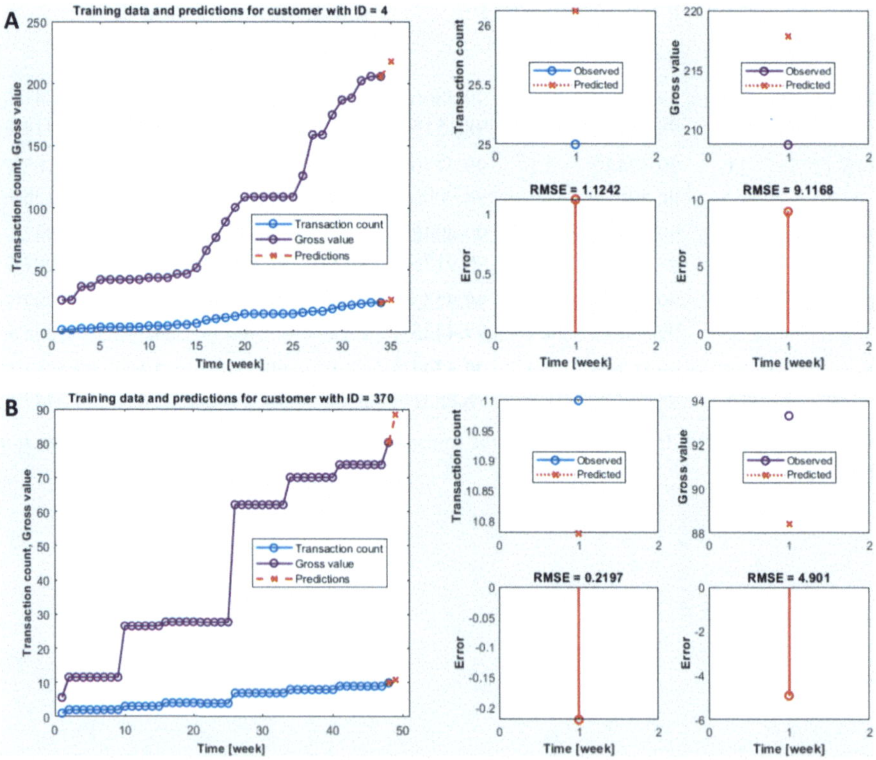

Fig. 1. Sample courses of historical transaction count and gross value (input) and predictions (output) for selected customer IDs 4 (A) and 370 (B) using the optimal LSTM deep neural network model.

size of 44%. This dataset allowed sequences as long as 3 weeks, which probably significantly worsened the generalization abilities of the network. Otherwise, as a result, the model trained using a 10% dataset exhibited the highest values of the determination coefficient (R^2) for the transaction count as well as the transaction gross value, reaching 99.9233% and 99.4425%, respectively (Table 7). This confirms the rightness of choosing an LSTM deep neural network for simultaneous prediction of both the transaction count and the gross value based on the purchasing history of customers.

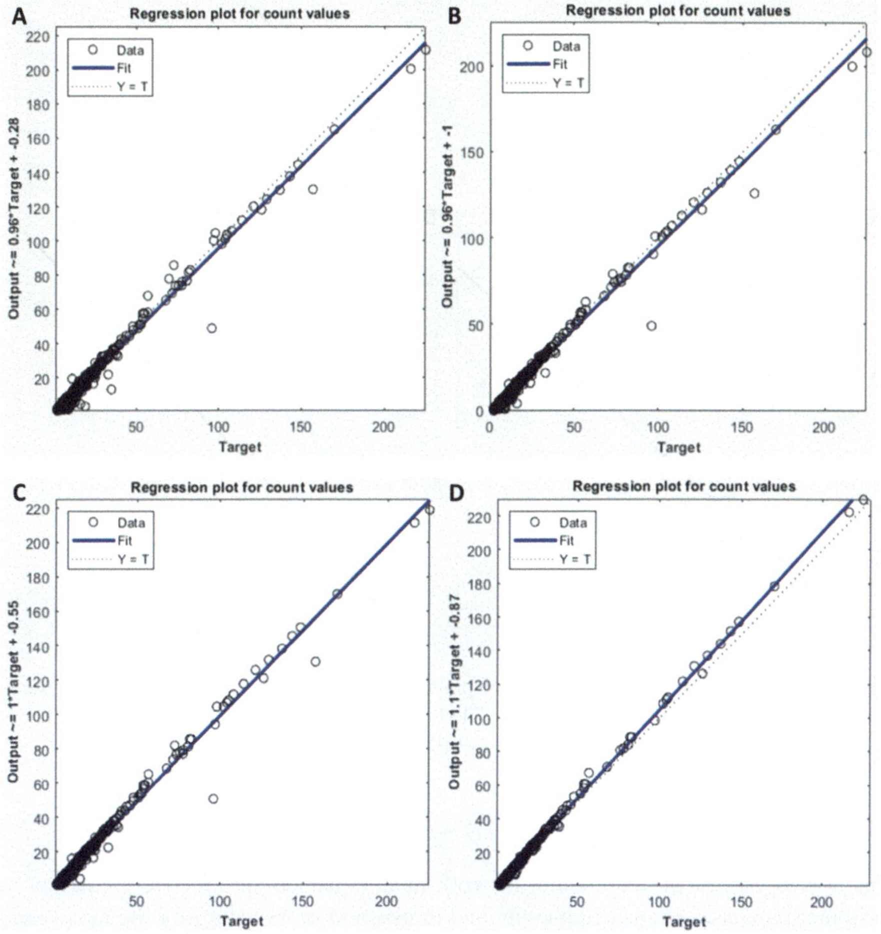

Fig. 2. Prediction performance as regression plot between observations and predictions for transaction count obtained by the optimal LSTM deep neural network model trained on different dataset sizes: 44% (A), 34% (B), 24% (C), and 10% (D).

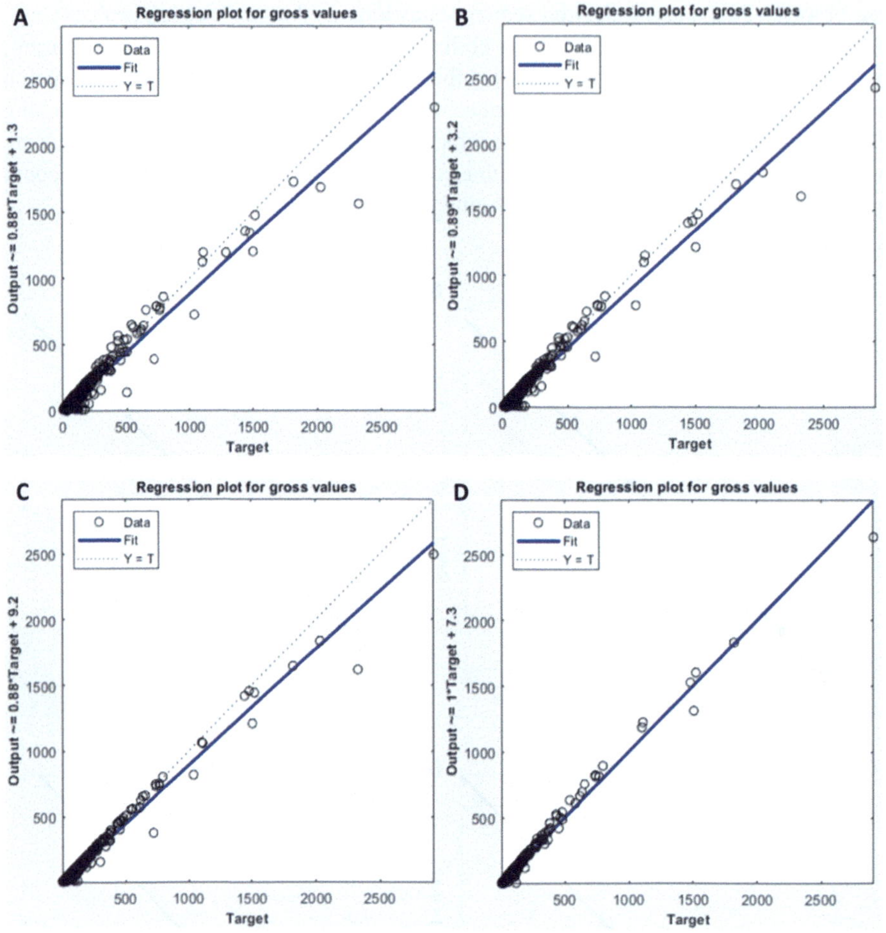

Fig. 3. Prediction performance as regression plot between observations and predictions for transaction gross value obtained by the optimal LSTM deep neural network model trained on different dataset sizes: 44% (A), 34% (B), 24% (C), and 10% (D).

5 Conclusion

The developed LSTM deep neural network model enabled an effective prediction of customer purchases (transaction count and transaction gross value) for the next week. However, as demonstrated through conducted research, its effectiveness was strictly dependent on the dataset size and data selection for training the model, hyperparameters of the applied learning algorithm, and the neural network structure.

Too large a training dataset deteriorated the quality of model training. As a result, the model trained using 44% (mini-batch size of 601 sequences) dataset of the customers' transaction sequences was characterized by approximately twice lower prediction quality for both the transaction count and the transaction gross value, compared to the model trained with 10% dataset (mini-batch size of 105 sequences).

The optimal model, trained using a 10% dataset (sequences with an observation window of at least 30 weeks), exhibited the highest prediction performance. For both transaction count and transaction gross value, it achieved the lowest RMSE values of 1.4 and 26.2, respectively, as well as the highest R^2 values of 99.9233% and 99.4425%, respectively.

To confirm the effectiveness of the proposed predictive model with the selected structure and hyperparameters of the learning process, it should be validated on another instance of the production dataset before implementation.

Founding. Project under the Intelligent Development Operational Program for 2014-2020, Axis 1 Support for R&D by enterprises, Measure 1.1 R&D projects of enterprises, Sub-measure 1.1.1 Industrial research and development works carried out by enterprises entitled: "Divante Open Pay" - development of an innovative, digital, loyalty payment card". Project co-financed by the European Regional Development Fund implemented under the competition of the National Center for Research and Development: Szybka Ścieżka.

References

1. Geiler, L., Affeldt, S., Nadif, M.: A survey on machine learning methods for churn prediction. Int. J. Data Sci. Anal. **14**(3), 217–242 (2022). https://doi.org/10.1007/s41060-022-00312-5
2. Singhal, R., et al.: Fast online 'next best offers' using deep learning. In: Proceedings of the ACM India Joint International Conference on Data Science and Management of Data, pp. 217–223. New York, USA (2019)
3. Vieira, A.: Predicting online user behaviour using deep learning algorithms. arXiv preprint arXiv:1511.06247 (2015). https://doi.org/10.48550/arXiv.1511.06247
4. Martínez, A., Schmuck, C., Pereverzyev, S., Jr., Pirker, C., Haltmeier, M.: A machine learning framework for customer purchase prediction in the non-contractual setting. Eur. J. Oper. Res. **281**(3), 588–596 (2020). https://doi.org/10.1016/j.ejor.2018.04.034
5. Fader, P.S., Hardie, B.G.: Probability models for customer-base analysis. J. Interact. Mark. **23**(1), 61–69 (2009). https://doi.org/10.1016/j.intmar.2008.11.003
6. Chaudhuri, N., Gupta, G., Vamsi, V., Bose, I.: On the platform but will they buy? Predicting customers' purchase behavior using deep learning. Decis. Support Syst. **149**, 113622 (2021). https://doi.org/10.1016/j.dss.2021.113622
7. Guo, L., et al.: Buying or browsing?: Predicting real-time purchasing intent using attention-based deep network with multiple behavior. In: Proceedings of the 25th ACM SIGKDD international conference on knowledge discovery & data mining, pp. 1984–1992 (2019). https://doi.org/10.1145/3292500.3330670
8. Korpusik, M., Sakaki, S., Chen, F., Chen, Y.Y.: Recurrent Neural Networks for Customer Purchase Prediction on Twitter. CBREcsys@ recsys **1673**, 47–50 (2016)
9. Zhang, Y., Pennacchiotti, M.: Predicting purchase behaviors from social media. In: Proceedings of the 22nd international conference on World Wide Web, pp. 1521–1532 (2013). https://doi.org/10.1145/2488388.2488521
10. Ramanand, J., Bhavsar, K., Pedanekar, N.: Wishful thinking-finding suggestions and'buy'wishes from product reviews. In: Proceedings of the NAACL HLT 2010 workshop on computational approaches to analysis and generation of emotion in text, pp. 54–61 (2010)
11. Pondel, M., et al.: Deep learning for customer churn prediction in e-commerce decision support. In Business Information Systems **1**, 3–12 (2021). https://doi.org/10.52825/bis.v1i.42

12. Farquad, M.A.H., Ravi, V., Raju, S.B.: Churn prediction using comprehensible support vector machine: An analytical CRM application. Appl. Soft Comput. **19**, 31–40 (2014). https://doi.org/10.1016/j.asoc.2014.01.031
13. Lu, J.: Predicting customer churn in the telecommunications industry—An application of survival analysis modeling using SAS. SAS User Group International (SUGI27) Online Proceedings **114**, 27 (2002)
14. Li, X., Li, Z.: A hybrid prediction model for E-commerce customer churn based on logistic regression and extreme gradient boosting algorithm. Ingénierie des Systèmes d'Information **24**(5), 525–530 (2019)
15. Gordini, N., Veglio, V.: Customers churn prediction and marketing retention strategies. An application of support vector machines based on the AUC parameter-selection technique in B2B e-commerce industry. Industrial Marketing Management **62**, 100–107 (2017). https://doi.org/10.1016/j.indmarman.2016.08.003
16. Amin, A., Al-Obeidat, F., Shah, B., Adnan, A., Loo, J., Anwar, S.: Customer churn prediction in telecommunication industry using data certainty. J. Bus. Res. **94**, 290–301 (2019). https://doi.org/10.1016/j.jbusres.2018.03.003
17. Çelik, O., Osmanoglu, U.O.: Comparing to techniques used in customer churn analysis. J. Multidiscip. Dev. **4**(1), 30–38 (2019)
18. Burez, J., Van den Poel, D.: CRM at a pay-TV company: Using analytical models to reduce customer attrition by targeted marketing for subscription services. Expert Syst. Appl. **32**(2), 277–288 (2007). https://doi.org/10.1016/j.eswa.2005.11.037
19. Gordini, N., Veglio, V.: Using neural networks for customer churn prediction modeling: preliminary findings from the Italian electricity industry. In: Atti del X° Convegno Annuale della Società Italiana Marketing dal titolo" Smart Life. Dall'Innovazione Tecnologica al Mercato, pp. 1–13. Università degli Studi di Milano-Bicocca, Italy (2013)
20. Gordini, N., Veglio, V.: Customer relationship management and data mining: A classification decision tree to predict customer purchasing behavior in global market. In Handbook of Research on Novel Soft Computing Intelligent Algorithms: Theory and Practical Applications **1**, 1–40 (2013). https://doi.org/10.4018/978-1-4666-4450-2.ch001
21. Verbeke, W., Martens, D., Mues, C., Baesens, B.: Building comprehensible customer churn prediction models with advanced rule induction techniques. Expert Syst. Appl. **38**(3), 2354–2364 (2011). https://doi.org/10.1016/j.eswa.2010.08.023
22. Verbeke, W., Dejaeger, K., Martens, D., Hur, J., Baesens, B.: New insights into churn prediction in the telecommunication sector: A profit driven data mining approach. Eur. J. Oper. Res. **218**(1), 211–229 (2012). https://doi.org/10.1016/j.ejor.2011.09.031
23. De Caigny, A., Coussement, K., De Bock, K.W.: A new hybrid classification algorithm for customer churn prediction based on logistic regression and decision trees. Eur. J. Oper. Res. **269**(2), 760–772 (2018). https://doi.org/10.1016/j.ejor.2018.02.009
24. Deng, Z., Lu, Y., Wei, K.K., Zhang, J.: Understanding customer satisfaction and loyalty: An empirical study of mobile instant messages in China. Int. J. Inf. Manage. **30**(4), 289–300 (2010). https://doi.org/10.1016/j.ijinfomgt.2009.10.001
25. Neslin, S.A., Gupta, S., Kamakura, W., Lu, J., Mason, C.H.: Defection detection: Measuring and understanding the predictive accuracy of customer churn models. J. Mark. Res. **43**(2), 204–211 (2006). https://doi.org/10.1509/jmkr.43.2.204
26. Xu, S., Lai, S., Qiu, M.: Privacy preserving churn prediction. In: Proceedings of the 2009 ACM symposium on Applied Computing, pp. 1610–1614 (2009). https://doi.org/10.1145/1529282.1529643
27. Sharma, A., Kumar Panigrahi, P.: A neural network based approach for predicting customer churn in cellular network services. Int. J. Comp. Appl. **27**(11), 26–31 (2011). https://doi.org/10.5120/3344-4605

28. Buckinx, W., Van den Poel, D.: Customer base analysis: partial defection of behaviourally loyal clients in a non-contractual FMCG retail setting. Eur. J. Oper. Res. **164**(1), 252–268 (2005). https://doi.org/10.1016/j.ejor.2003.12.010
29. Coussement, K., Van den Poel, D.: Churn prediction in subscription services: An application of support vector machines while comparing two parameter-selection techniques. Expert Syst. Appl. **34**(1), 313–327 (2008). https://doi.org/10.1016/j.eswa.2006.09.038
30. Xie, Y., Li, X., Ngai, E.W.T., Ying, W.: Customer churn prediction using improved balanced random forests. Expert Syst. Appl. **36**(3), 5445–5449 (2009). https://doi.org/10.1016/j.eswa.2008.06.121

Social Aspects of Artificial Intelligence in Management

Acceptance and Use of ChatGPT Among Accounting and Finance Higher Education Students

Ewa Wanda Ziemba(✉), Ewa Wanda Maruszewska, Dariusz Grabara, and Katarzyna Renik

University of Economics in Katowice, 40-287 Katowice, Poland
{ewa.ziemba,ewa.maruszewska}@ue.katowice.pl, {dariusz.grabara, katarzyna.renik}@uekat.pl

Abstract. This study investigates the acceptance and use of ChatGPT among higher education students specializing in accounting and finance. Employing and validating Venkatesh's Unified Theory of Acceptance and Use of Technology (UTAUT), a sample of 237 accounting and finance students was surveyed. The results confirm eight out of ten proposed hypotheses, highlighting that behavioral intention towards ChatGPT is shaped by effort expectancy, social influence, and performance expectancy. Furthermore, ChatGPT use is affected by behavioral intention and facilitating conditions. Interestingly, the findings reveal gender moderation in the impact of effort expectancy on behavioral intention, study level moderation in the influence of social factors on behavioral intention, and age moderation in the effect of facilitating conditions on ChatGPT use. These findings have significant potential to advance the understanding of ChatGPT acceptance and utilization within higher education, particularly in the domain of accounting and finance. Nonetheless, the use of ChatGPT for educational purposes is a new phenomenon that necessitates further exploration and repeated empirical validation across diverse research samples.

Keywords: ChatGPT · Accounting · Finance · UTAUT · Generative AI · Education

1 Introduction

Generative Pre-trained Transformer (ChatGPT) is considered a valuable artificial intelligence (AI) tool in research [1, 2] and business [3], and influences the labor market [4]. According to Eloundou et al. [4] analysis of a database containing information on 1,016 occupations, including work and worker characteristics and skill requirements, it was found that approximately 80% of the U.S. workforce could experience a minimum of 10% impact on their work tasks due to the introduction of GPTs. Additionally, around 19% of workers may see a significant impact, with at least 50% of their tasks being affected. Already in 2017, Frey and Osborne [5] predicted that the majority of workers in office and administrative support roles, including accountants and auditors, were among

M. Hernes and J. Wątróbski (Eds.): ECAI 2023, LNNS 1079, pp. 185–202, 2024.
https://doi.org/10.1007/978-3-031-66761-9_16

the occupations vulnerable to the advancements in computerization. In 2023, McGee [6] posed a set of inquiries to ChatGPT regarding its impact on the accounting profession as a whole and its various subfields. ChatGPT expressed the perspective that it would automate mundane tasks, enabling accountants to concentrate on more advanced and elevated responsibilities. Lately, accounting and finance professionals have confirmed that ChatGPT can understand and generate text like a human, making it a powerful tool for a variety of applications. These applications include answering accounting and finance questions, automating data entry, analyzing financial data and forecasting, generating financial reports, processing invoices and expenses, assisting with tax preparation, streamlining communication, and serving as a 'super assistant' for financial teams, as well as for training and education purposes [7–9]. Eloundou et al. [4] estimated that occupations such as financial quantitative analysts, accountants and auditors, and financial managers are among the professions where GPTs and GPT-powered software can significantly reduce the time required to complete a substantial portion of their tasks.

Emphasizes, however, that countries with well-trained, highly skilled populations have a higher chance of adopting artificial intelligence technologies, including ChatGPT [10, 11]. According to Eloundou et al. [4] analysis, individuals with Bachelor's, Master's, and professional degrees are found to have a higher level of exposure to GPTs and GPT-powered software compared to those without formal educational credentials. It is expected that accounting and financial professionals will need to adapt and acquire new skills to effectively collaborate with ChatGPT [9]. These skills should be acquired not only through work experience but also through Bachelor's, Master's, and professional education. Nonetheless, the use of ChatGPT in academia and higher education is a recent development that is still being actively explored [12, 13]. Many scholars have written about the potential benefits and concerns regarding the integration of ChatGPT in education [3, 12]. They have also provided educators with a set of strategies and techniques to ensure the responsible and successful implementation of ChatGPT in teaching [13]. The student context of utilizing ChatGPT was explored by Wood et al. [14], who compared ChatGPT's performance on accounting assessments to student averages on the same set of questions, analyzing where it performs better and worse. However, a comprehensive study is still lacking, one that can identify students' acceptance of ChatGPT and its use, a theme that has consistently emerged across research when new technologies appeared. All of these factors necessitate the examination and understanding of whether accounting and finance students are accepting of ChatGPT as a tool for learning. Therefore, this study aims to identify the acceptance and use of ChatGPT among accounting and finance higher education students.

The remainder of the paper is organized as follows. The next section discusses the utilization of ChatGPT in education and presents the theory of acceptance and use of new technology. Following that, the methodology employed in this study is presented. The subsequent sections report the research findings along with their discussion. Finally, the last section concludes our work.

2 Theoretical Background

2.1 ChatGPT in Higher Education

The application of AI in the field of natural language processing has led to the creation of intelligent chatbots and virtual assistants capable of comprehending and generating human language [15]. Among these, ChatGPT stands out as a powerful AI-powered chatbot [16]. It was launched in November 2022, and its subsequent release of GPT-4 in mid-March 2023 further solidified its capabilities [1]. ChatGPT is built upon the OpenAI language model and trained on an extensive dataset of human conversations. This enables it to perform intricate tasks and produce responses that closely resemble human language [16]. Leveraging deep learning techniques, ChatGPT exhibits a remarkable ability to understand, process, and generate natural language, ensuring a balance of accuracy and usability [12].

ChatGPT has the potential to reshape the educational landscape through its transformative capabilities [17, 18]. It can serve as an invaluable assistant to teachers, offering support in various tasks, and also function as a virtual tutor for students, providing personalized guidance and assistance [19]. ChatGPT can support teachers by generating course outlines, creating topic-specific content for lectures and presentations, generating questions and problem sets, and more. Similarly, students can benefit from ChatGPT by utilizing its assistance in solving complex problems, facilitating collaborative learning, answering questions, writing essays, explaining phenomena, and gaining a deeper understanding of specific topics, thereby enhancing their learning experience [17, 20].

Based on a review of studies that utilized tests, exams, and interviews to assess the performance of ChatGPT in higher education, Lo [19] revealed that its performance varies across different subject domains. Particularly, ChatGPT exhibited exceptional results in the field of economics and outperformed the average students. Wood et al. [14] presented findings regarding the performance of ChatGPT in the domain of accounting. According to their study, ChatGPT provided accurate responses for 56.5% of the questions and partially correct answers for an additional 9.4% of the questions. However, students demonstrated significantly higher performance, with an average of 76.7% on assessments compared to 56.5% for ChatGPT. Nevertheless, ChatGPT outperformed the student average in 15.8% of the assessments.

All the aforementioned findings confirm the high potential of ChatGPT in the field of accounting and finance education. However, it is crucial for students to receive education on the use of ChatGPT [19] and understand how they can effectively collaborate with ChatGPT to attain educational objectives [21]. As the utilization of ChatGPT in higher education is still in its early stages, knowledge is scarce regarding the acceptance and use of this novel technology among students.

2.2 New Technology Acceptance and Use Among Students

Since its development by Venkatesh et al. [22] UTAUT (Unified Theory of Acceptance and Use of Technology) model has been widely used for examining different technologies' acceptance and use explaining about 70% of the variance in behavioral intention to use technology and about 50% of the variance in technology use [23]. UTAUT is a definitive model that captures the essence of the complex nature of interactions as it encompasses four constructs that influence behavioral intention and technology use. The first construct is performance expectancy understood as the degree to which an individual believes that using the system will help him/her to attain gains in job performance and it is the strongest predictor of intention in past studies [24–26]. Effort expectancy, the second construct, is defined as the degree of ease associated with the use of the system, while social influence is the degree to which an individual perceives that important others believe he/she should use the new system. Finally, the fourth construct called facilitating conditions is associated with the beliefs of individuals that an organizational and technical infrastructure exists to support the use of the system. UTAUT states that the first three constructs directly affect the behavioral intention and thus the use of the technology, while the last one directly determines the use of the technology. All four constructs were incorporated in the past research with individual additions like system characteristics [27], trust [28, 29], or mobility understood as flexible access to the platform without any limitation of location or time [30].

Prior research conducted among students is mainly concentrated on different technologies that can supplement traditional learning. One of the early research starts with the investigation of the behavioral intent and use of different types of software [31] and nowadays it examines the chances to encompass artificial intelligence in the learning process [29]. At the beginning of this millennium, as researchers noticed the growing usage of instant messaging like Messenger or ICQ, the students in Singapore were examined showing no effect of performance expectancy on behavioral intent [27]. Contrary results were obtained when investigating the use of blogs [32], webinars [24–26], or e-learning websites [33] for blended learning purposes. Referring to the effort expectancy, also the results are not homogenous as Khechine [25] in his research conducted in 2018 as well as Pardamean and Susanto [32] did not confirm its influence on the behavior intention. Social influence was confirmed in all the above-mentioned studies except Khechine [25] while facilitating conditions gave uniform results in each study. All the above technologies are suggested to enhance learning through case-based teaching methods, their interactivity, as well as accessibility from any location, and enabling life-long learning, both asynchronous and synchronous. To achieve these advantages, the tools should answer the needs of the students in terms of each of the four constructs. Therefore, the authors interpret the insignificance of performance effort as a result of low-working conditions in the case of the use of instant messaging, while effort expectancy insignificance requires actions from webinars and blog producers making the usage more friendly for the students.

As internet technology opened a new era, a vast portion of UTAUT literature deals with mobile learning (m-learning) seen as a successor of e-learning [28, 34]. For example, Bakar et al. [35] investigated student portal system learning, Onaolapo, and Olawale [36] smartphone use, while Attuquayefio and Addo [37] students' information and communication technologies in a form of a mixture of hardware (computers), software (Microsoft Office tools) and telecommunication (Wi-Fi, e-mail, cellular phones, and the Internet). In most of the research, all four constructs had significantly affected the behavioral intention to use m-learning, although Bakar et al. [35] did not confirm performance effort, while Attuquayefio and Addo [37] obtained significant results only for effort expectancy. Low or average intention to use m-learning was explained due to limited functionalities of the student portal or user-unfriendly information and communication technologies with requisite technical support. It can be concluded that most of the students do not benefit fully from the opportunities associated with it, indicating a need for action to bring about a more favorable perception, enhance the quality of learning content, and provide training [38].

Other new technologies investigated in UTAUT research include mobile cloud storage services [39] and virtual classrooms [30]. In all the above-mentioned studies, students' behavioral intention or students' actual use of a certain technology served as a proxy for long usage prospects for a certain technology. To the best of our knowledge, no research has been dedicated to accounting and finance students who will become future professionals, whose knowledge and competencies in utilizing new technology will determine the quality of financial statements and tax calculations. Further, except for Long et al. [40] with a group of 35 students, and Raffaghelli et al. [29] research dedicated to students' early warning system, we did not find research dedicated to artificial intelligence in a form of a ChatGPT.

2.3 Research Hypotheses and Model Development

Thus, based on the above literature, we formulated the following hypotheses for Basic UTAUT (UTAUT1) model:

H1: Behavioral Intention (BI) is affected by Effort Expectancy (EE), Social Influence (SI), and Performance Expectancy (PE)

- H1.1: Behavioral Intention (BI) is affected by Performance Expectancy (PE)
- H1.2: Behavioral Intention (BI) is affected by Effort Expectancy (EE)
- H1.3: Behavioral Intention (BI) is affected by Social Influence (SI)

H2: ChatGPT use (USE) is affected by Behavioral Intention (BI) and Facilitating Conditions (FC)

- H2.1: ChatGPT use (USE) is affected by Behavioral Intention (BI)
- H2.2: ChatGPT use (USE) is affected by Facilitating Conditions (FC)

The UTAUT model was further modified with the addition of moderator variables. They were added to measure the moderation effect on Behavioral Intention and ChatGPT use (USE). Main hypotheses followed by detailed hypotheses (if needed) were developed:

H3: Age (AGE) and gender (GDR) moderate the effect of the influence of Performance Expectancy (PE) on Behavioral Intention (BI), such that the effect is stronger among younger students or males.

- H3.1: Age (AGE) moderates Performance Expectancy (PE) influence on Behavioral Intention (BI)
- H3.2: Gender (GDR) moderates Performance Expectancy (PE) influence on Behavioral Intention (BI)

H4: Age (AGE), gender (GDR), and level of study (LOS) moderate the effect of the influence of Effort Expectancy (EE) on Behavioral Intention (BI), such that the effect is stronger among younger students, or males, or students with a higher level of study.

- H4.1: Age (AGE) moderates Effort Expectancy (EE) influence on Behavioral Intention (BI)
- H4.2: Gender (GDR) moderates Effort Expectancy (EE) influence on Behavioral Intention (BI)
- H4.3: Level of study (LOS) moderates Effort Expectancy (EE) influence on Behavioral Intention (BI)

H5: Age (AGE), gender (GDR), and level of study (LOS) moderate the effect of Social Influence (SI) on Behavioral Intention (BI), such that the effect is stronger among younger students, males, or students with a higher level of study.

- H5.1: Age (AGE) moderates Social Influence (SI) effect on Behavioral Intention (BI)
- H5.2: Gender (GDR) moderates Social Influence (SI) effect on Behavioral Intention (BI)
- H5.3: Level of study (LOS) moderates Social Influence (SI) effect on Behavioral Intention (BI)

H6: Age (AGE) moderates the effect of Facilitated Conditions (FC) on ChatGPT use (USE), such that the effect is stronger among younger students.

H7: Level of study (LOS) moderates the effect of behavioral intention (BI) on ChatGPT use (USE), such that the effect is stronger among students who have a high appreciation for ChatGPT.

On this basis, UTAUT theoretical model was derived which is provided in Fig. 1.

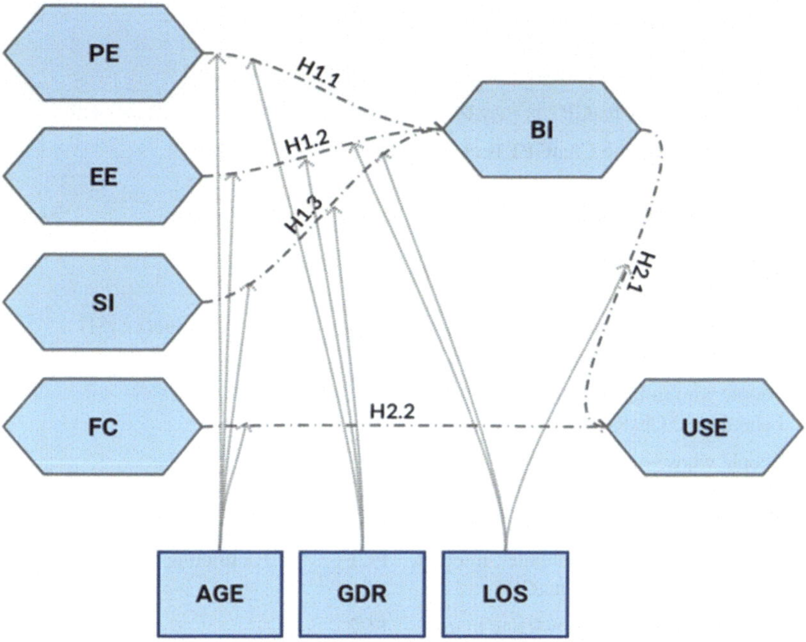

Fig. 1. Basic UTAUT model (UTAUT1) and addition of moderator variables for expanded model (UTAUT2). * Thin dotted lines represent the influence of moderators.

3 Material and Methods

3.1 Questionnaire

A questionnaire was developed through the method proposed by Venkatesh, Thong, and Xu [23]. Questions were organized to reflect theoretical constructs as shown in Table 1.

Table 1. Survey questions.

Id	Question	Variable	Construct
1	I find ChatGPT useful in gaining knowledge and skills	PE1	Performance Expectancy (PE)
2	Using ChatGPT increases my chances of gaining more knowledge and skills	PE2	
3	Using ChatGPT helps me gain knowledge and skills faster	PE3	
4	Using ChatGPT increases my productivity in gaining knowledge and skills	PE4	

(*continued*)

Table 1. (*continued*)

Id	Question	Variable	Construct
5	Learning to use ChatGPT is easy for me	EE1	Effort Expectancy (EE)
6	My interaction with ChatGPT is clear and understandable	EE2	
7	I find ChatGPT easy to use	EE3	
8	I easily become proficient in using ChatGPT	EE4	
9	People who are important to me believe that I should/should use ChatGPT	SI1	Social Influence (SI)
10	People who influence my behavior think I should use ChatGPT	SI2	
11	People whose opinions I value recommend that I use/should use ChatGPT	SI3	
12	I have the resources (computer, internet, etc.) necessary to use ChatGPT	FC1	Facilitating Conditions (FC)
13	I have the necessary knowledge to use ChatGPT	FC2	
14	ChatGPT is compatible with other technologies I use	FC3	
15	I can get help from others when I have difficulty using ChatGPT	FC4	
16	I plan to continue using ChatGPT in the future	BI1	Behavioral Intention (BI)
17	I will always try to use ChatGPT	BI2	
18	I plan to continue to use ChatGPT frequently	BI3	
19	I use ChatGPT to write my thesis (undergraduate/master's)	USE1	Use (USE)
20	I use ChatGPT while working on classes at the university	USE2	
21	I use ChatGPT for doing homework/preparing projects/writing dissertations/writing papers	USE3	
22	I use ChatGPT for self-study, gaining knowledge and skills on an interesting topic related to my studies	USE4	
23	Age	AGE	Age (AGE)

(*continued*)

Table 1. *(continued)*

Id	Question	Variable	Construct
24	Gender	GDR	Gender (GDR)
25	Level of study	LOS	Level of study (LOS)

Indicators for construct USE, Level of Study (LOS), and Field of Study (FOS) were coded as shown in Table 2.

Table 2. Constructs' coding and scoring.

Construct	Graduation	Code
USE	never	1
	only once	2
	several times	3
	once - twice a month	4
	once a week	5
	several times a week	6
	once a day or more often	7
Level of Study (LOS)	1st degree (bachelor)	1
	2nd degree (master) or higher	2
Gender*	Female	0
	Male	1
	Other	2

* Gender was coded according to Venkatesh et al. [23].

3.2 Sample and Data Collection

The questionnaire was distributed via Google Forms and sent directly by the Google Classroom platform to accounting and finance students at the University of Economics in Katowice, Poland. The data were collected between May 8th, 2023, and June 8th, 2023. According to Kock [41], a minimum sample size of 100–200 observations is required to ensure the accuracy and validity of the findings. A total of 237 valid responses were collected, resulting in a response rate of 14%. This means that 14% of all accounting and finance students at the university participated. The sampling error for an infinite population was approximately 6% at a confidence level of 95% (assuming $p = q = 0.5$). Hence, the study's sample of 237 students was considered appropriate.

3.3 Methods

The database has been processed to avoid incorrect answers. The process involved finding missing or incomplete values and automated answers. Automated answers such as choosing the same answer for all of the questions resulted in answer rejection.

For data analysis presentation basic descriptive statistics have been used, and for hypotheses testing Partial Least Squares (PLS-SEM) technique was used. PLS-SEM provides cause-effect relations between latent constructs. The technique analyzes path coefficients and indicators loadings to determine the strength and significance of the relationships between the latent constructs and is useful in hypotheses verification.

Developed PLS-SEM models were tested according to the Hair Jr et al. procedure [42]. The assessment included the evaluation of measurement and structural models using the SEMinR package of the R project.

4 Results

4.1 Respondents Profile

The basic descriptive profile of respondents is shown in Table 3.

Table 3. Respondents' profile.

Construct	Graduation	No of students	Percentage
Level of Study (LOS)	1st degree (bachelor)	134	56.5%
	2nd degree (master) or higher	103	43.5%
Gender	Female	166	70.0%
	Male	68	28.7%
	Other	3	1.3%

The results indicate that most answers were gathered from bachelor's degree students (55.6%), while the rest accounted for master's and higher degree students (44.4%). The group was also representative due to the greater involvement of females (70%) than males (28.7%) as it is registered that more women are attending in the presented field of study.

4.2 UTAUT1 and UTAUT2 Models Results

The baseline model testing was provided by analyzing the reliability of measurement indicators. The analysis revealed that besides 3 indicators all others have achieved medium to high values of indicator commonalities as shown in Table 4.

The FC1, FC4, and USE1 are below the threshold of 0.708, but they are greater than the rejection value of 0.4 and in the light of social research could be accepted.

Table 4. Reliability of measurement model indicator.

Indicator	Model/loadings	
	UTAUT1	UTAUT2
PE		
PE1	0.908	0.908
PE2	0.920	0.920
PE3	0.893	0.893
PE4	0.855	0.855
EE		
EE1	0.915	0.915
EE2	0.922	0.922
EE3	0.878	0.878
SI		
SI1	0.911	0.911
SI2	0.925	0.925
SI3	0.926	0.926
FC		
FC1	0.659	0.658
FC2	0.839	0.840
FC3	0.823	0.822
FC4	0.625	0.625
BI		
BI1	0.906	0.905
BI2	0.861	0.863
BI3	0.940	0.940
USE		
USE1	0.528	0.530
USE2	0.849	0.851
USE3	0.850	0.853
USE4	0.766	0.760

* Loading with moderate effect (more than 0.2 and less than 0.708) was greyed.

The reliability of the constructs is provided in Table 5. Analysis revealed that there could be redundancy between constructs (CR > 0.90), however, there was no construct that was greater than the rejection value of 0.95. CR with that value could indicate, that questions are not properly distinguished by respondents.

Table 5. Reliability of structural model constructs.

Model/ Construct	Cronbach's alpha		CR (rhoC)		AVE		rhoA	
	UTAUT1	UTAUT2	UTAUT1	UTAUT2	UTAUT1	UTAUT2	UTAUT1	UTAUT2
PE	0.916	0.916	0.941	0.941	0.800	0.800	0.918	0.918
EE	0.891	0.891	0.932	0.932	0.820	0.820	0.906	0.906
SI	0.911	0.911	0.944	0.944	0.848	0.848	0.917	0.917
BI	0.887	0.887	0.930	0.930	0.815	0.816	0.908	0.906
FC	0.726	0.726	0.829	0.829	0.551	0.551	0.766	0.766
USE	0.748	0.748	0.841	0.841	0.577	0.578	0.788	0.787

There were no detected collinearities between constructs. All values were well below the threshold of 3.

The model's discriminant validity was measured by the HTMT criterion. The criterion did not reveal that there are values more than 0.9, and thus models provided an adequate discriminant assessment. It was also confirmed by running further Fornell-Larcker procedures which didn't confirm discriminant issues.

Results for path coefficients are shown in Table 6.

5 Discussion

The basic model of UTAUT1 has confirmed that the UTAUT framework is adequate to analyze phenomena like ChatGPT. All fundamental constructs (PE, EE, SI, FC, and BI) play a significant role in the models, thus resulting in H1 and H2 hypotheses confirmation. As the literature presents limitations of UTAUT and thus develops an extension of this model, we also used UTAUT2 in order to improve the overall model fitting.

Hypotheses and their confirmations are presented in Table 7.

Statistically significant moderation effect confirmed that gender (GDR) influences Effort Expectations (EE) on Behavioral Intention (BI), thus confirming hypothesis H4.2. This finding is in line with Lin et al. [43]. Researchers argued that gender is one of the indirect moderators of the relationship between EE and BI. However, rather than direct inclusion of gender they established a more generalized concept of Learning-Teaching Fit. Moderation is also confirmed by Warsame and Ireri [44] in the analysis of new technology of mobile microfinance. In their work, gender was found to be a significant moderator between Effort Expectancy and Behavioral Intensions. In our work, significant moderation revealed that positive affection is showing greater influence among males than females.

The other positive impact of the influence of the moderator variable is noted for the moderator level of study (LOS) which affects the relationship between Social Influence (SI) and Behavioral Intention (BI). This finding suggests that the stronger the impact of Social Influence (SI), the higher the intention (BI) are toward using ChatGPT. Warsame and Ireri also confirmed that SI and BI relation is affected by factors like gender, aged, trust, or religious beliefs. While in our study we didn't confirm the moderator effect of

Table 6. ChatGPT model coefficient estimates.

Item	Model				Hypothesis **
	UTAUT1		UTAUT2		
R^2 (BI)	0.485		0.526		X
Adj. R^2 (BI)	0.479		0.496		
R^2 (USE)	0.418		0.441		
Adj. R^2 (USE)	0.413		0.426		
PE -> BI	0.440	*	0.449	*	H1.1
EE -> BI	0.207	*	0.224	*	H1.2
SI -> BI	0.227	*	0.220	*	H1.3
AGE -> BI			0.040		X
AGE -> USE			-0.141		X
GDR -> BI			0.060		X
LOS -> BI			-0.077		X
LOS -> USE			0.040		X
PE*GDR -> BI			-0.101		H3.2
EE*GDR -> BI			0.132	*	H4.2
SI*GDR -> BI			-0.067		H5.2
PE*AGE -> BI			0.085		H3.1
EE*AGE -> BI			0.028		H4.1
SI*AGE -> BI			-0.067		H5.1
EE*LOS -> BI			-0.002		H4.3
SI*LOS -> BI			0.115	*	H5.3
BI -> USE	0.571	*	0.574	*	H2.1
FC -> USE	0.149	*	0.139	*	H2.2
FC*AGE -> USE			-0.140	*	H6
BI*LOS -> USE			0.061		H7

* Path is significant with p-value<0.05; ** 'X' is provided if the hypothesis does not apply to the item.

gender and age, the significance of the Level of Study (LOS) could relate to trust as trust growth with knowledge development. Venkatesh et al. [23] also proposed experience as a moderator of Effort Expectancy (EE) and Behavioral Intensions (BI) relation. While experience also relates to knowledge development like leveling Level of Study (LOS) it confirms our findings. In conclusion, hypothesis H5.3 is then confirmed.

A negative significant impact is registered for moderator age (AGE) and the relationship between Facilitating Conditions (FC) and ChatGPT use (USE). This finding suggests that with increasing age, Facilitating Conditions (FC) have a lesser impact on

Table 7. Results of hypotheses verification.

Hypothesis		Description	Results*	
			UTAUT1	UTAUT2
H1	H1.1	BI is affected by PE	Confirmed	Confirmed
	H1.2	BI is affected by EE	Confirmed	Confirmed
	H1.3	BI is affected by SI	Confirmed	Confirmed
H2	H2.1	USE is affected by BI	Confirmed	Confirmed
	H2.2	USE is affected by FC	Confirmed	Confirmed
H3	H3.1	AGE moderates the effect of PE on BI		NC
	H3.2	GDR moderates the effect of PE on BI		NC
H4	H4.1	AGE moderates the effect of EE on BI		NC
	H4.2	GDR moderates the effect of EE on BI		Confirmed
	H4.3	LOS moderates the effect of EE on BI		NC
H5	H5.1	AGE moderates the effect of SI on BI		NC
	H5.2	GDR moderates the effect of SI on BI		NC
	H5.3	LOS moderates the effect of SI on BI		Confirmed
H6	H6	AGE moderates the effect of FC on USE		Confirmed
H7	H7	LOS moderates the effect of BI on USE		NC

* Confirmed – hypothesis is confirmed, NC – hypothesis is not confirmed.

ChatGPT use (USE). In the work of Alkhasawneh and Alanazy [45] on ICT adoption among academic staff, age wasn't found a significant moderator. In the works of Khechine et al. [25], Szyszka et al. [46], and Tomczyk et al. [47], researchers investigated the influence of age on the relationship of Facilitating Conditions (FC) on Behavioral Intention (BI) and found positive affection. However, they didn't examine included technology use in their model, which could significantly impact the role of FC on USE. In this context, our findings bring a new perspective on the moderation effect of LOS on FC and USE variables.

6 Conclusion

This study examines the acceptance and usage of ChatGPT among students in the field of accounting and finance in higher education. It adopts and verifies Venkatesh's theoretical model of acceptance and use of technology. The study confirms that ChatGPT use is influenced by behavioral intention and facilitating conditions. Moreover, behavioral intention towards ChatGPT is affected by effort expectancy, social influence, and performance expectancy. Additionally, the findings indicate that gender moderates the effect of effort expectancy on behavioral intention towards ChatGPT, with a stronger effect observed among younger or male students or with a higher level of study. Furthermore, the level of study moderates the effect of social influence on behavioral intention towards ChatGPT, with a stronger effect observed among younger students or male students, or those with a higher level of study. Lastly, age moderates the effect of facilitating conditions on ChatGPT use, with a stronger effect observed among younger students.

Our findings confirm results that are consistent with and build upon a recent study by Strzelecki [48], which also investigated students' acceptance and use of ChatGPT in Poland. We confirmed that the use of ChatGPT is positively associated with behavioral intention and facilitating conditions, as well as performance expectancy, effort expectancy, and social influence positively influencing behavioral intention. Our results reveal the moderating effect of gender on the relationship between effort expectancy and behavioral intention towards ChatGPT, the impact of the level of study on the relationship between social influence and behavioral intention, and the influence of age on the relationship between facilitating conditions and the use of ChatGPT. Strzelecki's study [48] did not reveal such roles for these moderators. The conclusions drawn from this comparison are twofold. First, our primary findings concerning the factors influencing the use of GPT and behavioral tendencies towards ChatGPT can likely be generalized to students specializing in economics, management, and computer science, and possibly to the broader context of Poland as well. Second, the moderating roles of the level of study, gender, and age demand further elucidation.

The findings obtained from this study hold significant potential in advancing the understanding of ChatGPT adoption and utilization within higher education, particularly in the accounting and finance domain. The outcomes of the UTAUT analysis reveal that this new AI technology is not only accepted but also likely to see increasing use by students for educational purposes. Therefore, both academia and practice cannot overlook the strengths and opportunities that ChatGPT offers for education, necessitating further studies in this domain. Educating not only students but also teachers about the benefits, limitations, and risks of chatbots, along with instructing them on how to effectively use ChatGPT in their work, while emphasizing the value of cultivating a broad range of knowledge and skills, holds the utmost importance. This aspect has also been highlighted in the existing literature [3, 14].

Like any exploratory study, this study has several limitations. First, ChatGPT and its usage in education purposes are new phenomena that have yet to be further explored and exposed to repeated empirical validation. Second, the sample included students from the University of Economics in Katowice in Poland. Caution should be taken when generalizing the findings to other fields of students' study, regions, and countries. Third, some statistical contraversions appeared. Discriminant validity problems were discovered in the basic model UTAUT model for PE, EE, SI, and BI constructs (CR > 0.9). This finding could imply, that further explanation of given questions for respondents should be provided. The novelty of the study includes the investigation of the level of study as the moderator variable on a set of relationships. In the presented case, the level of study was found significant negative moderator in the relationship between facilitating conditions and ChatGPT use. This finding should be further investigated especially with the conjunction of experience moderator.

Future studies should be conducted to gain a deeper understanding of the usage of ChatGPT in education including accounting and finance education and practice. It would be particularly valuable to explore the various methods through which such technologies can be integrated into accounting and finance higher education programs, enhancing the knowledge and experience of accounting and finance students. Furthermore, research investigating the adoption of these technologies by accounting and finance professionals should be a significant focus for future studies.

References

1. Chen, S., et al.: ChatGPT: Cross-cultural tourism research imperative. J. Econ. Manage. 45, 137–146 (2023). https://doi.org/10.22367/jem.2023.45.07
2. Korzyński, P., et al.: Generative artificial intelligence as a new context for management theories: analysis of ChatGPT. Centr. European Manage. J. **31**(1), 3–13 (2023). https://doi.org/10.1108/CEMJ-02-2023-0091
3. Dwivedi, Y.K., et al.: "So what if ChatGPT wrote it?" Multidisciplinary perspectives on opportunities, challenges and implications of generative conversational AI for research, practice and policy. Int. J. Inf. Manage. **71**, 1–61 (2023). https://doi.org/10.1016/j.ijinfomgt.2023.102642
4. Eloundou, T., Manning, S., Mishkin, P., Rock, D.: GPTs are GPTs: An Early Look at the Labor Market Impact Potential of Large Language Models. Working Paper. March, 27 (2023). https://doi.org/10.48550/arXiv.2303.10130. Last accessed 22 June 2023
5. Frey, C.B., Osborne, M.A.: The future of employment: How susceptible are jobs to computerisation? Technol. Forecast. Soc. Chang. **114**, 254–280 (2017). https://doi.org/10.1016/j.techfore.2016.08.019
6. McGee, R.W.: Will ChatGPT Destroy the Accounting Profession? ChatGPT Replies. Working Paper. April, 27 (2023). https://doi.org/10.13140/RG.2.2.12811.36644. Last accessed 22 June 2023
7. Dilmegani, C.: 12 GPT Accounting Use Cases & Benefits in 2023. April, 26 (2023). https://research.aimultiple.com/gpt-accounting/. Last accessed 22 June 2023
8. Guerra, M.: Chat GPT for Accounting: How AI is Transforming the Industry. March, 13 (2023). https://brandalytics.co/chat-gpt-for-accounting/. Last accessed 22 June 2023
9. Wattanajantra, A.: AI in finance: What ChatGPT and generative AI means for CFOs. May, 23 (2023). https://www.sage.com/en-gb/blog/ai-finance-chatgpt-generative-ai-cfos/. Last accessed 22 June 2023
10. Pahl, S.: An emerging divide: Who is benefiting from AI?. February (2023). https://iap.unido.org/articles/emerging-divide-who-benefiting-ai#fn-2303-0
11. Wach, K., et al.: The dark side of generative artificial intelligence: a critical analysis of controversies and risks of ChatGPT. Entreprene. Bus. Econ. Rev. **11**(2), 7–24 (2023). https://doi.org/10.15678/EBER.2023.110201
12. Farrokhnia, M., Banihashem, S. K., Noroozi, O., Wals, A.: A SWOT analysis of ChatGPT: implications for educational practice and research. Innovat. Edu. Teach. Int. 1–15 (2023). https://doi.org/10.1080/14703297.2023.2195846
13. Halaweh, M.: ChatGPT in education: Strategies for responsible implementation. Contemp. Edu. Technol. **15**(2), 1–11 (2023). https://doi.org/10.30935/cedtech/13036
14. Wood, D.A., et al.: The ChatGPT artificial intelligence chatbot: how well does it answer accounting assessment questions? Issues Account. Educ. **38**(4), 1–28 (2023). https://doi.org/10.2308/ISSUES-2023-013
15. Caldarini, G., Jaf, S., McGarry, K.: A literature survey of recent advances in chatbots. Information **13**(41), 1–22 (2022). https://doi.org/10.3390/info13010041
16. Sunjak, T.: ChatGPT: The End of Online Exam Integrity? Working Paper. https://doi.org/10.48550/arXiv.2212.09292. Last accessed 22 June 2023

17. Haleem, A., Javaid, M., Singh, R.P.: An era of ChatGPT as a significant futuristic support tool: A study on features, abilities, and challenges. BenchCouncil Transactions on Benchmarks, Standards and Evaluations **2**(4), 1–8 (2022). https://doi.org/10.1016/j.tbench.2023.100089
18. Mhlanga, D.: Open AI in Education, the Responsible and Ethical Use of ChatGPT Towards Lifelong Learning. Working paper. February, 11 (2023). https://doi.org/10.2139/ssrn.4354422. Last accessed 22 June 2023
19. Lo, C.K.: What is the impact of ChatGPT on education? a rapid review of the literature. Education Sciences **13**(410), 1–15 (2023). https://doi.org/10.3390/educsci13040410
20. Rahman, M.M., Watanobe, Y.: ChatGPT for education and research: opportunities, threats, and strategies. Appl. Sci. **13**(5783), 1–23 (2023). https://doi.org/10.3390/app13095783
21. Tlili, A., et al.: What if the devil is my guardian angel: ChatGPT as a case study of using chatbots in education. Smart Learning Environments **10**(15), 1–24 (2023). https://doi.org/10.1186/s40561-023-00237-x
22. Venkatesh, V., Morris, M.G., Davis, G.B., Davis, F.D.: User acceptance of information technology: toward a unified view. MIS Q. **27**(3), 425–478 (2003)
23. Venkatesh, V., Thong, J.Y.L., Xu, X.: Consumer acceptance and use of information technology: extending the unified theory of acceptance and use of technology. MIS Q. **36**(1), 157–178 (2012). https://doi.org/10.2307/41410412
24. Lakhal, S., Khechine, H., Pascot, D.: Influence of business students ' psychological factors on behavioral intentions to use webinar in distance education : integration of autonomy to the Unified Theory of Acceptance and Use of Technology (UTAUT) model. J. Comput. High. Educ. **25**, 93–121 (2013). https://doi.org/10.1007/s12528-013-9069-3
25. Khechine, H., Lakhal, S., Pascot, D., Bytha, A.: UTAUT Model for Blended Learning : The Role of Gender and Age in the Intention to Use Webinars. Interdiscipl. J. E-Learning and Learn. Obj. **10**, 33–52 (2014). http://www.ijello.org/Volume10/IJELLOv10p033-052Khechine0876.pdf. Last accessed 22 June 2023
26. Khechine, H., Lakhal, S.: Technology as a double-edged sword: From behavior prediction with UTAUT to students' outcomes considering personal characteristics. J. Info. Technol. Edu. Res. **17**, 63–102 (2018). https://doi.org/10.28945/4022
27. Lin, J., Chan, H.: Instant Messaging Acceptance and Use Among College Students. Pacific Asia Conference on Information Systems PACIS 2004 Proceedings. 15, pp. 181–194. Communications of the Association for Information Systems (CAIS), Shanghai (2004). https://aisel.aisnet.org/pacis2004/15. Last accessed 22 June 2023
28. Chao, C.: Factors Determining the Behavioral Intention to Use Mobile Learning: An Application and Extension of the UTAUT Model. Frontiers in Psychology **10**, 1–14 (2019). https://doi.org/10.3389/fpsyg.2019.01652
29. Raffaghelli, J.E., Rodríguez, M.E., Bañeres, D.: Computers & Education Applying the UTAUT model to explain the students' acceptance of an early warning system in Higher Education. Comput. Educ. **182**, 1–14 (2022). https://doi.org/10.1016/j.compedu.2022.104468
30. Alshammari, S.H.: Determining the factors that affect the use of virtual classrooms: A modification of the UTAUT Model. J. Info. Technol. Edu. Res. **20**, 117–135 (2021). https://doi.org/10.28945/4709
31. Marchewka, J.T., Kostiwa, K.: Communications of the IIMA An Application of the UTAUT Model for Understanding Student Perceptions Using Course Management Software An Application of the UTAUT Model for Understanding Student Perceptions Using Course Management Software. Communications of the IIMA **7**(2), 93–104 (2007). https://doi.org/10.58729/1941-6687.1038
32. Pardamean, B., Susanto, M.: Assessing User Acceptance toward Blog Technology Using the UTAUT Model. Int. J. Math. nd Comp. Simul. **6**(1), 203–212 (2012). http://citeseerx.ist.psu.edu/viewdoc/download?doi=10.1.1.457.4427&rep=rep1&type=pdf. Last accessed 22 June 2023

33. Tan, P.J.B.: Applying the UTAUT to Understand Factors Affecting the Use of English E-Learning Websites in Taiwan. SAGE Open 1–12 (2013). https://doi.org/10.1177/2158244013503837
34. Izkair, A.S., Lakulu, M.M.: The moderating effects of gender on factors affecting the intention to use mobile learning. J. Info. Technol. Edu. Res. **22**, 199–233 (2023). https://doi.org/10.28945/5094
35. Bakar, A.A., Zaidi, F., Razak, A., Salihin, W., Abdullah, W.: Assessing the effects of UTAUT and self-determination predictor on students continuance intention to use student portal. World Applied Sci. J. **21**(10), 1484–1489 (2013). http://www.idosi.org/wasj/wasj21(10)13/13.pdf. Last accessed 22 June 2023
36. Onaolapo, S., Oyewole, O.: Performance expectancy, effort expectancy, and facilitating conditions as factors influencing smart phones use for mobile learning by postgraduate students of the University of Ibadan, Nigeria. Interdisc. J. e-Skills and Lifelong Learn. **14**, 95–115 (2019). https://doi.org/10.28945/4085
37. Attuquayefio, S.N., Addo, H.: Using the UTAUT model to analyze students' ICT adoption. Int. J. Educ. Dev. Using Inf. Commun. Technol. **10**, 75–86 (2014)
38. Almaiah, M.A., Alamri, M.M., Al-Rahmi, W.: Applying the UTAUT model to explain the students' acceptance of mobile learning system in higher education. IEEE Access, 1–9 (2019). https://doi.org/10.1109/ACCESS.2019.2957206
39. Arpaci, I.: Computers in human behavior understanding and predicting students ' intention to use mobile cloud storage services. Comput. Hum. Behav. **58**, 150–157 (2016). https://doi.org/10.1016/j.chb.2015.12.067
40. Long, J., Yuan, M.J., Lee, H.M.: How to program a chatbot – An introductory project and student perceptions. Issu. Info. Sci. Info. Technol. **16**, 1–31 (2019). https://doi.org/10.28945/4282
41. Kock, N.: Minimum sample size estimation in PLS-SEM: An application in tourism and hospitality research. In: Applying partial least squares in tourism and hospitality research, 1–16. Emerald Publishing Limited (2018). https://doi.org/10.1108/978-1-78756-699-620181001
42. Hair Jr., J.F., et al.: Partial Least Squares Structural Equation Modeling (PLS-SEM) Using R: A Workbook. Springer, Switzerland (2021). https://doi.org/10.1007/978-3-030-80519-7
43. Lin, P.C., Lu, H.K., Liu, S.C.: Towards an education behavioral intention model for e-learning systems: An extension of UTAUT. J. Theor. Appl. Inf. Technol. **47**(3), 1200–1207 (2013)
44. Warsame, M.H., Ireri, E.M.: Moderation effect on mobile microfinance services in Kenya: An extended UTAUT model. J. Behav. Exp. Financ. **18**, 67–75 (2018). https://doi.org/10.1016/j.jbef.2018.01.008
45. Alkhasawneh, S., Alanazy, S.: Adopt ICT among academic staff in Aljoufuniversity: Using UTAUT model. Mediterr. J. Soc. Sci. **6**(1), 490–494 (2015). https://doi.org/10.5901/mjss.2015.v6n1p490
46. Szyszka, M., Tomczyk, Ł, Kochanowicz, A.M.: Digitalisation of Schools from the Perspective of Teachers' Opinions and Experiences: The Frequency of ICT Use in Education, Attitudes towards New Media, and Support from Management. Sustainability **14**(14), 8339 (2022). https://doi.org/10.3390/su14148339
47. Tomczyk, Ł., Potyrała, K., Włoch, A., Wnęk-Gozdek, J., Demeshkant, N.: Evaluation of the Functionality of a New E-Learning Platform vs. Previous Experiences in E-Learning and the Self-Assessment of Own Digital Literacy. Sustainability **12**(23), 10219 (2020). https://doi.org/10.3390/su122310219
48. Strzelecki, A.: To use or not to use ChatGPT in higher education? A study of students' acceptance and use of technology. Interact. Learn. Environ. 1–14 (2023). https://doi.org/10.1080/10494820.2023.2209881

Virtual Reality Games in Promoting Waste-Sorting Behaviour

Anna Borawska(✉), Małgorzata Łatuszyńska, and Mariusz Borawski

University of Szczecin, Cukrowa 8, 71-004 Szczecin, Poland
anna.borawska@usz.edu.pl

Abstract. Virtual reality is a rapidly expanding topic, both in terms of technological advances and the domains of its applications – also in promoting pro-environmental behaviours or increasing environmental awareness. This study aims to present the research procedure that could enable a comparison of immersive virtual reality and desktop games concerning promoting waste-sorting behaviour in terms of their effectiveness with the use of cognitive neuroscience techniques. The article provides a contribution to the field of waste management by proposing a methodology for investigating the most effective ways of promoting waste-sorting behaviour. While there are some limitations to this approach, the potential benefits can be significant.

Keywords: Virtual Reality · serious games · cognitive neuroscience · waste-sorting

1 Introduction

Waste-sorting behavior is one of the many pro-environmental actions that individuals can take to reduce their impact on the environment. It has been recognized as one of the most effective pro-environmental actions that individuals can take, as it directly reduces the amount of waste sent to landfills and helps to conserve natural resources. In general terms, waste-sorting behavior refers to the actions individuals take to separate different types of waste materials, such as recyclables, compostables, and non-recyclables [1].

Improving the source separation of waste is essential for managing waste in an environmentally friendly manner. One way to achieve this is through citizen education on the proper ways to source-separate waste at home by education and awareness-raising campaigns. By educating citizens about the importance of waste sorting and the benefits it brings to the environment, individuals can be motivated to participate in the waste sorting process. Various channels can be used to educate citizens, including social media, public service announcements, and educational workshops. In recent years, more and more often virtual reality games have also been used for this purpose.

Virtual reality (VR) is a rapidly expanding topic, both in terms of technological advances and the domains of its applications – also in promoting pro-environmental behaviors or increasing environmental awareness. According to literature, VR systems (including VR games) can be categorized into three groups: immersive, semi-immersive,

M. Hernes and J. Wątróbski (Eds.): ECAI 2023, LNNS 1079, pp. 203–214, 2024.
https://doi.org/10.1007/978-3-031-66761-9_17

and non-immersive [2]. An immersive VR system is capable of simulating the sensation of completely exploring a virtual world, whereas a semi-immersive system (e.g., utilizing multiple projection screens rather than goggles) or a non-immersive system (e.g., computer monitors) can offer some degree of realism, albeit without an extensively heightened impression of presence [3]. Games can be created in all three types of VR systems - therefore, the question arises to what extent the level of immersion determines the effectiveness of the game in the aspect of promoting waste-sorting behavior.

The effectiveness of messages promoting socially desirable attitudes, regardless of the communication channels (media) used, can be examined in many different contexts using common methods for gathering data such as questionnaires, interviews, and observations [4]. This kind of research typically focuses on metrics such as brand recognition and recall [5], attitudes and behavior change [6], behavioral intentions [7], self-efficacy [8], knowledge, motivation, and persuasion [9], or observed behavior [10]. These measures are also relevant from the perspective of waste-sorting behavior promotion. Although such methods and metrics are commonly used to gather information, they may not always yield reliable knowledge. This is because people may have difficulty or be unwilling to accurately express their true preferences when asked directly [11]. To address these issues, cognitive neuroscience methods can be employed, which involve collecting psychophysiological data to determine objective metrics that define the subconscious reactions and opinions of participants towards presented stimuli. These methods encompass a range of tools and techniques used to measure, map, and record activity within the nervous system in real time. According to Harris, Ciorciari, and Gountas [12], the most commonly used techniques in this area include electroencephalography (EEG), functional magnetic resonance imaging (fMRI), measurements of galvanic skin response (GSR) and heart rate (HR), as well as eye-tracking.

It should be emphasized that there is a great interest in conducting experimental research using virtual reality and simultaneous registration of nervous system activity using various tools and methods of cognitive neuroscience but mainly in the context of advertising (e.g. [13]). There are also studies in the literature on comparing immersive and non-immersive virtual reality games [14], however, there are very few of them using neuroscience (e.g. [15]) and not in the context of games designed to promote pro-environmental attitudes.

Considering the above, in this study, we aim to present the research procedure that could enable a comparison of different games concerning promoting waste-sorting behavior in terms of their effectiveness. Due to the implementation of the suggested procedure, it would become feasible to make a comparative analysis of the efficacy of immersive and non-immersive games, and subsequently ascertain the optimal methodology to be employed in such games.

The reminder of the study is organized as follows. Section 2 provides an overview of available games promoting waste-sorting behavior and additional literature review looking for research studies with a focus on games focusing on waste management. Section 3 presents two versions of VR game (immersive and non-immersive), which will be used for conducting experimental research and Sect. 4 describes the proposed research procedure. Finally, Sect. 5 presents the conclusions and presentation of limitations, as well as an indication of the direction of future research.

2 Virtual Reality Games Promoting Waste Sorting Behaviour

2.1 Applications Review

The topic of waste sorting is prevalent in many games, particularly those designed for children. The mechanics of the sorting game are straightforward and thus, relatively easy to incorporate into game design. Additionally, an effectively implemented game can have significant educational benefits. However, due to the repetitive nature of simply sorting waste, game designers must use their creativity to maintain the game's entertainment value over time.

The most basic waste sorting games involve the disposal of waste into appropriately labeled bins. Such games are so simplistic that they are also found in non-digital form, as exemplified by the games Material Sorting Recycling Truck by Masterkidz [16] and Ecohero [17] by Adamigo. Similar games in the computer version can be found on websites that provide free online games. An example is the Recycling game available on Jeja website [18]. It involves throwing garbage into the appropriate containers. The graphics of the game are quite simple, due to the fact that it is quite an old game, still written on the basis of Flash technology. For this reason, it is practically unavailable on most computers (it requires violating security rules to run). The game Trash Sorting For Kids, available on the online gaming site Fog [19], is identical in concept. Rubbish is scattered randomly on various types of background graphics. The player's task is to classify the garbage into the appropriate category by moving it to the correct bin. The main disadvantage of the game is the lack of any point system and the monotony of the stages, in which only the background graphics and garbage categories change. The Garbage Sorting Truck puzzle game [20] is on the same gaming site. It is a puzzle game in which the player has to make the garbage fall into the right truck by opening the appropriate "passages". The game has a lot of potential as it could theoretically scale the difficulty level properly, but the levels are poorly balanced, and the difficulty of the levels fluctuates a lot. In addition, the large number of advertising cutscenes in the game is very tiring.

Games about sorting garbage can also be found on websites with educational games for children. An example is the game Litter Critters from the ABCYa website [21]. In the game, garbage moves on a conveyor belt. The player must move them to the bin if they are recyclable (easy version) or decide what type of waste it is (hard version). The player has 1 min to do so. The game scores points for correct choices, but there are only two levels (easy and hard) and the game becomes boring very quickly. A similar game is offered by TurtleDiary, a website with educational games for children. The game is called Recycling for Kids [22]. In the game, garbage arrives on the conveyor belt, which you have to move the mouse to one of three bins: compost, recyclable garbage, non-recyclable garbage. The big advantage of the game is that the garbage has verbal descriptions, so it can contain more problematic garbage, such as greasy paper. The player has unlimited time to make decisions and does not earn points or praise. In the game, nothing changes over time except garbage, so it's very monotonous. On National Geographic Kids there is a game Recycle Roundup [23]. The player controls a superhero who collects garbage falling from the sky and throws it into one of three containers. He has a certain amount of time for this. The game is nicely refined in terms of graphics, and at the end there is information on how many kilograms of garbage have been recycled and composted. The game has only one stage, which makes it unsuitable for longer games, and the control of the superhero is not well implemented.

Garbage sorting games can also be found on the websites of recycling companies. For example, on the website of the company Routeware there is a game called Recycle Right Alexandria [24]. It is a game in which the player indicates where the garbage appearing in turn is to be placed. Despite having inferior graphics, the game offers a unique feature that entices players to continue playing - the ability to expand the park in the city by selecting appropriate attractions upon completion of each stage. This addition significantly enhances the gameplay experience and makes it much more engaging for the player. Finally, after completing the game, it is possible to print a certificate.

In addition to online browser games, there are also desktop games. An example of such a game is SORTED! [25]. The game offers much better graphics than browser games. The garbage moves on a special conveyor belt, and the player must sort it by dragging it to the appropriate containers. The degree of difficulty of the game is regulated by the speed of the belt. Players can earn extra points by finding valuable items among the garbage. The big advantage of the game is the ability to play with two players on a shared screen. For now, the game only exists in a demo version.

Another example of a desktop game is Forest Ranger Simulator [26], in which garbage sorting is only one element of the gameplay. The player's task is to take care of the forest: cleaning it, feeding animals, scaring off poachers, etc. The player can sell garbage on the Internet or throw it into one of 6 types of containers. A comprehensive approach to the issue, rather than focusing only on sorting, makes the game more attractive and suitable for longer gameplay. For now, it is in early access, and the release date is planned for April 24, 2023.

The only immersive virtual reality game included in this review is the game Trash Time [27], which was created thanks to a grant from the Slovak Arts Council. In the game, the player takes on the role of a robot whose task is to sort garbage. To make the game more attractive, the possibility of improving the robot has been introduced, thanks to which you can increase the efficiency of its work and thus also profits.

The presented review of generally available games that fit into the subject of waste management shows that the vast majority of them are 2D or 3D games, which are non-immersive games – according to the division of VR systems adopted for the purposes of this article. There is no game that was created in two versions. Therefore, in order to carry out the research proposed in this article, a game will be constructed that will be able to run in an immersive and non-immersive environment. This will allow for the examination of how the level of immersion can influence the effectiveness of the message conveyed by the game.

2.2 Literature Review

In addition to analyzing the available games related to waste sorting/management, a literature review was conducted to investigate the type of research carried out on subjects that piqued our interest, with particular emphasis on research that utilized cognitive neuroscience techniques in virtual reality.

The review focused mainly on academic literature relating to serious games, virtual reality, waste sorting and cognitive neuroscience. When selecting studies for the review the one of the largest multidisciplinary database was used – SCOPUS. The search was performed in April 2023 and was executed with the use of basic combined keywords, relating to research question, as: ("serious game" AND waste AND (sorting OR management)), ("virtual reality" OR VR), (neuroscience OR psychometrics OR biometrics). Additionally, some general filters were applied to improve the search results for the aim of this study. These filters include: (1) search fields: title, abstract, keywords; (2) conference papers and articles; (3) papers written in English.

The search was conducted in two phases. The first used the phrase: ("serious game") AND waste AND (sorting OR management). A total of 33 publications were obtained from 2011–2023. Upon analyzing the summaries of the publications, it was found that only 19 of them were somewhat related to the topic, with six of them being in a context other than the one considered in this article. Therefore, 13 other publications were selected for a deeper analysis, and the summarized findings are presented in Table 1.

Table 1. Research on waste sorting-games

Author(s)	Type of game	Article goal	Publishing domain	Empirical research	Country
Bifulco et al. [28]	Virtual World Serious Game	Presenting the main concepts of waste collection and garbage recycling to primary school students	Distributed Multimedia Systems	No report of scientific testing	Italy
Sreelakshmi et al. [29]	Serious Digital Game (Unity 2D)	Highlight success of game-based learning through a waste sorting game	Computing Communication and Networking Technologies	20 participants	India
Menon et al. [30]	Serious 2D Digital Game using Microsoft Kinect	Presenting a prototype of the a serious game on how to sort depending on the various categories of waste	Serious Games and Applications for Health	9 participants	India
Idrobo et al. [31]	Serious 3D Digital Game	Teach correct waste sorting	Telematics and Computing	5 participants	Colombia

(*continued*)

Table 1. (*continued*)

Author(s)	Type of game	Article goal	Publishing domain	Empirical research	Country
Luo et al. [32]	Serious 2D Digital Game	Promoting recycling and composting accuracy via a digital sorting game that provides immediate feedback and evaluating its effectiveness	Environmental Management	Totally 458 participants	Canada
Castellano et al. [33]	Serious 2D Digital Game using the social robot as player	Presenting a serious game in which the social robot plays with a child to teach him to correctly recycle the materials	Systems, Man and Cybernetics	Totally 21 participants	Italy
Ambarwati and Ardi [34]	Serious Simulation Serious Simulation 2D Game	Developing a preliminary model of a serious game for educating about waste electrical and electronic equipment system management	Industrial and Systems Engineering	No report of scientific testing	Indonesia
Gaggi et al. [35]	Educational 2D Game for mobile devices	The creation of a serious game able to teach people of all ages how to match the various trash cans with the corresponding type of waste	Smart Objects and Technologies for Social Good	20 participants	Italy

(*continued*)

Table 1. (*continued*)

Author(s)	Type of game	Article goal	Publishing domain	Empirical research	Country
Ishoj-Paris et al. [36]	3D Video Game	Teaching preteens how to recycle by sorting and identifying common household waste	Computer-Human Interaction in Play	No report of scientific testing	Canada
Theethum et al. [37]	Immersive Virtual Reality Serious Game	Improving the players' waste sorting skills by the Virtual Reality game	Computer Science and Software Engineering	20 participants	Thailand
Bekoum and Robinot [38]	Serious 2D Digital Game	Investigating what a role might play video game on plastic pollution in changing ones' daily habits	Sustainability	12 participants	Canada
Hoffmann and Pfeiffer [39]	Mobile game 2D	Investigation the effect of game elements design on long-term learning outcomes in correct waste sorting	Business and Information Systems Engineering	266 participants	Germany
Stenberdt and Makransky [40]	Immersive Virtual Reality Serious Game	Exploring the use of a novel IVR game on pro-environmental waste-sorting behavior for climate change education	Computers & Education	Totally 300 participants	Denmark

In the second phase, the originally obtained 33 results were searched for the phrase: ("virtual reality" OR VR), yielding 5 results [28, 30, 37, 41, 42], of which only three that are included in the table [28, 37, 40] refer to immersive virtual games about waste management. The originally obtained 33 results were also searched for the phrase (neuroscience OR psychometrics OR biometrics) obtaining 3 publications [33, 39, 41]. However, none of them presents studies of the effectiveness of games using cognitive neuroscience techniques.

The review of the literature therefore indicates the existence of a research gap in the field of experimental research on the effectiveness of immersive virtual games promoting pro-environmental behavior (including waste sorting) using cognitive neuroscience techniques. This article is a contribution to filling this gap, as it presents a proposal for a research procedure along with a description of the author's serious game made in two versions: immersive (using a head mounted device) and non-immersive (desktop).

3 Description of the Game

The game that will be used in the planned study is a tower defense game. Games of this type consist in placing defensive towers along the roads that opponents are moving. They are partly of a strategic nature due to the need to properly plan the arrangement of defensive towers with limited financial resources allocated for this purpose. This type of game was chosen because of the great potential in gradating the difficulty of individual stages. Most games that teach garbage sorting are quite monotonous, because it is difficult to add many potentially interesting elements to throwing garbage into containers, and the segregation process can be hindered mainly by the pace of garbage inflow.

In the game, defensive towers will be replaced with garbage cans, and enemies with garbage. Garbage will flow down rivers to the sea, and garbage cans placed by the player will catch them. Five types of bins will be used: a bin for plastics and metals, a bin for paper, a bin for glass, a bin for biodegradable waste and a bin for mixed waste. In addition, there will be a device for catching waste intended for ecoports (electronics, paint cans, etc.). Each of the dumpsters will automatically catch garbage and convert it into cash, which can be used to buy more garbage cans or upgrade them. The exception will be the mixed waste bin, which will catch all rubbish, but it will cost the player a certain amount. Placing each dumpster will require you to spend a certain amount of money.

In addition to the proceeds from the sale of garbage, the player will also have periodic (every specific unit of time) income. These revenues will result from the work of fisheries enterprises that can operate when the river is clean. However, the more garbage flows into the sea, the lower these revenues will be. In addition, these revenues will be reduced by the costs of maintaining the landfill. The bigger it will be, i.e., the more garbage the player will deliver there, the higher these costs will be. The player loses when too much garbage flows into the sea or when the landfill grows beyond a certain threshold. The player wins when the amount of garbage assumed for a given stage is released into the rivers and they are caught or reach the sea, but their number in the sea and in the landfill will not be too large.

The next stages will differ in the number of garbage discharge points, the complexity of the river network, the amount of available cash, limits on the amount of garbage in the sea and in the landfill, and the amount of garbage released. This gives great opportunities to vary the difficulty scale of individual levels, which should make the game not monotonous. For comparison purposes, two versions of the game will be used: a desktop game and an immersive VR game. This will, among other things, determine whether the VR environment will increase the player's level of immersion.

4 Experimental Procedure Plan

The experimental procedure for comparing the effectiveness of games with different levels of immersion was developed on the basis of the authors' previous experience in conducting similar studies and a literature review of publications related to the subject of the work. The proposal includes two phases - designing and conducting relevant research. At each of these stages, specific steps must be followed to ensure the proper conduct of the research.

At the beginning of the procedure, an experimental plan is prepared. At this stage, all factors deemed relevant to the study are selected. They result mainly from the characteristics of the games that were created for the purpose of the study. Then, the appropriate measurement methods (traditional or neuroscientific) should be selected depending on the needs resulting from the nature of the games under study and the aspects we are interested in. For the purpose of our specific research, these methods would include questionnaires from the traditional side and EEG, GSR and HR measurements, and eye-tracking as neuroscientific ones. In this phase, appropriate metrics are also selected – depending on the methods that are used. Survey will measure recognition and recall of the game and its components and the intended behavior of the participants towards the waste-sorting. In addition, it allows to assess the degree of immersion and presence of subjects in virtual reality using tools such as The Igroup Presence Questionnaire (IPQ). Neurometrics will allow to register the memorization (EEG), emotions (GSR and HR) and the attention (eye-tracking) of the players during the game.

Based on the choices of methods and metrics, the number and characteristics of the test persons should also be determined and divided into appropriate subgroups according to the considered variants of the tested materials. At the end of the preparatory stage, a detailed experiment scenario is developed, taking into account the presentation of the tested materials and the interaction of users with the game.

Once the study has been designed, the next step is to recruit participants and conduct the experiment. If necessary, a preliminary selection of volunteers is carried out to ensure that the participants meet certain conditions required by the study, such as familiarity with VR technology or specific attitudes towards waste sorting. After completing the pool of study participants, each individual is randomly assigned to test a selected game, whether immersive or not, and then invited to the research laboratory. The experiment should be conducted during optimal hours, typically between 9 and 11 am, when participants' cognitive abilities are at their peak.

To ensure accurate data collection, it is crucial to take appropriate preparation steps before commencing the experiment. Firstly, the research laboratory room should maintain appropriately set and monitored environmental parameters such as temperature and humidity, which is recommended to minimize their impact on the gathered data. Before starting the research, it is important to familiarize the subjects with the experimental environment and discuss the course of the study with them. After familiarizing themselves with the procedure, the research participants should give their written consent to participate in the research project. If necessary, demographic data are also collected after obtaining consent. The next step is to assemble and calibrate the VR goggles (for participants assigned to the immersive group) and other research equipment selected for the study.

At this stage, training may also be carried out to make the respondents acquainted with navigation in a virtual environment and the use of games. Moreover, this step allows you to exclude people who are not comfortable in the virtual environment and experience dizziness, headaches, fatigue and negative emotional activity such as anxiety and stress. After starting the experiment, the measuring apparatus records the basic course of psychophysiological signals of the subject at rest. In the further part of the study, a game begins during which psychophysiological data is collected. The recorded signals will be supplemented with data from questionnaires. At this stage, the practical part of the study ends, followed by an analysis of the collected data, taking into account both subjective and objective measures. The procedure ends with the interpretation and presentation of the results.

5 Conclusion

The purpose of the article was to introduce a research methodology that can be used to compare the effectiveness of different games in promoting waste-sorting behaviour. The article emphasizes the recommended course of action for the study and the characteristics of games in both non-immersive and immersive virtual reality that can be utilized for the experiment. This approach has the potential to provide valuable insights into the most effective methods for encouraging waste-sorting behaviour.

However, there are some limitations that must be taken into consideration. One potential issue is simulator sickness, which some participants may experience during the virtual reality experiments. This could exclude certain individuals from participating in the study. Additionally, creating a virtual environment that would be sufficiently immersive and realistic may pose a challenge. The registration of psychophysiological signals could also be problematic due to the sensitivity of cognitive neuroscience tools to external noise.

Despite these potential challenges, the proposed approach has the potential to offer significant benefits for effectively promoting waste-sorting behaviour. The next step of the research is to conduct an experiment using the recommended procedure and virtual reality games. This will help to determine the suitability of the procedure and games and enable refinement of the approach.

In summary, the article provides a valuable contribution to the field of waste management by proposing a novel methodology for investigating the most effective ways of promoting waste-sorting behaviour. While there are some limitations to this approach, the potential benefits are significant and justify further research into this promising area.

References

1. Jamro, I.A., Chen, G., Mangi, S.A., Ma, W., Allehyani, S.: Management of university solid waste in China through gasification technology: an analysis of waste composition and energy potential. Environ. Sci. Pollut. Res. **29**, 84191–84205 (2022)
2. Bamodu, O., Ye, X.: Virtual reality and virtual reality system components. In: Proceedings of the 2nd International Conference on Systems Engineering and Modeling. Atlantis Press, China (2013).

3. Martín-Gutiérrez, J., et al.: Virtual technologies trends in education. EURASIA J. Math. Sci. Technol. Educ. **13** (2017)
4. Borawski, M.: Use of computer game as an element of social campaign focusing attention on reliability of information in the internet. In: Nermend, K., Łatuszyńska, M. (eds.) CMEE 2017. SPBE, pp. 127–139. Springer, Cham (2018). https://doi.org/10.1007/978-3-319-99187-0_10
5. Chung, S., Sparks, J.V.: Motivated processing of peripheral advertising information in video games. Commun. Res. **43**, 518–541 (2016)
6. Yoo, S.-C., Peña, J.: Do violent video games impair the effectiveness of in-game advertisements? The impact of gaming environment on brand recall, brand attitude, and purchase intention. Cyberpsychol. Behav. Soc. Netw. **14**, 439–446 (2011)
7. Russell-Bennett, R., et al.: A hierarchy-of-effects approach to designing a social marketing game. J. Nonprofit Public Sect. Mark. **28**, 105–128 (2016)
8. Buller, M.K., et al.: Marketing fruit and vegetable intake with interactive games on the internet. Soc. Mark. Q. **15**, 136–154 (2009)
9. Yam, A., Russell-Bennett, R., Foth, M., Mulcahy, R.: How does serious M-game technology encourage low-income households to perform socially responsible behaviors?: M-game technology and socially responsible behaviors. Psychol. Mark. **34**, 394–409 (2017)
10. Pempek, T.A., Calvert, S.L.: Tipping the balance: use of advergames to promote consumption of nutritious foods and beverages by low-income African American children. Arch. Pediatr. Adolesc. Med. **163**, 633 (2009)
11. Calvert, G.A., Brammer, M.J.: Predicting consumer behavior: using novel mind-reading approaches. IEEE Pulse **3**, 38–41 (2012)
12. Harris, J.M., Ciorciari, J., Gountas, J.: Consumer neuroscience for marketing researchers. J. Consum. Behav. **17**, 239–252 (2018)
13. Russo, V., Bilucaglia, M., Zito, M.: From virtual reality to augmented reality: a neuromarketing perspective. Front. Psychol. **13**, 965499 (2022)
14. Barrett, R.C.A., et al.: Comparing virtual reality, desktop-based 3D, and 2D versions of a category learning experiment. PLoS ONE **17**, e0275119 (2022)
15. Zumbach, J., Seitz, C., Bluemke, M.: Impact of violent video game realism on the self-concept of aggressiveness assessed with explicit and implicit measures. Comput. Hum. Behav. **53**, 278–288 (2015)
16. Masterkidz: Material Sorting Recycling Truck. https://www.masterkidztoys.com/. Accessed 14 Apr 2023
17. Adamigo: Ekobohater. https://www.adamigo.pl/. Accessed 14 Apr 2023
18. Jeja.pl: Recykling. https://gry.jeja.pl/. Accessed 14 Apr 2023
19. FOG: Trash Sorting for Kids. https://html5.gamedistribution.com/. Accessed 14 Apr 2023
20. FOG: Garbage Sorting Truck. http://www.freeonlinegames.com/. Accessed 14 Apr 2023
21. ABCYa: Litter Critters - Learn to Sort and Recycle. https://www.abcya.com/. Accessed 14 Apr 2023
22. Turtle Diary: Reduce Reuse Recycle Game. https://www.turtlediary.com/. Accessed 15 Apr 2023
23. National Geographic Kids: Recycle Roundup. https://kids.nationalgeographic.com/. Accessed 15 Apr 2023
24. City of Alexandria: Recycle Right Alexandria. https://recollect.net/sorting-game/alexandriava/. Accessed 15 Apr 2023
25. Steam: SORTED!. https://store.steampowered.com/. Accessed 15 Apr 2023
26. Steam: Forest Ranger Simulator. https://store.steampowered.com/. Accessed 15 Apr 2023
27. Steam: Trash Time. https://store.steampowered.com/. Accessed 15 Apr 2023

28. Bifulco, I., et al.: The TIE project: agile development of a virtual world serious game on waste disposal. In: The 17th International Conference on Distributed Multimedia Systems Knowledge Systems Institute, USA, Firenze, 18–20 August 2011, pp. 204–209 (2011)
29. Sreelakshmi, R., McLain, M., Rajeshwaran, A., Rao, B., Jayakrishnan, R., Bijlani, K.: Gamification to enhance learning using Gagne's learning model. In: 2015 6th International Conference on Computing, Communication and Networking Technologies (ICCCNT). pp. 1–6. IEEE, Dallas-Fortworth (2015)
30. Menon, B.M., Unnikrishnan R, Muir, A., Bhavani, R.R.: Serious game on recognizing categories of waste, to support a zero waste recycling program. In: 2017 IEEE 5th International Conference on Serious Games and Applications for Health (SeGAH), pp. 1–8. IEEE, Perth (2017)
31. Idrobo, M.L., Fernanda Saenz, M., Márceles, K., Chanchí, G.E., Vidal, M.I., Burbano, C.L.: Recycling: a serious game focused on the classification of waste. In: Mata-Rivera, M.F., Zagal-Flores, R. (eds.) WITCOM. CCIS, vol. 944, pp. 234–245. Springer, Cham (2018). https://doi.org/10.1007/978-3-030-03763-5_20
32. Luo, Y., Zelenika, I., Zhao, J.: Providing immediate feedback improves recycling and composting accuracy. J. Environ. Manage. **232**, 445–454 (2019)
33. Castellano, G., et al.: Learning waste recycling by playing with a social robot. In: 2019 IEEE International Conference on Systems, Man and Cybernetics (SMC), pp. 3805–3810. IEEE, Bari (2019)
34. Ambarwati, L., Ardi, R.: Serious simulation game preliminary design as education tool for waste electrical and electronic equipment management system in Indonesia. In: 2019 Asia Pacific Conference on Research in Industrial and Systems Engineering (APCoRISE), pp. 1–6. IEEE, Depok (2019)
35. Gaggi, O., et.al.: Learning how to recycle waste using a game. In: Proceedings of the 6th EAI International Conference on Smart Objects and Technologies for Social Good, pp. 144–149. ACM, Antwerp (2020)
36. Ishoj-Paris, Y. et al.: AXO: a video game that encourages recycling to preteens. In: Extended Abstracts of the 2021 Annual Symposium on Computer-Human Interaction in Play, pp. 350–355. ACM, Virtual Event (2021)
37. Theethum, T., et al.: SEPBO: trash separator bot VR game. In: 2021 18th International Joint Conference on Computer Science and Software Engineering (JCSSE), pp. 1–6. IEEE, Lampang (2021)
38. Bekoum Essokolo, V.-L., Robinot, E.: «Let's go deep into the game to save our planet!» How an immersive and educational video game reduces psychological distance and raises awareness. Sustainability. **14**, 5774 (2022)
39. Hoffmann, G., Pfeiffer, J.: Gameful learning for a more sustainable world: measuring the effect of design elements on long-term learning outcomes in correct waste sorting. Bus. Inf. Syst. Eng. **64**, 459–482 (2022)
40. Stenberdt, V., Makransky, G.: Mastery experiences in immersive virtual reality promote pro-environmental waste-sorting behavior. Comput. Educ. **198**, 104760 (2023)
41. Panagiotopoulou, L., et al.: Design of a serious game for children to raise awareness on plastic pollution and promoting pro-environmental behaviors. J. Comput. Inf. Sci. Eng. **21**, 064502 (2021)
42. Papamichael, I., et al.: Unified waste metrics: a gamified tool in next-generation strategic planning. Sci. Total. Environ. **833**, 154835 (2022)

Psychophysiological Research in the Design of Useful Interfaces for e-Learning Materials

Mateusz Piwowarski, Małgorzata Nermend(✉), and Kesra Nermend

University of Szczecin, Cukrowa 8, 71-004 Szczecin, Poland
{mateusz.piwowarski,malgorzata.nermend,kesra.nermend}@usz.edu.pl

Abstract. Designing effective human-computer interfaces comes with significant challenges and this applies to interfaces used in e-learning as well. One way to design effective interfaces is by studying human reactions at the level of psychophysiological processes. This enables the analysis of system components' impact on cognitive and emotional human functions. The aim of this publication is to present a methodological approach for designing useful e-learning materials based on the measurement and analysis of eye movement activity and facial expression. The research results and conclusions are both methodological and applied in nature.

Keywords: Human-Computer Interfaces · E-learning Materials · Eye Tracking · Facial Expression Analysis

1 Introduction to the Role of Interfaces in e-Learning

Users engage with diverse information systems through user interfaces, making the design, implementation and testing a crucial aspect that significantly impacts the user experience [1]. The main characteristics that determine the quality of user interfaces include ease of learning how to use them and their overall usability. Related to the first characteristic is the time to familiarize oneself with the interface and the subsequent support from the system provider. As per for the second characteristic, it pertains to the functionality and convenience of use. Usability relates to the effectiveness, efficiency of the interface and is linked to user feedback [2]. By analyzing both of these characteristics (while performing specific activities) and additionally combining them with knowledge of users' cognitive, emotional capabilities and limitations, one can obtain information about the properties of a good quality interface [3]. This aligns with established methodologies that connect affective and emotional aspects of human behaviour to the field of human-computer interaction (HCI). These concepts include the areas of emotional design, hedonomics, kansei engineering and affective computing [4].

Users of the systems have different tastes, preferences, as well as the physical abilities to operate them. When designing a useful interface, it is necessary to take into account not only the aspect of its actual production (technical correctness), but also the broader user experience. Without understanding the impact of the interface on its perception by users, it is difficult to talk about the usefulness and pleasure of using systems. Users who

M. Hernes and J. Wątróbski (Eds.): ECAI 2023, LNNS 1079, pp. 215–227, 2024.
https://doi.org/10.1007/978-3-031-66761-9_18

use interfaces that elicit positive reactions are more creative, are more engaged. They rate systems as easier to use friendlier, even if they have some limitations, drawbacks [5–7]. This is particularly important in e-learning systems, whose main purpose is to effectively transfer knowledge. If an e-learning system, including modules responsible for providing learning resources, is improperly designed, users may not be sufficiently motivated to use it such systems should be easy to use, intuitive, engaging, realistic, interactive, and impose minimal cognitive load on the user [8, 9].

The learning process of a specific course is heavily influenced by the design of the content and structure and the organization of teaching resources. Knowing and understanding the level of user interaction with the materials and their impact on cognitive and emotional functions will also enable the development of design guidelines for developers. Consequently, there is a need to modify e-learning design paradigms. In addition to the evident user-centered design approach, the learning-centered design paradigm should also be considered, taking into account three learning theories: behaviorist, cognitive and constructivist [10, 11]. By integrating both methodological approaches, it is possible to develop an interface that will assist users in achieving their learning goals. Such an interface will feature well-structured content, good functionality, seamless navigation and a pleasant user experience [2, 10, 12]. In the interface design process, several models (e.g., mental, designer, etc.) stand out, each with its own set of goals including predicting user behavior, ensuring consistency between design levels, or evaluating different aspects of the design [13, 14].

Various methods are used to study the usability of interfaces or user experience. Some of these methods are declarative involving the participants expressing informed opinions or judgments. However, these methods do not allow for an understanding of the biological basis of human functioning or the body's reactions to the received stimuli (concerning cognitive, emotional processes). In the context of user interfaces, this understanding is of great importance. On the other hand, methods based on physiological measurement provide insight into the automatic cognitive processes that occur rapididly, in a simplified, unconscious and often emotion-based manner [15].

The aim of the article is to present the possibility of using psychophysiological research (eye tracking, face reading) in the process of designing useful interfaces for e-learning materials. The presented research results, which include analyses of selected cognitive and emotional functions, expand the knowledge of the impact of various forms of visualization of e-learning resources on the audience.

2 Materials and Methods

2.1 Protocol and Stimuli

The experiment followed the structure of short e-learning course, comprising of two thematic lessons consisting of a total of 112 screens (92 with content and 20 with quizzes). Each screen was treated as a separate stimulus and analyzed in detail. The topics covered a specific area of medicine. To fulfil the research objective, three versions of training materials with identical content were prepared. The structure of each lesson included an introduction (general information about the topic), the contents of the issue, a quiz to test knowledge, and a conclusion (Fig. 1).

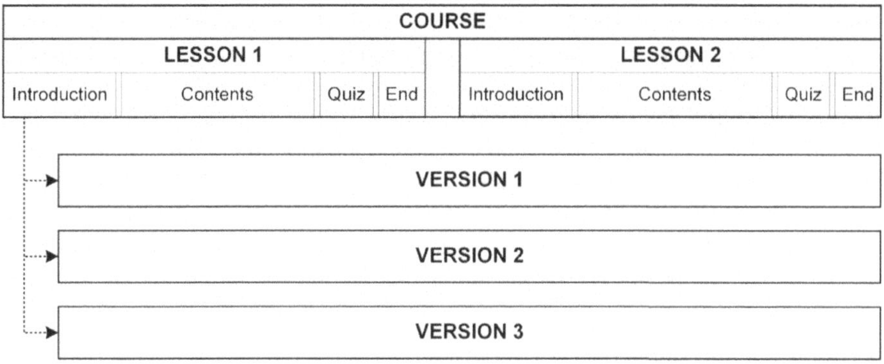

Fig. 1. Structure of the analyzed training materials.

The interfaces of each version of the course were developed in the specialized iSpring Suite of software (dedicated tools for creating courses, quizzes, simulations and e-learning lessons). Each version of the interface has slight differences in terms of visualization, color scheme and the content layout. Version 1 of the interface, was the most classic, looking similar to typical multimedia presentations. Version 2 was enriched with graphic elements, aiming to better distinguish specific content elements (numbered lists, bulleted lists, object captions, parts of the screen content). Version 3 was the most minimalist, featuring the fewest graphic elements.

2.2 Apparatus and Research Techniques

The study used eye tracking and face reading techniques. Eye tracking is a research technique for studying gaze focus and determining metrics based on fixations, saccades or gaze points. Fixations are moments when human eyes essentially stop scanning an area, holding the gaze in place (for a specified period of time). Saccades are a type of eye movement used to quickly move the gaze from one point of interest to another. Gaze points, are the reading points of focusing the gaze, according to the frequency of the measuring device [16]. Eye tracking analyses most often involve gaze paths, heat maps, and Area of Interest (AOI) – based metrics. In eye tracking analyses, gaze paths, heat maps and metrics – based Area of Interest (AOI) are most commonly analyzed. These are used in cognitive studies and in observations of oculomotor behavior [17].

The stationary (screen-based) Tobii Pro X3–120 eye tracker was used for the study, and the data acquired for analysis was processed using the I-HMM (Hidden Markov Model) fixation filtering algorithm. The settings for the minimum fixation duration were set at 60 ms, the maximum inter-fixation interval at 75 ms, noise reduction using the Moving Average algorithm, and gap-filling interpolation with a maximum length of 75 ms.

Face reading is a technique that enables expression recognition and emotion identification based on facial muscle movement. It uses the Facial Action Coding System (FACS) facial assessment method developed by Ekman and Friesen. FACS codes muscle movements in the form of Action Units (AUs). These reflect the contraction or relaxation of muscles (one or more). The 2012 FACS defines 27 AUs, as well as eye positions and movements, head movements, Visibility Codes, Action Descriptors and Gross Behavior [18, 19].

Affectiva's AFFDEX 2.0 algorithm [20], EMFACS [21] and Logitech BRIO Webcam 4K were used to recognize facial expressions and identify emotions. It was assumed that classification of expression measures would be made for moderately strong facial reactions. Threshold values for valence identification were set at 50%. When aggregating the data, the length of the time window in which the data were analyzed for groups of study participants was set to 500 ms.

2.3 Analyzes and Indicators of Evaluation

Visual attention analyses (eye tracking) were based on fixations. Eye tracks, heat maps and AOI-based metrics were used. A description of the metrics is provided in Table 1. Heat maps and metrics values were aggregated for the study groups considered.

Table 1. AOI metrics used in the study.

AOI metrics	Metric description
Respondent base	Number of respondents who were presented with the stimulus (screen)
Respondent count	Number of respondents who gazed at the AOI
Dwells with fixations	Average of how often respondents gazed at the AOI at least once
Fixation count	Average number of fixations detected inside the AOI
TTFF AOI (ms)	Average time elapsed before the first fixation was detected inside the AOI
Dwell time (%)	The average amount of time respondents fixated on the AOI relative to the time the AOI was active
First fixation duration (ms)	Average duration of respondents' first fixation inside the AOI

The AFFDEX classifier was used to test the valence of emotions (face reading). The following FACS-compliant AUs were used as measures of positive valence: Smile and Cheek Raise. For negative valence AU: Inner Brow Raise, Brow Furrow, Nose Wrinkle, Upper Lip Raise, Lip Corner Depressor, Chin Raise, Lip Press, Lip Suck [20, 22]. Measure values were aggregated for each study group.

2.4 Participants

Thirty respondents (16 women and 14 men) participated in the study. They were young people who were studying and working. The mean age and standard deviation of the study participants were: M = 21.80; SD = 4.31. The subjects were divided into three groups, with 10 people in each. There were 5 women and 5 men in group one (M =

22.10; SD = 4.46), 4 women and 6 men in group two (M = 21.80; SD = 4.83), and 5 women and 5 men in group three (M = 21.50; SD = 4.06). All participated in the study in an informed and voluntary manner. They signed consent to participate in the experiment and to process personal data. Prior to the start of the research procedure, the participants were informed about the course of the experiment and the research and measurement equipment used. Details about the content of the study (the stimuli provided) were not disclosed.

The experiment was conducted in accordance with the principles outlined in the Declaration of Helsinki (2013 version) [23]. It was reviewed by the Bioethics Committees at Regional Chamber of Physicians in Szczecin.

3 Research Results

The results of the analyses were grouped according to characteristic content presentation layouts (interfaces). They included screens of training materials with the following structures (predominant share): bulleted list, numbered lists, mixed numbered lists, textual content, graphic elements, tables with text content, videos and quizzes.

The visualization of the course screens was standardized for each structure, separately for each version. This standardization facilitated the analysis of each structure and improved the graphical presentation of the course. Despite the unified visualization of screens with identical structures, each screen had partial differences from the others (e.g. different length of text, different number of items in lists, different graphic elements, etc.). Therefore, each screen was analyzed individually. Cumulative analyses were then conducted for each structure and generalized conclusions were drawn. Due to editorial constraints (number of pages), only selected results of the analyses are presented in this publication. However, the conclusions, apply to the entire study.

3.1 Bulleted List

Interfaces with bulleted content contained, in most cases, 5 to 12 items and 1 or 2 short paragraphs of text. The same analyses were performed for each version of the interface. The eye tracking analyses focused on gaze paths, heat maps and AOI metrics. Heat maps and marked AOIs for each version are shown in Figs. 2, 3 and 4, while calculated AOI metrics are shown in Tables 2, 3 and 4. Emotion valence values determined from facial expression analysis are included in Table 4. The considered versions of content with bulleting (one of the screens) differ in the bullet points used, the highlighting of text from the list, and the distances between items.

Version 1 used, among other things, a small bullet marker, standard line spacing, a highlighted screen header and a graphic element (man). The areas of interest analyzed included the screen header, the text with the question, the bullet point, and the man figure (Fig. 2). Averaged values of AOI metrics (for the study group) are included in Table 2.

Fig. 2. Heat map and areas of interest for version 1 of the interface (bulleted list).

Table 2. Averaged values of AOI metrics for version 1 of the interface (bulleted list).

AOI metrics	Header (Nagłówek)	Questions (Pytania)	Bullets (Wypunktowanie)	Man (Postać)
Respondent base	10	10	10	10
Respondent count	9	10	10	8
Dwells with fixations	2,2	3,5	3,0	1,1
Fixation count	19,9	31,6	116,3	13,3
TTFF AOI (ms)	865,9	1135,9	19,7	4549,9
Dwell time (%)	10,3	11,7	19,7	3,2
First fixation duration (ms)	79,6	42,4	17,6	21,9

Version 2 of the interface used a larger bullet point, highlighted text, and partially modified the graphic design. The graphic element and AOI remained the same as in version 1 (Fig. 3). The calculated AOI metrics for version 2 of the interface are included in Table 3.

Fig. 3. Heat map and areas of interest for version 2 of the interface (bulleted list).

Table 3. Averaged values of AOI metrics for version 2 of the interface (bulleted list).

AOI metrics	Header (Nagłówek)	Questions (Pytania)	Bullets (Wypunktowanie)	Man (Postać)
Respondent base	10	10	10	10
Respondent count	10	10	10	6
Dwells with fixations	1,7	2,8	2,4	1,0
Fixation count	11,8	28,3	131,9	13,0
TTFF AOI (ms)	689,2	1261,2	1350,4	8172,0
Dwell time (%)	6,2	8,3	22,2	3,2
First fixation duration (ms)	72,8	50,0	28,3	14,6

Version 3 of the interface contained the least graphic and highlighting elements. The bullet point was changed. The number of AOIs was reduced due to the absence of the man figure (Fig. 4). The values of the determined AOI metrics are shown in Table 4.

Fig. 4. Heat map and areas of interest for version 3 of the interface (bulleted list).

Table 4. Averaged values of AOI metrics for version 3 of the interface (bulleted list).

AOI metrics	Header (Nagłówek)	Questions (Pytania)	Bullets (Wypunktowanie)
Respondent base	10	10	10
Respondent count	10	10	10
Dwells with fixations	1,7	2,3	2,1
Fixation count	8,1	22,9	138,9
TTFF AOI (ms)	865,3	1076,3	1608,3
Dwell time (%)	4,4	5,4	22,7
First fixation duration (ms)	82,5	28,2	14,5

The considerable highlighting of the bulleted list item is reflected in a more even focus on its content (heat maps). This is particularly evident in version 3. The direction of scanning of bulleted text (whole sentences) is more in line with the natural reading of this type of content (top–bottom, left–right) for versions 3 and 2 (in that order). In the case of version 1 (for the majority of respondents), gaze paths showed gaze returns in the vertical direction. This could indicate either a misunderstanding of the content or that the content was not scanned very accurately with the eyes.

The added man figure (a decorative element) did not significantly affect the reading of the relevant content. Analysis of AOI metrics shows that this element did not attract enough attention to distract from the rest of the content. The averaged TTFF value (AOI – man) shows that the gaze was focused on this element further than for the other AOIs. In the case of limited reading time of the stimulus (screen), some respondents skipped it. In other cases, eye scanning was too short, making the time not classified as fixation. The layout of the screen structure also shows that reading the questions took the least amount of time for version 3, and the most for version 1. Respondents spent the most time reading the contents of the bulleted list in versions 2 and 3, much less in version 1.

Averaged numerical values indicating valence levels (direction of emotions) for the bulleted interface are included in Table 5. These values are presented as a percentage of the duration of detected positive, negative or neutral emotions (no identified facial expressions), relative to the total presentation time of the stimulus.

Table 5. Averaged values of emotion valence for version 1 of the interface (bulleted list).

Interface variant	Positive	Negative	Neutral
Variant 1	14,28%	10,04%	75,68%
Variant 2	23,65%	6,39%	69,96%
Variant 3	21,97%	5,15%	72,88%

As can be seen, the automated analysis of facial expressions showed a mostly neutral emotional reception. The most positive emotions were identified for variant 2 (23.65%), a little less for variant 3 (21.97%) and the least for variant 1 (14.28%). Negative emotion valence was highest for version 1 and lowest for version 3.

3.2 Numbered Lists

The numbered list versions of the content analyzed differed in the numbering character used, the highlighting of the text in the list (version 2) and the distances between items. In addition, version 1 and version 2 added a graphic element (man figure). Versions 1 and 3 were similar, differing in the numbering character used and the distances between them. Version 2 included a clear graphic character for consecutive list numbers and highlighting the text on each line.

Heat maps indicated a more even focus of visual attention on list items for version 2. Gaze paths confirmed a smoother reading of numbered content for version 2 (according to the order of list items). The averaged time of focusing visual attention in the AOI (Numbering) was the longest for version 2, the shortest for version 1. As with bulleted lists, the added graphic element (figure man) did not interfere with the reading of the rest of the screen content (especially in version 2). The visual presentation of the numbered list in version 2 was quite effective in drawing the eye. Other content elements did not attract the eye enough to introduce an element of reading disorder (as is the case with the header in version 1). The averaged values of positive emotion valence for screens of this structure, for the different versions of the interfaces, were as follows (version 1 – 11.34%, version 2 – 16.82%, version 3 – 17.12%).

3.3 Mixed Numbered Lists

Nested numbered lists were visually similar in versions 1 and 3. They differed in the numbers used and their size. Version 2 used (similar to numbering and bulleting) content highlighting and reduced the number of items per screen. This solution made it easier to visually scan the entire content.

Version 2 organized the numbered content to a much greater degree than both other versions. This was reflected in the way it was read. It forced the eye to focus on the numbered list (heat maps, gaze paths. The placed man figure was not too distracting. This may be due to the fact that the same graphic has been shown many times before (AOI analysis). The averaged values of positive emotion valence for the screens of this structure were as follows (version 1 – 10.71%, version 2 – 18.28%, version 3 – 13.69%).

3.4 Textual Content

The text in all versions was placed in several paragraphs. In addition, in version 1 and version 2, a graphic element (figure man) was added. Analyzing heat maps and selected gaze paths, it can be seen that a more even focus of gaze on the text content was achieved in versions 2 and 3. Paragraphs of text highlighted by color and placement relative to other paragraphs (version 3) focused on visual attention more. Arranging the text in visibly separated blocks resulted in smoother content scanning (gaze path analysis).

When considering the marked areas of interest (AOIs), it is important to note the relatively long time to observe the screen header (relative to the other AOIs) and the returns of gazing at it (particularly in version 1). These are unfavorable proportions. Shorter blocks of text resulted in relatively shorter eye-scan time for its contents. This is true for version 3 and to a lesser extent for version 2. The averaged values of positive emotion valence for the screens of this structure were as follows (version 1 – 14.63%, version 2 – 25.81%, version 3 – 13.17%).

3.5 Graphic Elements

In the course analyzed, the posted graphics are similar in form, being mostly photos. The arrangement of the photos on the screen tended to have a similar layout (graphic element and caption). They differed visually in the signature and minor graphic elements of the screen template. The heat maps obtained for versions 1, 2 and 3 were similar. Analysis of gaze paths provided information that a slightly more orderly way of gazing (photo, caption) was obtained for version 2. Bolding the photo caption and subtle separation by means of a line had a favorable effect on the eye-scanning process. The applied highlighting of the signature in version 3 interfered with the fluidity of reading.

Analysis of the areas of interest (photos, captions) indicates that the visualization in version 2 maintained the best balance between looking at the photo and reading the caption below. In the course studied, graphics were also presented in a different layout. For example, the photo on the left and the caption on the right. Although visually such a screen may look attractive, scanning such content with the eye is less effective. It is more natural for human vision to scan from top to bottom than from left to right. The averaged values of positive emotion valence for screens of this structure were as follows (version 1 – 15.08%, version 2 – 20.19%, version 3 – 11.93%).

3.6 Tables with Text Content

Tables with text content, especially with more text, are quite difficult to read. The versions considered differed in text placement, font highlighting and the graphic design of the table.

Analysis of heat maps and gaze paths indicated that for larger tables, the important area for focusing visual attention is the center vertically of the table. In versions 2 and 3, section headings inside the table were centered and the font was bolded. This resulted in preserving the natural way of scanning the content with the eye from top to bottom. In the case of smaller tables, it was noticed that their contents were scanned more accurately. In addition, the simplification of the table's graphical appearance resulted in more focused attention on its contents. The averaged values of positive emotion valence for the screens of this structure were as follows (version 1 – 9.46%, version 2 – 11.38%, version 3 – 11.22%).

3.7 Videos

The screens displaying the videos had a similar appearance. They varied across versions in the template utilized to present the entire course, including the header. With such a screen design, the header assumed special significance as it provided information about the subject of the video.

In each version, the screen header was read fairly accurately (reading time), with the longest for version 1, the shortest for version 2 (AOI). The gaze paths showed that the title was read accurately before launching the video. The effect of providing a preview in the movie window on the overall screen reading was also analyzed. For one screen, the preview showed the initial frame of the video, while for another there was black content instead. In the latter case, it provided a distracting element, forcing the eyes to focus on other elements of the screen. The averaged values of positive emotion valence for screens of this structure were as follows (version 1 – 16.88%, version 2 – 23.41%, version 3 – 20.11%).

3.8 Quizzes

The quizzes for each version had a similar design. There was the same layout (typical for any quiz), and the difference was of the graphic element. In version 1, it was a little larger and more conspicuous, in version 2 it was a little different, smaller. In version 3 it was absent, which made the distribution of text (question, answers) change a bit.

The heatmaps showed a similar distribution of focusing attention on the question and answers (in all versions). The graphic element in versions 1 and 2 also attracted attention. In version 1, a little more than in version 2 (AOI analysis). This is due to its size, color scheme, and design. Based on the analysis, it can be concluded that in the case of quizzes, additional elements (graphics type) are unnecessary. They can be a distracting factor. However, an analysis of successive quiz questions (with the same design) shows that the use of non-intensive graphics (the same on all quiz screens) causes them to cease to be a distraction over time. The averaged values of positive emotion valence for the screens of this structure were as follows (version 1–7.37%, version 2–9.18%, version 3–10.24%).

4 Conclusion

The study aimed to verify how human beings perceive the interfaces of e-learning materials (prepared in 3 versions). The visualization of training materials, available in digital form, is often considered a crucial element of their effectiveness. Visual attractiveness plays a significant role in this aspect. It is important to design the content in a way that facilitates maximum absorption and recall. Therefore, it is crucial to follow design guidelines that are derived from direct observations of how content is read, scanned and emotionally received. A number of important conclusions can be drawn from the research.

The content of individual screens should be concise (without excess information). Uniformity of content presentation throughout the course has a positive effect on emotional reception and visual focus. Individual screens should have headings, but they should not be too visually prominent. The font used must be legible and of appropriate size (not too large). Screens with text content should be clearly separated into paragraphs, and paragraphs can be distinguished in different ways (e.g. offset, font color). When using numbered or bulleted lists, it is advisable that they are visually appealing and clearly highlight individual items (they cannot give the effect of a single block of text). When using several types of elements on one screen (bullets, numbering), they should be distinguished from each other. Graphics that are important to the content should be of good quality and appropriately captioned (e.g., in bold, small font size, without using unnecessary background highlighting). Graphic elements of a decorative nature repeated on successive screens are not a distraction. It is recommended to simplify tables, and in the case of large tables, divide them into smaller ones and place them on subsequent screens. For quizzes, decorative elements are not recommended.

More precise effects of graphical user interfaces on users' cognitive and emotional functions can be studied using more advanced physiological measurement methods, such as the study of electrodermal activity or electrical brain activity (EEG).

References

1. Brey, P., Søraker, J.H.: Philosophy of computing and information technology. In: Meijers, A. (ed.) Philosophy of Technology and Engineering Sciences, pp. 1341–1407. North-Holland, Amsterdam (2009). https://doi.org/10.1016/B978-0-444-51667-1.50051-3
2. Uden, L.: Human–computer interaction and learning. In: Seel, N.M. (ed.) Encyclopedia of the Sciences of Learning, pp. 1464–1469. Springer US, Boston, MA (2012). https://doi.org/10.1007/978-1-4419-1428-6_363
3. Vera, A.H.: Understanding users: the knowledge-level of analysis. In: Marsh, J.P., Gorayska, B., Mey, J.L. (eds.) Human Factors in Information Technology, pp. 89–103. North-Holland (1999). https://doi.org/10.1016/S0923-8433(99)80009-4
4. Jeon, M.: Emotions and affect in human factors and human–computer interaction: taxonomy, theories, approaches, and methods. In: Jeon, M. (ed.) Emotions and Affect in Human Factors and Human-Computer Interaction, pp. 3–26. Academic Press, San Diego (2017). https://doi.org/10.1016/B978-0-12-801851-4.00001-X
5. Lidwell, W., Holden, K., Butler, J.: Universal principles of design, revised and updated: 125 ways to enhance usability, influence perception, increase appeal, make better design decisions, and teach through design. Rockport Pub (2010)
6. Fishwick, M.: Emotional design: why we love (or hate) everyday things. The Journal of American Culture **27**, 234 (2004)
7. Van Gorp, T., Adams, E.: Design for emotion. Elsevier, Boston (2012)
8. Shen, L., Wang, M., Shen, R.: Affective e-learning: using "Emotional" data to improve learning in pervasive learning environment. J. Educ. Technol. Soc. **12**, 176–189 (2009)
9. Lewis, R., Stoney, S., Wild, M.: Motivation and interface design: maximising learning opportunities. J. Comput. Assist. Learn. **14**, 40–50 (1998). https://doi.org/10.1046/j.1365-2729.1998.1410040.x
10. Pillai, S., Lucas, K., Mello, A.: A Learner-Centered Design Method for Educational Technology. Presented at the (2014)
11. Uden, L.: Editorial: learning technology. Int. J. Learn. Technol. **1**, 1–15 (2004)
12. Cullen, R.M., Harris, M., Hill, R.R.: The learner-centered curriculum: design and implementation. Jossey-Bass, San Francisco, CA (2012)
13. Moore, R.J., An, S., Ren, G.-J.: The IBM natural conversation framework: a new paradigm for conversational UX design. Human-Computer Interaction **38**, 168–193 (2023). https://doi.org/10.1080/07370024.2022.2081571
14. Trætteberg, H.: Model-based User Interface Design (2004)
15. Cacioppo, J., Tassinary, L., Berntson, G.: Handbook of psychophysiology (2007). https://doi.org/10.13140/2.1.2871.1369

16. Lobão-Neto, R., Brilhault, A., Neuenschwander, S., Rios, R.: Real-time identification of eye fixations and saccades using radial basis function networks and Markov chains. Pattern Recogn. Lett. **162**, 63–70 (2022). https://doi.org/10.1016/j.patrec.2022.08.013
17. Hu, T., Wang, X., Xu, H.: Eye-tracking in interpreting studies: a review of four decades of empirical studies. Frontiers in Psychology **13** (2022)
18. Ekman, P., Friesen, W., Hager, J.: Facial action coding system [E-book]. Research Nexus, Salt Lake City, UT (2002)
19. Ekman, P., Friesen, W.V.: Measuring facial movement. J Nonverbal Behav. **1**, 56–75 (1976). https://doi.org/10.1007/BF01115465
20. Bishay, M., et al.: AFFDEX 2.0: A Real-Time Facial Expression Analysis Toolkit, http://arxiv.org/abs/2202.12059 (2022)
21. Friesen, W., Ekman, P.: EMFACS-7: Emotional Facial Action Coding System (1983)
22. Piwowarski, M., Wlekły, P.: Factors disrupting the effectiveness of facial expression analysis in automated emotion detection. Procedia Computer Science. **207**, 4296–4305 (2022). https://doi.org/10.1016/j.procs.2022.09.493
23. World Medical Association: World Medical Association Declaration of Helsinki: Ethical Principles for Medical Research Involving Human Subjects. JAMA **310**, 2191 (2013). https://doi.org/10.1001/jama.2013.281053

Analysis of the Successful and Bankrupt Digital Currency Exchanges Based on Open Data

Piotr Stolarski(✉) and Włodzimierz Lewoniewski

Poznań University of Economics and Business, al. Niepodległości 10, 61-875 Poznań, Poland
piotr.stolarski@ue.poznan.pl
https://ue.poznan.pl

Abstract. In this work we proposed two models on prediction of the successful and bankrupt digital currency exchanges based on financial data and measures collected from Wikipedia and Wikidata. The purpose of the article is to indicate how intermediary entities providing services related to innovative digital asset markets are characterized in the dominant open online sources. At the same time, this information is related to the knowledge flowing from portals and knowledge bases specializing in crypto-assets. Such a comparison of data and knowledge from two of the aforementioned groups of sources is intended to identify important aspects that objectify the risk analysis process for the selected group of entities.

Risk analysis requires obtaining data and knowledge from a variety of sources, as objective as possible. This is the main precondition for having satisfactory and reliable results. Such analysis is necessary in order to make a better selection of intermediary entities for transactions carried out with digital means of exchange of value, and thus should significantly increase the awareness of entities operating in these markets and give the opportunity to reduce specific risks.

Keywords: Crypto-assets · Wikipedia · Wikidata · Open Data · Blockchain Security

1 Introduction

The emergence of Bitcoin in 2008 and the subsequent boom in the crypto-asset economy is a phenomenon on an unprecedented scale in modern world economic history. The growing interest in cryptocurrency technology also makes it an attractive target for potential attacks and manipulation. The technology itself, although secure, is the source of the emergence of entirely new and unprecedented threats. This is causing an increase in demand for ways and techniques that will be able to keep users, cryptocurrency entities and information stored on

M. Hernes and J. Wątróbski (Eds.): ECAI 2023, LNNS 1079, pp. 228–239, 2024.
https://doi.org/10.1007/978-3-031-66761-9_19

the blockchain network safe. These solutions, in order to be effective, must be prepared for frequent threats.

One of the most frequently identified and actual scenarios of cryptocurrency-denominated cash loss situations are incidents related to wallet storage services and exchanges [3]. The infamous history of fraud and theft associated with intermediation in crypto-asset markets began with the first exchange, Mt.Gox. We are undoubtedly reminded of the reality of the threats by the recent events in 2022 related to the collapse of the FBX exchange. It is also highly likely that this is not the last chapter that will be written in the infamous history of misappropriation of capital of investors who invest funds in innovative digital currency technologies.

The basic mechanism to protect against the discussed risk cases is to build awareness and knowledge among potential investors about the operation of crypto-assets, the role of intermediaries and the dangers of their interactions. In principle, there are two options for protection here: resignation from the services of intermediaries and risk management based on proper analysis of the entities selected for cooperation. Due to the degree of development of the market and the need for easy swapping of assets between crypto-fiat domains, the first solution is unsatisfactory and applicable only to a very limited extent.

Risk analysis requires obtaining data and knowledge from a variety of sources, as objective as possible. This is a prerequisite for obtaining satisfactory and reliable results, which can then translate into the right choice of service providers and, consequently, increased security of capital. In the age of the digital economy, there are essentially two primary sources that investors can use.

The first group is, of course, industry-specific and specialized portals and databases devoted to investment topics, especially those related to the crypto-asset market.

The second group is open Internet sources, among which Wikipedia, as the largest and most popular open encyclopedia and multi-temporal knowledge base, undoubtedly leads the way.

The third group is media with a particular emphasis on social networks. While the order of listing the first group is indisputable, the other two may compete for the palm of primacy. Nevertheless, prudent investors, emphasizing independence and object-ism of the analyzed information, should choose the order indicated in their analytical strategy. This is dictated by the fact that news from the media and the accuracy of the representation of the state of market players may not be free from intentional bias or even manipulation.

In this article, we focus on comparing data and knowledge from two of the aforementioned groups of sources in the context of intermediary entities providing services related to innovative digital asset markets. In this regard, the purpose of the article is to indicate how these entities are depicted in Wikipedia and to relate the information recorded there to the knowledge flowing from portals and knowledge bases specializing in crypto-assets. Such an analysis is necessary for a better selection of intermediary entities for transactions carried out with digital means of value exchange, and thus should significantly increase the awareness of entities operating in these markets and cause a reduction in the listed risks.

2 Methodology and Literature Review

The risk management process consists of the following steps: identification, measurement, control and monitoring. The risks associated with the use of intermediaries in crypto-asset markets include the following situations:

- risk of bankruptcy of the intermediary,
- the risk of fraud or misappropriation of part or all of the entrusted capital,
- the risk of unreliable execution of orders,
- the risk of a hacking attack.

Cross-exchange trading induces risk spillover in the crypto market, especially for centralized exchanges, which compound crypto volatility and counterparty risk [15].

Within the framework of the adopted management method, an attempt will be made to determine the magnitude of risks within the framework of a qualitative approach. An attempt will then be made to verify the available information on the analyzed entities and to identify potentially relevant factors affecting the control of individual risks.

2.1 Literature Review

Sonia Arsi et al. [1] provided a detailed overview on the main risks to be considered by crypto-traders, namely, technology, fraud, legal, market, liquidity, and COVID-19 pandemic risks. The main findings showed that the occurrence of any technological failure tends to rise insecurity and distrust in the cryptocurrency technology. This fact can be even further spoiled through frauds' schemes and fake trading volumes. Additionally, the legal framework of the cryptocurrencies is still inconclusive. In terms of market risk, these crypto-assets are riskier than fiat currencies and there is a significant risk contagion across large-cap cryptocurrencies.

George Milunovich and Seung Ah Lee [15] constructed a database containing eight key characteristics on 238 cryptocurrency exchanges and employ machine learning techniques to predict whether a cryptocurrency market will remain active or whether it will go out of business. Four models have been developed. While all give good results, the best model is one that employs random forest classifier. It produces results with accuracy of about 90 % on training and 86 % on test data. The identified important predictors include exchange lifetime, transacted volume and cyber-security measures.

Yifu Wang et al. [13] proposed a Multivariate Heterogeneous AutoRegression for Crypto Market (MHAR-CM) to specifically investigate interconnectedness among 9 different exchanges on Bitcoin in a high-frequency resolution via dynamic partial correlation networks. The authors found that the scale of an exchange determines its influence towards the others, and the interconnectedness among these exchanges is stronger during a crisis. They also concluded with some remarks on FTX due to its recent bankruptcy. Furthermore, they demonstrated that a portfolio considering the dynamics of partial correlation networks offers a promising result in terms of risk diversification.

Oskar Liljeström [11] provided a research on the market efficiency of cryptocurrencies, by studying the risk exposure of identical investments between different cryptocurrency exchanges. The study included four cryptocurrencies and nine different exchanges. The study concluded that there are significant market inefficiencies on the cryptocurrency market which diminishes over time.

Dean Fantazzini and Raffaella Calabrese [7] considered a unique dataset of 144 exchanges, active from the first quarter of 2018 to the first quarter of 2021 and analyzed the determinants surrounding the decision to close an exchange using credit scoring and machine learning techniques. This is a robust research and the main finding is that having a public developer team is by far the most important (negative) determinant of closure, followed by the CER cybersecurity grade, the age of the exchange, and the number of traded cryptocurrencies available on the exchange. Contr-intuitively, such variables as country of registration and exchange type are not reported as important. This is consistent with other researches [16].

Patrick McCorry et al. [14] presented extensively a survey of possible exchanges heists. The approach of the authors differs from many other publications as they proposed a reactive measure instead of theft prevention. The measure is a fail-safe mechanism to counteract the heist.

Myungim Lim and Hangbae Jang [12] proposed a plan to reduce the operational risk of Korean cryptocurrency exchanges. Moreover, in their work the authors analyzed the current situation of Korean cryptocurrency exchanges and presented the direction of development of each exchange platform in order to protect their users. They also noticed that the risks associated with cryptocurrency exchanges are greater than the price risk itself.

Tyler Moore and Nicolas Christin [16] focused on risk posed by failures of currency exchanges giving a large number of examples at the beginning. The authors noticed that while decentralization is a key feature of cryptocurrencies, an extensive ecosystem of 3rd-party intermediaries supports transactions. The ecosystem includes digital exchanges. Most risk crypto-assets holders face stems from the interactions with these intermediaries as they act as central authorities. The paper focused on Bitcoin solely, however authors provided very interesting analysis of 40 exchanges. In order to improve the equality of data used, a number of sources is checked. A Bitcoin Wiki is among them. In the last part two statistical models were conceptualized. The first one is a survival model for exchanges and the other is a breach model. The models use 3 variables as their input.

Chang Yeon Kim and Kyungho Lee [8] analyzed vulnerabilities of potential cryptocurrency exchanges and individual user wallets. Moreover, this paper suggested policy risk management methods using international standards such as from NIST and ISO. Blockchain, weaknesses of countermeasures management system, countermeasures to deal with them, management vulnerability of investors and management plan.

2.2 Research Gap

The mentioned works extensively explore the risks associated with cryptocurrency trading, market efficiency, the determinants of exchange closures, and various methods for predicting the survival or failure of exchanges, among other topics. However, a noticeable gap is the lack of research combining objective open online sources like Wikipedia and Wikidata with specialized portals to holistically assess the characterization of intermediary entities in the crypto-assets market. Most works either focused on specific attributes of exchanges (e.g., cybersecurity, public development teams) or used machine learning and statistical models to predict exchange outcomes, but there was no cohesive attempt to juxtapose open online data with specialized crypto sources for a comprehensive risk analysis.

This study fills this gap by creating prediction models for the success and bankruptcy of digital currency exchanges using financial data from open data from Wikipedia and Wikidata. By comparing information from open online sources with specialized crypto knowledge bases, our work aims to provide a more objective and comprehensive risk analysis process for these intermediary entities. This new approach not only leverages diverse data sources for enhanced reliability but also serves the broader goal of educating entities in the crypto market. Consequently, by offering a multifaceted analysis, this study is poised to bolster market participants' awareness and help reduce associated risks.

3 Crypto-Assets Economy

3.1 Market Characteristics

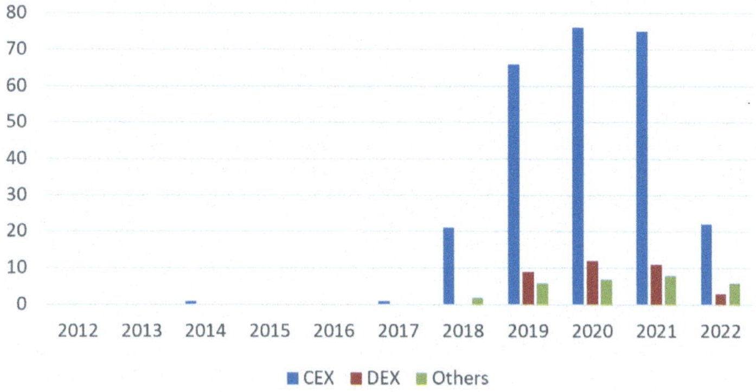

Fig. 1. Closed or bankrupted crypto exchanges.

There are in total 529 crypto-assets intermediaries exchanges as of 30 November 2022 [2]. Only to a little surprise the majority of them, i.e. 282 decentralized exchanges (dex) while only little smaller group of 243 of them are spot ones. Another 3 subjects are lending enterprises. 43 entities out of the set also offer products analogous to derivatives. The number of entities that have ceased to operate in the considered market can be estimated at about 390. The reasons for the cessation of operations vary, but they are usually related to security breaches or fraud.

A detailed distribution of the number of entities that ceased operations as intermediaries by year is presented on Fig. 1. The exponential growth of such events in years 2019–2021 is quite characteristic here. Consequently, the reasons leading to the cessation of offering services are presented on Fig. 2.

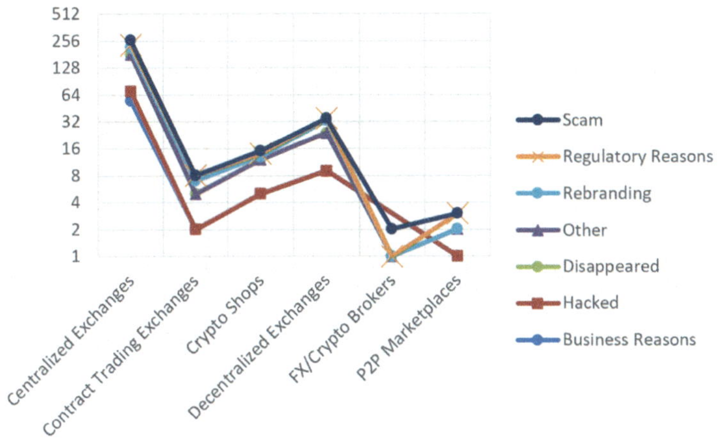

Fig. 2. Reasons of crypto exchanges failure.

3.2 Model Dataset

For the purpose of modeling the bankruptcy of crypto-assets intermediaries, a special dataset was prepared. The collection describes various entities operating in the market in question. The collection was divided into 2 subsets: operating entities and closed entities. Both subsets were created on the basis of the information collected and presented in the Sect. 3.1. However, this information was critically analyzed, cleaned and supplemented. If reliable and complete data on a specific entity could not be obtained, the data was removed from the collection as defective or incomplete. The collection prepared in this way formed the basis for building a number of models for predicting the continuity of the analyzed entities. Some of the models built additionally took into account information from an alternative source, which in this case was Wikipedia.

4 Wikipedia

Wikipedia articles often appear in the first positions in search engine results. Therefore popularity of search query in search engine on particular topic can correlate with page views of separate Wikipedia article on such topic. Depending on language versions popularity of particular topic may vary [9]. In order to identify Wikipedia articles about digital currency exchanges we can use different approaches. One of the simplest - to use category (or categories) of one of the Wikipedia articles which is included in considered topic (in our case - digital currency exchanges). Let's take for example article on FTX in English Wikipedia [6]. The Fig. 3 shows the list of the article categories. As we can see, this article aligned to over 10 different categories, but the most relevant for this research are following: 'Digital currency exchanges' [5] and 'Bitcoin exchanges' [4].

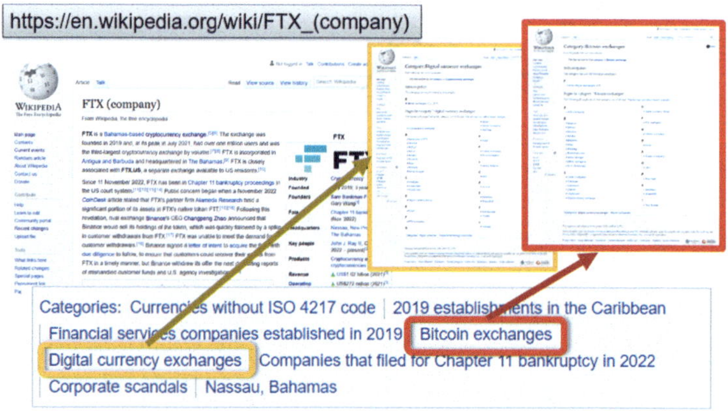

Fig. 3. Categories of article about FTX (company) in English Wikipedia.

Semantic knowledge base Wikidata contains information about over 100 million objects. Each object is presented in separate page with various statements. The Fig. 4 presents Scheme of the Wikidata item about FTX.

In order to find Wikidata items related to cryptocurrency or digital currency exchangers the following values of property "instance of" (P31) were searched in over 100 millions items:

- digital currency exchanger (Q3511068)
- bitcoin exchange (Q20107444)
- cryptocurrency exchange (Q25401607)

Based on the 3 mentioned statements we were able to find 91 Wikidata items[1]. However not all of them have related Wikipedia article in at least one

[1] Wikidata SPARQL query and the result are available on the web page: https://w.wiki/5$DP.

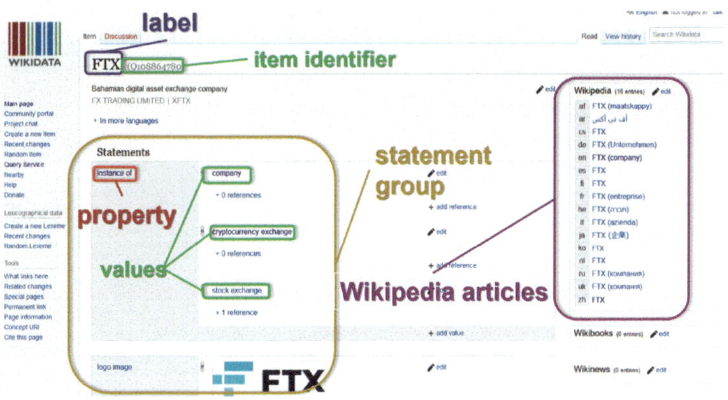

Fig. 4. Scheme of the Wikidata item about FTX based on [17]

language version. We used the obtained Wikipedia articles names in different languages for pageviews analysis. After adding up the number of views for all language versions of a particular digital currency exchangers described on Wikipedia, we were able to draw up a ranking of the most popular exchangers in separate months. The Fig. 5 presents monthly popularity ranking of articles about digital currency exchanges on Multilingual Wikipedia.

It is important to note, that digital currency exchanges, which in the surveyed month were not described in Wikipedia as a separate article in at least one language version, do not have any data on the number of views, and therefore do not have any position in the ranking. However, we can also notice that more and more articles about digital currency exchanges appear over time.

For Wikipedia articles about digital currency exchanges we collected different quality and popularity measures. Some of them are widely used in models on automatic quality assessment of Wikipedia articles and their sources [9,10]. One of collected measures is a number of authors of the Wikipedia articles. Such measures were obtained in two ways: by counting all unique authors of the Wikipedia articles and (separately) only registered authors (non-bots). They named as "allauthors" and "regauthors" respectively. Another extracted measure is a number of internal citations of Wikipedia articles (called "citations"). This number shows how many other Wikipedia articles contains link to selected one.

5 Results

5.1 Basic Dataset Models

The Random Forest family of algorithms was used for modeling. These algorithms are widely used and offer great potential for quality and flexibility. Models were created using a classifier from Scikit. Several scenarios of the model construction were considered. Initially, different hyper-parameters were taken into account. This includes depth of trees and density of forest.

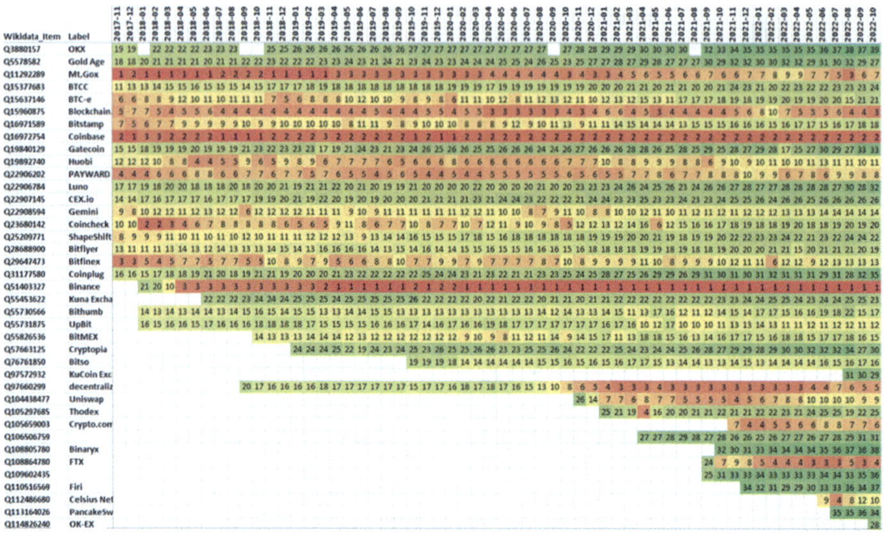

Fig. 5. Monthly popularity ranking of articles about digital currency exchanges on Multilingual Wikipedia.

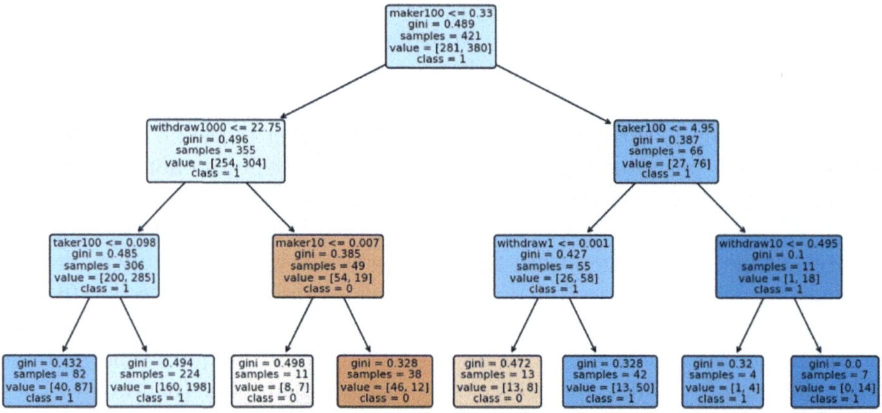

Fig. 6. Exemplary 3-levels deep tree generated as part of the forest model.

Firstly, in case of the number of tree levels, two options were tested. Models that consisted of 3 layers trees (Fig. 6) and 2 layers trees. In this case the smaller number of levels gave minimally better results.

Secondly, regarding the forest density, two approaches were explored. One with a small number of trees (n = 3) and the other with a large number of trees (n = 100). In this case the larger number of trees provided more accurate models. Results for models with 3 layers trees and 2 layers trees are presented on Fig. 7.

2 levels	precision	recall	f1-score
0	0.59	0.48	0.53
1	0.69	0.78	0.73
accuracy			0.66
macro avg	0.64	0.63	0.63
weighted avg	0.65	0.66	0.65

3 levels	precision	recall	f1-score
0	0.57	0.48	0.52
1	0.68	0.76	0.72
accuracy			0.64
macro avg	0.63	0.62	0.62
weighted avg	0.64	0.64	0.64

Fig. 7. Evaluation metrics of 2-level and 3-level tree models.

5.2 Wiki Extension Dataset Model

Finally, a more advanced dataset extended with the information provided from Wikipedia and Wikidata were also used to build another batch of models. The extension data consisted of different information quality and popularity indicators related to the appropriate entries of business entities. The whole dataset was traditionally divided into a train subset and a test subset during the models creation.

	precision	recall	f1-score
0	0.60	1.00	0.75
1	1.00	0.78	0.88
accuracy			0.83
macro avg	0.80	0.89	0.81
weighted avg	0.90	0.83	0.84

Fig. 8. Evaluation metrics of the model based on wiki information.

The results for the obtained model are demonstrated on Fig. 8. These results are up to expectations in terms of the typically used measures that describe the quality of such models. Nevertheless, they are in line with the results presented by some other researchers in the above mentioned works at Sect. 2.1. Moreover, we can also receive information about importance of particular parameters in the obtained models. Figure 9 shows ranking of parameters importance for two selected models.

Basic

features	importances
withdraw1000	0.150546
withdraw100	0.124753
withdraw1	0.119998
withdraw10	0.119161
taker1000	0.092055
taker100	0.077936
taker10	0.077634
taker1	0.076259
maker1	0.049528
maker10	0.046066
maker100	0.030129
maker1000	0.024305
usallowed	0.004618
credit	0.003740
wire	0.003272

+ wiki

features	importances
citations	0.164572
allauthors	0.130734
regauthors	0.113692
langs	0.111706
enlang	0.100574
pageviews	0.078625
withdraw100	0.038458
maker10	0.032971
maker100	0.032808
maker1000	0.028149
taker100	0.025132
withdraw1000	0.025092
withdraw1	0.023256
taker1000	0.021458
maker1	0.019097
taker1	0.017879
withdraw10	0.012962
taker10	0.009600
usallowed	0.007463
credit	0.005771
wire	0.000000

Fig. 9. The importance of particular parameters of the 2-level trees model (on the left) and of the model based on wiki extension (on the right).

6 Conclusions

This study delved into the analysis of both successful and bankrupt digital currency exchanges, leveraging financial data and information from open sources such as Wikipedia and Wikidata. The primary objective was to discern the characteristics of intermediary entities operating within the dynamic landscape of digital asset markets.

The authors highlighted the importance of comprehensive risk analysis in the selection of intermediary entities for digital asset transactions. In order to support this process forecasting models was built using machine learning techniques. The models use data from diverse sources. Firstly, the crypto-asset specialized portals and secondly, knowledge bases such as Wikipedia and Wikidata.

The resulted models are more accurate when using the enhanced information taken from mentioned knowledge bases. However, there are some limitations. The primary limitation of this study is the reliance on open data sources. One need to be aware that these sources may not always provide complete, up-to-date or accurate information. Moreover, the information about digital currencies on the Wikipedia is fragmented. The external factors such as regulatory changes, market sentiment, or unexpected events are other issues which limit the approach taken. The presented models does not take into account the mentioned elements and so may not be universally applicable regardless of circumstances.

The main theoretical implication is the highlighting of the importance of considering multiple sources of data and knowledge in risk analysis, especially in emerging and rapidly evolving markets like digital currencies. For practitioners the research emphasizes the necessity of conducting thorough due diligence when selecting intermediaries. A tool that may be useful in such activity is provided.

The presented tool in order to be more valuable need to be upgraded. This is a task for planned future research. The plans include data augmentation, incorporation of more advanced machine learning techniques, longitudinal studies to track the performance of digital currency exchanges over time as well as inclusion of regulatory aspects.

References

1. Arsi, S., Ben Khelifa, S., Ghabri, Y., Mzoughi, H.: Cryptocurrencies: key risks and challenges. In: Cryptofinance: A New Currency for a New Economy, pp. 121–145. World Scientific (2022)
2. CoinMarketCap: Cryptocurrency Prices, Charts And Market Capitalizations. coinmarketcap.com (2022)
3. Disparte, Dante Alighieri: Category: Bitcoin exchanges (2018). https://www.forbes.com/sites/dantedisparte/2018/07/21/beware-of-crypto-risks-10-risks-to-watch/?sh=4c1dd1be5f17
4. English Wikipedia: Category: Bitcoin exchanges (2022). https://en.wikipedia.org/wiki/Category:Bitcoin_exchanges
5. English Wikipedia: Category: Digital currency exchanges (2022). https://en.wikipedia.org/wiki/Category:Digital_currency_exchanges
6. English Wikipedia: FTX (2022). https://en.wikipedia.org/wiki/FTX
7. Fantazzini, D., Calabrese, R.: Crypto exchanges and credit risk: modeling and forecasting the probability of closure. J. Risk Financ. Manag. **14**(11), 516 (2021)
8. Kim, C.Y., Lee, K.H.: Risk management to cryptocurrency exchange and investors guidelines to prevent potential threats. In: 2018 International Conference on Platform Technology and Service (PlatCon), pp. 1–6 (2018)
9. Lewoniewski, W., Węcel, K., Abramowicz, W.: Multilingual ranking of wikipedia articles with quality and popularity assessment in different topics. Computers **8**(3), 60 (2019)
10. Lewoniewski, W., Węcel, K., Abramowicz, W.: Identifying reliable sources of information about companies in multilingual wikipedia. In: 2022 17th Conference on Computer Science and Intelligence Systems (FedCSIS), pp. 705–714. IEEE (2022)
11. Liljeström, O.: Efficiency of cryptocurrency exchanges: risk exposure analysis of identical assets (2019)
12. Lim, M., Jang, H.: A study on the risk reduction plan of cryptocurrency exchange. J. Platform Technol. **8**(4), 29–37 (2020)
13. Lu, W., Wang, Y., Lin, M.B., Ren, R., Härdle, W.K.: Cross-exchange crypto risk: a high-frequency dynamic network perspective (2022). SSRN 4308825
14. McCorry, P., Moser, M., Ali, S.T.: Why preventing a cryptocurrency exchange heist isn't good enough. In: Matyas, V., Švenda, P., Stajano, F., Christianson, B., Anderson, J. (eds.) Security Protocols XXVI. Security Protocols 2018. LNCS, vol. 11286, pp. 225–233. Springer, Cham (2018). https://doi.org/10.1007/978-3-030-03251-7_27
15. Milunovich, G., Lee, S.A.: Cryptocurrency exchanges: predicting which markets will remain active. J. Forecast. **41**(5), 945–955 (2022)
16. Moore, T., Christin, N.: Beware the middleman: empirical analysis of bitcoin-exchange risk. In: Sadeghi, A.R. (eds.) Financial Cryptography and Data Security. FC 2013. LNCS, vol. 7859, pp. 25–33. Springer, Berlin, Heidelberg (2013). https://doi.org/10.1007/978-3-642-39884-1_3
17. Wikidata: Q108864780 (2022). https://www.wikidata.org/wiki/Q108864780

Intelligent Multiple Criteria Decision Analysis Methods

The New Algorithm for Effective Reducing the Number of Pairwise Comparisons in the Decision Support Methods

Andrii Shekhovtsov[1] and Wojciech Sałabun[1,2(✉)]

[1] National Telecommunications Institute, ul. Szachowa 1, 04-894 Warsaw, Poland
{a.shekhovtsov,w.salabun}@il-pib.pl
[2] West Pomeranian University of Technology in Szczecin, ul. Żołnierska 49, 71-210 Szczecin, Poland
wojciech.salabun@zut.edu.pl

Abstract. The pairwise comparison matrix is often used in different multi-criteria decision-making methods. The matrix identification process is related to the curse of dimensionality because as the size of the matrix increases, the number of queries to experts increases exponentially. One such method is a Characteristic Objects METhod (COMET) which requires building a Matrix of Expert Judgments (MEJ) in order to evaluate alternatives. However, to build MEJ, a decision maker or an expert should answer multiple pairwise comparison questions, which can be expensive in both financial and time terms. In this paper, we propose an algorithm that efficiently reduces the number of pairwise comparison questions to an expert. The algorithm is based on the triads' consistency and achieves more than ninety percent reduction in questions for big expert judgment matrices. Although the original algorithm is proposed for the COMET method, it could be generalized and used with such methods as RANking COMparison (RANCOM) and Analytic Hierarchy Process (AHP). The efficiency of the algorithm is proved using a simulation approach. The results of those simulations, as well as the pseudocodes, are presented in the paper.

Keywords: Triads · Pairwise Comparison · COMET

1 Introduction

Multi-criteria decision analysis (MCDA) is a branch of operational research that aims to help decision-makers solve complex multi-criteria problems. Methods and approaches provided by the MCDA domain are widely used for solving many practical decision-making problems in energetic [2,4], healthcare [10,17], and other domains [9,19]. Very often, decision-making methods require the identification of the pairwise comparison matrix. Among such methods, we can distinguish Analytic Hierarchy Process (AHP) [13], which is a very popular methodology in the MCDA domain. However, the AHP method is often criticized because

M. Hernes and J. Wątróbski (Eds.): ECAI 2023, LNNS 1079, pp. 243–254, 2024.
https://doi.org/10.1007/978-3-031-66761-9_20

of rank-reversing behavior [1,16] and issues with consistency of the pairwise comparison matrix, which was investigated in many research work [7,11,12].

Another method that uses the pairwise comparison matrix is the Characteristic Objects METhod (COMET), which is completely resistant to the rank reversal paradox [14]. However, this method is susceptible to the curse of dimensionality and requires building a large pairwise comparison matrix in order to process complex decision-making problems. To address the problem of inconsistency in the COMET, the new consistency coefficient for the decision matrices was proposed by Sałabun et al. [15]. This research was based on the works by [5] and [8]. Sałabun's work distinguishes four groups of triads. One of them is called consistent triads and is particularly interesting for us because, based on them, it is possible to predict some part of the pairwise comparison matrix with a limited number of questions to an expert.

The main contribution of this paper is the proposition of the new algorithm, which is based on consistent triads and can efficiently reduce the number of pairwise comparisons needed to identify a pairwise comparison matrix in such algorithms as COMET. We also present simulation experiments that prove the stability and usability of the proposed algorithm.

The rest of the paper is structured as follows: in Sect. 2, we explain how the MEJ is built and how triad analysis could be used to measure the consistency of the MEJ. In Sect. 3, we describe how triad analysis could help to reduce the number of expert queries in algorithms that use pairwise comparison matrices such as COMET and RANCOM [20]. Next, in Sects. 4.1 and 4.2, we show two experiments that show how the proposed approaches behave when the expert provides consistent answers and under the assumption that the expert will make a mistake with some probability. In Sect. 5, we discuss the results presented in the previous sections and the limitations of the approach. Finally, in Sect. 6, we conclude our work and provide some directions for future works.

2 Preliminaries

2.1 Matrix of Expert Judgments in COMET

One of the steps of the Characteristic Objects METhod is identifying the MEJ matrix. The full algorithm of this method is described in other works [14], and the implementation of this method can be found in the Python library of the MCDA methods called pymcdm [6]. In this paper, we will show the part of COMET's algorithm related to the identification of the Matrix of Expert Judgements (MEJ).

As one of the first steps in the COMET method, the domain expert should determine the Matrix of Expert Judgment (MEJ). This matrix is used to define the preferences of the characteristic objects (CO). During the identification step, the expert should make pairwise comparisons of defined CO. The structure of the MEJ matrix is defined as follows (1):

$$\begin{pmatrix} \alpha_{11} & \alpha_{12} & \dots & \alpha_{1n} \\ \alpha_{21} & \alpha_{22} & \dots & \alpha_{2n} \\ \dots & \dots & \dots & \dots \\ \alpha_{n1} & \alpha_{n2} & \dots & \alpha_{tn} \end{pmatrix} \quad (1)$$

where α_{ij} is the result of comparing CO_i and CO_j by the expert. Possible values are described by (2) and depend solely on the knowledge of the expert:

$$\alpha_{ij} = \begin{cases} 0.0, & f_{exp}(CO_i) < f_{exp}(CO_j) \\ 0.5, & f_{exp}(CO_i) = f_{exp}(CO_j) \\ 1.0, & f_{exp}(CO_i) > f_{exp}(CO_j) \end{cases} \quad (2)$$

where f_{exp} is the expert mental judgment function. It is worth noting that all values on the diagonal are equal to $\alpha_{ii} = 0.5$, and the upper and lower triangle part of the MEJ is connected by the Equation $\alpha_{ji} = 1 - \alpha_{ij}$. Based on these two properties, we can define the number of comparisons required as (3):

$$p = \binom{n}{2} = \frac{n(n-1)}{2} \quad (3)$$

2.2 Triads Analysis

In 2021 Sałabun et al. proposed a new consistency coefficient for the COMET method, which uses triad analysis to check how consistent the MEJ matrix is [15]. This coefficient was designed based on [5] and [8] works.

Suppose that we have defined n characteristic objects for the specific decision-making problem $\mathbf{CO} = \{CO_1, CO_2, \dots CO_n\}$. The subset contains three characteristic objects (CO_i, CO_j, CO_k) is called a triad. The number of all possible triads for specific n can be calculated using Eq. (4).

$$T = \frac{n!}{(n-3)!3!} \quad (4)$$

Assuming that each characteristic object has a certain unknown evaluation which is constant over time, and the expert made no mistakes in their judgments, the expert's preferences must be in a transitive relation.

Therefore, based on (1) and (2), we can determined a set of rules (5). If the triplet α_{ij}, α_{jk}, α_{ik} is match one of the rules, then the triad (CO_i, CO_j, CO_k) is consider consistent.

$$\begin{array}{llllllll} if & \alpha_{ij} = 1.0 & and & \alpha_{jk} = 1.0 & then & \alpha_{ik} = 1.0 \\ if & \alpha_{ij} = 1.0 & and & \alpha_{jk} = 0.5 & then & \alpha_{ik} = 1.0 \\ if & \alpha_{ij} = 0.5 & and & \alpha_{jk} = 1.0 & then & \alpha_{ik} = 1.0 \\ if & \alpha_{ij} = 0.5 & and & \alpha_{jk} = 0.0 & then & \alpha_{ik} = 0.0 \\ if & \alpha_{ij} = 0.0 & and & \alpha_{jk} = 0.0 & then & \alpha_{ik} = 0.0 \\ if & \alpha_{ij} = 0.0 & and & \alpha_{jk} = 0.5 & then & \alpha_{ik} = 0.0 \\ if & \alpha_{ij} = 0.5 & and & \alpha_{jk} = 0.5 & then & \alpha_{ik} = 0.5 \end{array} \quad (5)$$

The paper [15] also describes unknown T_{unk} weak inconsistent T_{inc}^{weak} and strong inconsistent triads T_{inc}^{strong}, resulted in 20 inconsistent triad rules. We omit the definitions of those rules in this paper because they can be found in the original paper and are unnecessary for understanding the algorithm proposed in this paper.

The coefficient of consistency based on the inconsistent triads is called C_{Tr}. It can be computed as (6):

$$C_{Tr} = 1 - \frac{T_{inc}^{weak} + T_{inc}^{strong}}{T} \tag{6}$$

2.3 Weighted Spearman's Rank Correlation Coefficient

For a sample of size N, the rank values x_i and y_i we define Weighted Spearman's rank correlation coefficient r_w by (7). In this approach, the positions at the top of both rankings are more important to the final correlation value. The weight of significance is calculated for each comparison. Therefore, this coefficient examines whether the differences appeared and not where they appeared [3].

$$r_w = 1 - \frac{6\sum_{i=1}^{N}(x_i - y_i)^2((N - x_i + 1) + (N - y_i + 1))}{N^4 + N^3 - N^2 - N} \tag{7}$$

3 Proposed Approach

Analyzing how the C_{Tr} consistency coefficient works, we came up with how the consistent triads could be used to help fill the MEJ matrix and successfully reduce the number of expert queries. In the case of the MEJ matrix with size n, the number of expert queries which is required to fill it is $\frac{n(n-1)}{2}$, e.g., for the $n = 10$, we require 45 queries and for $n = 20$ we require 190. However, the number of queries is susceptible to the curse of dimensionality and increases fast. For $n = 200$, we need to answer 19900 pairwise comparison questions, which is a very big number. Even if the expert answers one question per second, it will still require several hours to answer them all, and it can negatively influence the quality of the answers. To give a proper response, we need much more time. This indicates a need for an algorithm that can reduce the number of questions to answer.

The proposed approach is based on the fact that we can use a partially filled MEJ matrix to fill in the rest. For example, if we asked the expert and got the values $\alpha_{12} = 1$ and $\alpha_{25} = 1$. Therefore, we could tell that $\alpha_{15} = 1$ because we have a consistent triad like this. If the triad we try to use is inconsistent, we could use values α_{13} and α_{35} to determine the value of α_{15}. This means, that for filling up α_{ik} we could use values α_{ij} and α_{jk}, where i and k should be chose a priory and the $j \in \{i+1, i+2, i+3, \ldots, k\}$.

Based on that fact, we propose the algorithm called Full Triad Support (FTS) that allows filling in the MEJ matrix based on checking every value of j possible for given i and k. Algorithm 1 presents the pseudo-code of the method. The entry requirements for this algorithm are defining the size of the MEJ n and the empty matrix X of size n, which will be the MEJ matrix to fill. The function $query(i, j)$ indicates the moment in which we need to ask an expert to compare CO_i and CO_j and therefore determine the values of α_{ij}. We count the number of $query()$ function calls and use it later to determine the reduction level. This approach fills the matrix diagonal by diagonal, and the variable s in the loop on line 2 decides which diagonal will be filled at the moment. Then the loop on the line 3 and the formula on the line 4 determines the values of the i and k indices. Then, we iterate over every possible value of the j using the loop on the line 5. In this loop, we first check if α_{jk} is already defined. If not, we should ask the expert for its value (lines 6-8). On lines 9-14, we try to find the consistent triad which matches the values of α_{ij} and α_{jk} for current j. If a triad like this exits, we use it to fill in the value of α_{ik} and break the loop. If not, we do the same thing for the next iteration for another j value. Finally, if there is no suitable triad and α_{ik} is undefined. We should ask an expert about its value (lines 16-18). The last thing, we need to use the filled-in upper triangle part of the X to fill the lower triangle part.

Algorithm 1 Full Triad Support (FTS) algorithm.

Require: Size of MEJ n
Require: Expert query function $query()$
Require: Empty matrix $X = [\alpha_{ij}]$ of size n
1: $\alpha_{1,2} \leftarrow query(1, 2)$
2: **for** s **in** $2, 3, 4, \ldots, n-1$ **do**
3:   **for** i **in** $1, 2, 3, \ldots, n-s-1$ **do**
4:     $k = i + s + 1$
5:     **for** j **in** $i+1, i+2, i+3, \ldots, k-1$ **do**
6:       **if** α_{jk} **is undefined then**
7:         $\alpha_{jk} \leftarrow query(j, k)$
8:       **end if**
9:       **if** α_{ik} **is undefined then**
10:         **if** Consistent triad T for α_{ij} and α_{jk} exists **then**
11:           $\alpha_{ik} \leftarrow$ conclusion of T.
12:           **break**
13:         **end if**
14:       **end if**
15:     **end for**
16:     **if** α_{ik} **is undefined then**
17:       $\alpha_{ik} \leftarrow query(i, k)$
18:     **end if**
19:   **end for**
20: **end for**
21: Compute lower triangle of X based on upper triangle

4 Study Case

To prove the proposed methodology's efficiency and robustness, we design two research experiments that will demonstrate the method's superiorities and limitations.

4.1 Experiment 1

The first experiment was designed to show the proposed algorithms' efficiency in reducing the needed expert queries. The procedure of the experiment is described by Algorithm 2. The procedure was repeated ten thousand times for different sizes of matrices $n \in \{25, 50, 75, 100, 125, 150, 175, 200\}$. The algorithm is based on a randomly generated MEJ matrix X, which is consistent. On line 1, we calculate the preferred vector P from random MEJ X, and then we apply every of three methods for building MEJ to reconstruct MEJ X based on values in P. Notice that NTS stands for No Triad Support and FTS for Full Triad Support. Notation *method*(P) means that we use vector P to emulate the expert in the *method*. Therere, in lines 3-4 we reconstruct matrix $X' = X$ using the P vector. Then, we write the number of queries actually used in order to process them later.

Algorithm 2 The experiment procedure repeated 10000 times for $n \in \{25, 50, 75, 100, 125, 150, 175, 200\}$.

Require: Size of MEJ n
Require: Random consistent MEJ X
1: $P \leftarrow MEJ\_to\_P(X)$
2: **for** *method* **in** $\{NTS, FTS\}$ **do**
3: $X' \leftarrow method(P)$
4: **assert** $X = X'$
5: **write** Number of queries to expert made for *method* and MEJ size n.
6: **end for**

Figure 1 (left side) shows how the number of expert queries changes with the MEJ size. The NTS line shows the function $f(n) = \frac{n(n-1)}{2}$, or the number of expert queries needed to complete the MEJ matrix in the most straightforward way. For the FTS, the reduction is drastic, and for $n = 200$ with FTS support, the expert needs to answer approximately 2500 questions.

In Fig. 1 (right side), we show the reduction in queries in percents calculated as $r = (1 - q/f(n)) \cdot 100\%$, where q stands for an actual number of queries and $f(n) = \frac{n(n-1)}{2}$ which is a formula for a number of queries without triad support. If we analyze Fig. 1 (right side), we can see a reduction of the expert queries with the FTS algorithm achieves up to 90% for the matrix with size $n = 200$. The reduction is increasing slowly with the size of the MEJ instead of staying on the same level.

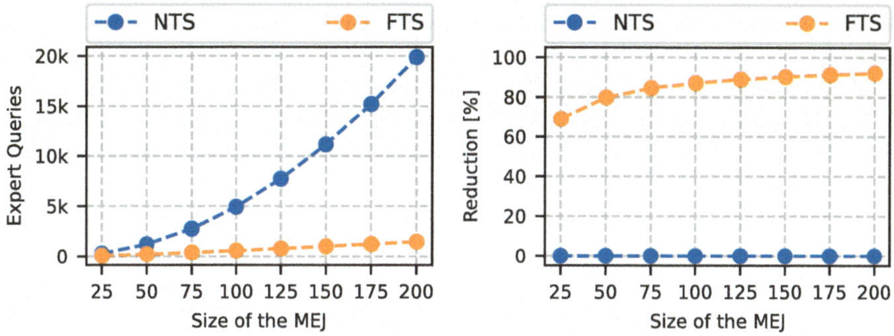

Fig. 1. Average number of expert queries needed (left) and average reduction in queries (right) with triad support and without it (left). Average value calculated from 10000 repeats.

4.2 Experiment 2

The presented results show the efficiency of the proposed algorithms in decreasing the number of expert queries needed to fill the MEJ matrix. However, in the first experiment, we assume that the expert's answers are consistent and constant. In some cases, this assumption can be broken because an expert can make a mistake in their answers.

Therefore, in the second experiment, we want to investigate the accuracy of the obtained results under the assumption that an expert made a mistake with some probability. With given probability Pr_{error}, instead of true value α_{ij}, expert will provide a value α'_{ij}, which is randomly chosen from the set of possible values $\alpha'_{ij} \in \{0, 0.5, 1\}$.

Algorithm 3 The experiment procedure repeated 3000 times for different values of n and Pr_{error}.

Require: Size of MEJ n
Require: Probability of the error Pr_{error}
Require: Random consistent MEJ X
1: $P \leftarrow MEJ\_to\_P(X)$
2: $R \leftarrow rank(P)$
3: **for** *method* **in** $\{NTS, RTS, FTS\}$ **do**
4: $\quad X' \leftarrow method(P, Pr_{error})$
5: $\quad P' \leftarrow MEJ\_to\_P(X')$
6: $\quad R' \leftarrow rank(P')$
7: $\quad$ **write** Number of queries to the expert, $MAE(P, P')$, $C_{Tr}(X')$, $r_w(R, R')$
8: **end for**

We design the simulation experiment similar to the first one but with the addition of the probability of an error in the expert's judgments. The algorithm of the experiment is presented on Algorithm 3. It was repeated 3000 times for a given probability of the error and different sizes of the MEJ. The start of the procedure is similar to the first experiment. In lines 1-2, we create preferences vector P

from the MEJ matrix X and then calculate the rank of the characteristic objects R based on P. Next, on lines 3-6, we apply one of three methods for building the MEJ and calculate P' and R'. In this case, the property $X = X'$ is wrong, so $P' \neq P$ and we can measure the difference of it. For this experiment, we write mean absolute error between P and P', consistency of the reconstructed MEJ, and the values of correlation coefficient r_w for rankings of characteristic objects.

As said previously, we repeat the procedure for the second experiment 3000 times for different values of n and Pr_{error}, simulating the experts who evaluate different characteristic objects with a certain probability of mistakes. The simulation results are presented in several visualizations, which will be described in the following part of the paper.

In Fig. 2, we present the visualization which shows how the mean MAE value between the original P vector and reconstructed P' vector depends on the size of MEJ for NTS or FTS algorithms and for the different probability of the error which is shown by the color. The general rule here is that the MAE value increases with the increase of the probability error. However, when we analyze it closer, for the NTS algorithm, the MAE error decreases with the increase of the size of the MEJ. However, for the FTS algorithm, it is another way around. The maximum MAE we could observe for NTS is approximately 0.18 for the biggest error probability and for the smaller matrix, which requires a small number of queries to be filled. However, for the FTS, the maximum MAE is approximately 0.3 and is observed for the biggest investigated matrices and for the highest probability error.

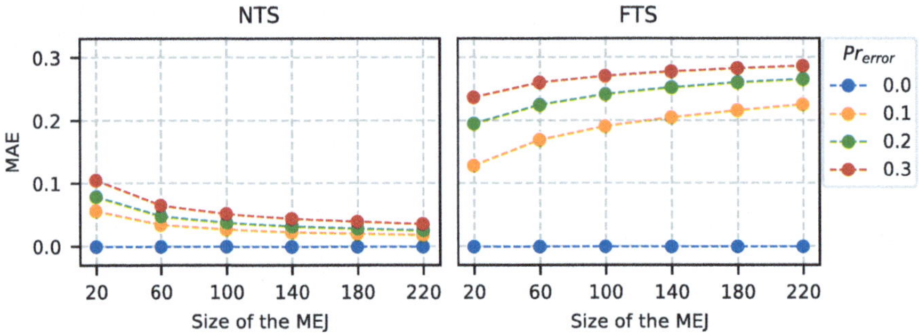

Fig. 2. Dependency of mean MAE value between P and P' vectors from size of the MEJ for different algorithms, and different probability of the error (color).

However, because of how the COMET method works, the more interesting measure will be the r_w rank correlation coefficient between rankings of the characteristic objects. The values of r_w are in the range $[-1, 1]$, where 1 means strong positive correlation and 0 means uncorrelated rankings. The visualization of r_w values is presented in Fig. 3. If we do not use triad support (NTS), we can obtain almost perfect correlation even when experts are tended to make errors often. The smallest value $r_w \approx 0.9$ was obtained for the smallest investigated MEJ matrix for the biggest investigated expert probability error (0.3). However, it should be said that the value $r_w = 0.9$ is a very good correlation between two ranking vectors. On the other hand, when we use FTS, for $n = 20$, we have $r_w \approx 0.5$, and when the expert is more uncertain ($Pr_{error} = 0.3$), we obtain a preference vector that barely correlated with the original one ($r_w \approx 0.2$).

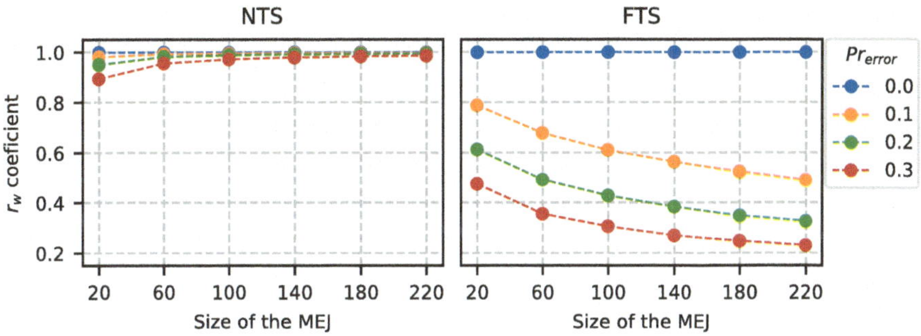

Fig. 3. Dependency of mean r_w value between R and R' vectors from size of the MEJ for different algorithms, and different probability of the error (color).

Another analyzed measurement was a C_{Tr} consistency coefficient described in Sect. 2. The results on consistency are very interesting. If we do not use any triad support (NTS), the consistency decreases with decreasing of Pr_{error}, with no regard to the size of the MEJ. However, for the FTS algorithm, the mean value of C_{Tr} equals 1 with no regard to the size of the MEJ or error probability, which is a really good result. Because its values are mostly constant, we do not present a visualization of C_{Tr} values.

5 Discussion

The results shown in Sect. 4.1 prove that the proposed algorithms based on the triad support can efficiently reduce the number of expert queries needed to build a pairwise comparison matrix. Figure 1 shows that when using the FTS algorithm, the expert needs to answer only approximately 2000 instead of almost 20000. It means that we can achieve more than a 90% reduction in the expert queries for the big MEJ matrices. Nevertheless, two thousand questions are still expensive in terms of an expert's time. Even if an expert makes one comparison in 30 s, doing 2000 will still require more than sixteen hours. Such complicated decision problems require a combination of different approaches to solve efficiently. For example, the FTS algorithm can potentially be combined with the structural approach shown in [18]. In that case, we get a huge reduction, and an expert may take more time to provide a reliable and correct answer.

Theoretically possible maximum reduction in the expert queries using the FTS algorithm is a reduction to $n-1$ queries instead of $\frac{n(n-1)}{2}$ queries for MEJ matrix with size n. Such reduction is only possible in simple cases when all of the values α_{ij} above the main diagonal fulfill the transitivity rules. In this case, the expert will need to answer only $n-1$ question, and the rest of the MEJ matrix will be filled using triad support. The reverse situation, when the expert will be required to fill all of the matrices manually, is very unlikely to occur because of the high stability of the FTS algorithm.

That said, the FTS algorithms allow the same MEJ matrix to be built, with no triad support, with a smaller number of expert queries required. In Experiment 1, we have assumed that an expert makes no error and that their responses are consistent over time. It is hard to ensure this, therefore we suggest another experiment that shows how those algorithms will behave when the expert could make mistakes in their judgments. The second experiment is described in Sect. 4.2 and provides interesting results. Because of how the FTS algorithm fills the MEJ, errors made when answering the first questions accumulate, resulting in big differences in identified MEJ. Visualizations on Figs. 2 and 3 show that with the FTS algorithm, average r_w correlation drops even to 0.2 when an expert is very uncertain ($Pr_{error} = 0.3$). On the other hand, when we use it for small matrices and the expert is more consistent in their answers ($Pr_{error} = 0.1$), the average r_w correlation is approximately 0.8, which is a very good result. This fact underlines that we should use the FTS algorithm only with confident domain experts who will be consistent in their answers. It will ensure that the MEJ will be consistent from the beginning, otherwise, large differences between constructed MEJ matrices are possible.

6 Conclusion

In this paper, we propose the Full Triad Support algorithm that allows decreasing the time of identifying pairwise comparison matrices by reducing the number of required comparisons. The algorithm is based on consistent triads and allows us to reduce up to 90% expert queries when identifying the MEJ matrix. Application of this algorithm allows recreating the same MEJ matrix as the MEJ matrix built without support algorithm. Combining this approach with the structural approach for reducing the number of queries will reduce it even more, drastically reducing the time costs of identifying the MEJ matrix. The only limitation is that experts should be consistent in their answers because the FTS algorithms tend to accumulate errors in the preliminary steps, and the final preference vector can differ significantly from the original one.

Proposed algorithm also create some directions for future research and improvements, for example:

- adaptation of the FTS algorithm for the RANCOM and AHP methods;
- investigation of how the probability of making a mistake by an expert will influence the final ranking on chosen practical study case;
- proposes the approach which will try to eliminate expert's mistakes in comparisons, so the resulting MEJ will be more similar to the original one;

- combines the algorithm proposed in this paper with a structured approach to lower the number of queries [18] to decrease the number of required comparisons even more.

Acknowledgements. The work was supported by the National Science Centre 2021/41/B/HS4/01296 (A.S. and W.S.).

References

1. Barzilai, J., Golany, B.: AHP rank reversal, normalization and aggregation rules. INFOR: Inf. Syst. Oper. Res. **32**(2), 57–64 (1994)
2. Cinelli, M., Burgherr, P., Kadziński, M., Słowiński, R.: Proper and improper uses of MCDA methods in energy systems analysis. Decis. Support Syst. **163**, 113848 (2022)
3. Pinto da Costa, J., Soares, C.: A weighted rank measure of correlation. Aust. N. Z. J. Stat. **47**(4), 515–529 (2005)
4. Deveci, M., Pamucar, D., Oguz, E.: Floating photovoltaic site selection using fuzzy rough numbers based LAAW and RAFSI model. Appl. Energy **324**, 119597 (2022)
5. Kendall, M.G., Smith, B.B.: On the method of paired comparisons. Biometrika **31**(3/4), 324–345 (1940)
6. Kizielewicz, B., Shekhovtsov, A., Sałabun, W.: pymcdm-the universal library for solving multi-criteria decision-making problems. SoftwareX **22**, 101368 (2023)
7. Kułakowski, K.: Notes on order preservation and consistency in AHP. Eur. J. Oper. Res. **245**(1), 333–337 (2015)
8. Kułakowski, K.: Inconsistency in the ordinal pairwise comparisons method with and without ties. Eur. J. Oper. Res. **270**(1), 314–327 (2018)
9. Li, X.Y.: Algorithms for multiple attribute decision making with dual hesitant pythagorean fuzzy information and their application to supplier selection. Int. J. Knowl.-based Intell. Eng. Syst. **25**(4), 465–475 (2021)
10. Memarpour Ghiaci, A., Garg, H., Jafarzadeh Ghoushchi, S.: Improving emergency departments during COVID-19 pandemic: a simulation and MCDM approach with MARCOS methodology in an uncertain environment. Comput. Appl. Math. **41**(8), 368 (2022)
11. Pant, S., Kumar, A., Ram, M., Klochkov, Y., Sharma, H.K.: Consistency indices in analytic hierarchy process: a review. Mathematics **10**(8), 1206 (2022)
12. Saaty, T.L.: A scaling method for priorities in hierarchical structures. J. Math. Psychol. **15**(3), 234–281 (1977)
13. Saaty, T.L.: Decision making with the analytic hierarchy process. Int. J. Serv. Sci. **1**(1), 83–98 (2008)
14. Sałabun, W.: The characteristic objects method: a new distance-based approach to multicriteria decision-making problems. J. Multi-Criteria Decis. Anal. **22**(1–2), 37–50 (2015)
15. Sałabun, W., Shekhovtsov, A., Kizielewicz, B.: A new consistency coefficient in the multi-criteria decision analysis domain. In: Paszynski, M., Kranzlmüller, D., Krzhizhanovskaya, V.V., Dongarra, J.J., Sloot, P.M.A. (eds.) Computational Science – ICCS 2021. ICCS 2021. LNCS, vol.12742, pp. 715–727. Springer, Cham (2021). https://doi.org/10.1007/978-3-030-77961-0_57

16. Sałabun, W., Ziemba, P., Wątróbski, J.: The rank reversals paradox in management decisions: the comparison of the AHP and COMET methods. In: Czarnowski, I., Caballero, A., Howlett, R., Jain, L. (eds.) Intelligent Decision Technologies 2016. IDT 2016. SIST, vol. 56, pp. 181–191. Springer, Cham (2016). https://doi.org/10.1007/978-3-319-39630-9_15
17. Seker, S., Bağlan, F.B., Aydin, N., Deveci, M., Ding, W.: Risk assessment approach for analyzing risk factors to overcome pandemic using interval-valued q-rung orthopair fuzzy decision making method. Appl. Soft Comput. **132**, 109891 (2023)
18. Shekhovtsov, A., Kołodziejczyk, J., Sałabun, W.: Fuzzy model identification using monolithic and structured approaches in decision problems with partially incomplete data. Symmetry **12**(9), 1541 (2020)
19. Wang, H.J.: Interval-valued intuitionistic fuzzy multiple attribute decision making and their applications. Int. J. Knowl.-based Intell. Eng. Syst. **25**(2), 251–277 (2021)
20. Więckowski, J., Kizielewicz, B., Shekhovtsov, A., Sałabun, W.: RANCOM: a novel approach to identifying criteria relevance based on inaccuracy expert judgments. Eng. Appl. Artif. Intell. **122**, 106114 (2023)

Multi-criteria Assessment of the Sustainable Share of Renewable Energy Sources in European Countries Using the SSP-TOPSIS Method

Aleksandra Bączkiewicz[1(✉)] and Jarosław Wątróbski[1,2]

[1] Institute of Management, University of Szczecin, ul. Cukrowa 8, 71-004 Szczecin, Poland

{aleksandra.baczkiewicz,jaroslaw.watrobski}@usz.edu.pl

[2] National Institute of Telecommunications, ul. Szachowa 1, 04-894 Warsaw, Poland

Abstract. The dynamic development of decision support systems is resulting in the emergence of new decision support tools. Additional attributes of these instruments making available broader analytical capabilities require more complex algorithms of applied methods than popular and simple compensation methods derived from American school. Adaptations based on existing methods can provide new analytical capabilities while maintaining the simplicity of a known algorithm. This paper presents a newly developed Strong Sustainability Paradigm Based Technique for Order Preference by Similarity to Ideal Solution (SSP-TOPSIS) method incorporating criteria compensation reduction modeling. The proposed method is based on the well-known Technique for Order Preference by Similarity to Ideal Solution (TOPSIS) algorithm, and the innovative criterion compensation reduction step makes the technique suitable for assessments that require consideration of the strong sustainability paradigm.

The developed method was applied to the practical case of multi-criteria evaluation of the sustainable use of renewable energy sources in the economies of selected European countries. Attributes of the SHARES (Short assessment of renewable energy sources) model proposed in Eurostat (the statistical office of the European Union) covering energy generation from renewable energy sources (RES) and its consumption in various branches of the economy were used as criteria for the research framework. The research was performed for multiple cases considering criteria compensation reduction modeling. The results identified Sweden as the most balanced country regarding RES exploitation. Other countries that scored well were Austria, Germany, and the Netherlands.

Keywords: Strong sustainability paradigm · SSP-TOPSIS · MCDA · Renewable energy sources

M. Hernes and J. Wątróbski (Eds.): ECAI 2023, LNNS 1079, pp. 255–267, 2024.
https://doi.org/10.1007/978-3-031-66761-9_21

1 Introduction

The extensive development of decision support has conditioned the growth in the number and types of tools designed to help decision-makers and stakeholders with higher-quality multi-criteria evaluation to analyze the fulfillment of sustainability goals [12]. An essential attribute of decision support tools is their usability for stakeholders. Decision support tools differ in complexity. In some instances, decision-makers prefer simple-to-use tools. Among these tools, multi-criteria methods derived from the American school can be mentioned, such as Technique for Order Preference by Similarity to Ideal Solution (TOPSIS) [11], Analytic Hierarchy Process (AHP) [3,9], and Vlsekriterijumska Optimizacija I Kompromisno Resenje (VIKOR) [6]. However, some situations require consideration of particular aspects like the strong sustainability paradigm in sustainability assessment [8]. The strong sustainability paradigm implies limited compensation of evaluation criteria. In addition, decision-support tools should allow exploration of the effects of alternative models providing stakeholders with more in-depth analyses relevant to the local context [14].

The above observations motivated the development of a new multi-criteria method based on the TOPSIS method. It allows modeling of criterion compensation reduction, a major weakness of popular and easy-to-use multi-criteria methods from the American school [5]. The proposed method is named Strong Sustainability Paradigm Based Technique for Order Preference by Similarity to Ideal Solution (SSP-TOPSIS). In this research, the SSP-TOPSIS method was used to solve the multi-criteria problem of assessing the sustainable share of renewable energy sources (RES) in the economy of selected European countries.

The European Union considers the sustainable development paradigm of regional, national, and local economies as a priority for inclusion in long-term growth and development strategies, in line with the Sustainable Development Goals (SDGs), with the requirements set by the 2030 Agenda [4]. New ambitious targets set for achievement in 2050 in response to international commitments to reduce greenhouse gas emissions have enabled the EU to take the global lead in renewable energy development [1,7]. Solar, wind, and photovoltaic energy support the EU's low-carbon economy and transition to a climate-neutral stage [10]. Incorporating RES to generate clean electricity (RES-E) is undoubtedly a major pillar of the energy sector's decarbonization strategy [10]. However, other energy balance components are also important, such as the use of RES in transportation (RES-T), where biofuels replace fossil fuels, or the utilization of RES for heating or cooling (RES-H&C), which contributes to reducing greenhouse gas emissions [13].

In assessing sustainability relative to the share of RES in the economy of countries, it is important to take an approach that will prevent the possibility of compensating the poor performances of some assessment criteria by the exceptionally favorable values of other few criteria [2]. Indeed, it is essential to balance good performances across all aspects of sustainability [9].

The above facts motivated to investigate the sustainable share of RES in European countries regarding the evaluation criteria included in the SHARES

(Short assessment of renewable energy sources) database provided by Eurostat (the statistical office of the European Union). SHARES tool focuses on the harmonized calculation of the share of energy from renewable sources. The main benefit of the SHARES tool is that EU countries are committed to following the same method to calculate the indicators related to the share of energy from renewable sources. Using this tool prevents irregularities that may arise from varying parameters and rules used in different calculation methods. SHARES tool considers, among other attributes, four criteria groups interesting from the perspective of this research: Electricity generation from RES, Electricity consumption in transport from RES, Electricity consumption in heating and cooling from RES, and Gross final consumption of energy from RES [10]. The abovementioned criteria were used to construct a model for assessing European countries regarding RES participation in the economy and industry. The countries were then evaluated using the newly developed SSP-TOPSIS method that allows the modeling of criteria compensation reduction. The evaluation was carried out for different levels of compensation reduction within the four considered groups of criteria of the research framework.

The rest of the paper is organized as follows. Section 2 provides the basics, assumptions, and mathematical formulas of the newly proposed SSP-TOPSIS method. Section 3 presents and describes the dataset used for this research. Results are provided and discussed in Sect. 4, while conclusions and future work directions are drawn in Sect. 5.

2 Methodology

The SSP-TOPSIS method is a newly developed method based on the assumptions of the popular and well-recognized TOPSIS method, which considers the distance of alternatives from two reference solutions: ideal and anti-ideal, in the evaluation. The method uses an innovative step that makes it possible to model the reduction of criteria compensation, which is a drawback of multi-criteria methods derived from American school like TOPSIS. This step involves reducing the compensation by subtracting the MD (Mean Deviation) from the performance value. MD is the difference between the performance of the alternative for a given criterion and the averaged performance within the criterion across all alternatives multiplied by the sustainability coefficient. The sustainability coefficient s is an element for modeling compensation reduction and takes values in the range of 0 to 1. By default, it can be set as the standard deviation from the normalized decision matrix within each evaluation criterion. Detailed Steps of the Proposed SSP-TOPSIS Method are provided below.

In addition to SSP-TOPSIS, an objective criteria weighting method named Criteria Importance Through Inter-criteria Correlation (CRITIC) was applied for the research.

Step 1. Calculate the Mean Deviation MD_{ij} for each performance value included in the decision matrix by subtraction the mean value $\overline{x_j}$ from performance values x_{ij} for each C_j criterion. Next, multiply the result by s_j coefficient.

Coefficient s_j defines the sustainability coefficient, which represents the level of criteria performance compensation reduction, and its values are real numbers in the range from 0 to 1. Criteria are numbered by $j = 1, 2, \ldots, m$. Higher values of s_j indicate a more significant reduction in the compensation of a j-th criterion performance value. In turn, low values of s_j represent a low degree of reduction in the compensation of a given j-th criterion. The whole procedure of Mean Deviation calculation is performed using Eq. (1).

$$MD_{ij} = (x_{ij} - \overline{x_j})s_j \tag{1}$$

Step 2. Assign 0 values to these MD_{+ij} for profit criteria C_{+j} that are lower than 0. When MD_{+ij} is lower than 0 it denotes that x_{+ij} is lower than $\overline{x_{+j}}$. Set 0 values for these MD_{-ij} for cost criteria C_{-j} that are higher than 0. It means that r_{-ij} are higher than $\overline{x_{-j}}$. The procedure described in this step is performed with Eq. (2)

$$MD_{ij} = 0 \, \forall \, MD_{+ij} < 0 \, \vee \, MD_{-ij} > 0 \tag{2}$$

where MD_{ij} denotes Mean Deviation values calculated for criteria C_j. This step is very significant because it aims to prevent improving unmentionable performance values that are outliers from the average toward the worse.

Step 3. Subtract MD_{ij} values from performance values x_{ij} contained in decision matrix x_{ij} as Eq. (3) shows.

$$t_{ij} = x_{ij} - MD_{ij} \tag{3}$$

The rest of the steps are the same as the classic TOPSIS method.

Step 4. Normalize the decision matrix shown in Eq. (4) using the selected normalization method, such as the Minimum-Maximum normalization or Vector normalization, the default normalization for the TOPSIS method. In Minimum-Maximum normalization, normalized values denoted by r_{ij}^{+} for profit criteria and r_{ij}^{-} for cost criteria are received applying Eq. (5). After this normalization procedure, all criteria are already converted to profit criteria.

$$T = [t_{ij}]_{m \times n} = \begin{bmatrix} t_{11} & t_{12} & \cdots & t_{1n} \\ t_{21} & t_{22} & \cdots & t_{2n} \\ \vdots & \vdots & \vdots & \vdots \\ t_{m1} & t_{m2} & \cdots & t_{mn} \end{bmatrix} \tag{4}$$

$$r_{ij}^{+} = \frac{t_{ij} - min_j(t_{ij})}{max_j(t_{ij}) - min_j(t_{ij})}, \; r_{ij}^{-} = \frac{max_j(t_{ij}) - t_{ij}}{max_j(t_{ij}) - min_j(t_{ij})} \tag{5}$$

Step 5. Calculate the weighted normalized decision matrix. For this purpose, multiply values in the normalized decision matrix by the respective criteria weights w_j using Eq. (6).

$$v_{ij} = r_{ij}w_j \tag{6}$$

Criteria weights were determined using the objective Criteria Importance Through Inter-criteria Correlation (CRITIC) method [12].

Step 6. Determine the Positive Ideal Solution (PIS) using Eq. (7) and Negative Ideal Solution (NIS) by Eq. (8). PIS includes the maximum values of the weighted normalized decision matrix, while NIS includes its minimum values. Due to the previous normalization of the decision matrix, splitting criteria into profit and cost is unnecessary.

$$v_j^+ = \{v_1^+, v_2^+, \ldots, v_n^+\} = \{max_j(v_{ij})\} \quad (7)$$

$$v_j^- = \{v_1^-, v_2^-, \ldots, v_n^-\} = \{min_j(v_{ij})\} \quad (8)$$

Step 7. Calculate distance from PIS D_i^+ and NIS D_i^- for each alternative using Eq. (9). The default metric for distance calculation in the TOPSIS algorithm is Euclidean distance.

$$D_i^+ = \sqrt{\sum_{j=1}^{n}(v_{ij} - v_j^+)^2},\ D_i^- = \sqrt{\sum_{j=1}^{n}(v_{ij} - v_j^-)^2} \quad (9)$$

Step 8. Calculate the score for each considered alternative as Eq. (10) demonstrates. The C_i value is always in the range from 0 to 1. The alternative that has the highest C_i value is the best scored. The ranking is generated by sorting alternatives according to preference values in descending order.

$$C_i = \frac{D_i^-}{D_i^- + D_i^+} \quad (10)$$

3 The Dataset

For the purpose of assessing the sustainable contribution of individual RES to the economies of European countries, it was decided to use a dataset provided by Eurostat called SHARES (Short assessment of renewable energy sources), made available on the Eurostat website [10]. The most up-to-date dataset containing data collected for 2021 was downloaded from https://ec.europa.eu/eurostat/web/energy/database/additional-data (accessed on 8 June 2023). For the analysis, 29 European countries were selected, comprising 27 European Union countries and additionally Iceland and Norway. The choice of these countries is justified by the fact that sustainable RES is a policy priority of the European Union, while Iceland and Norway are countries where RES is essential due to domestic politics and the availability of natural resources, especially Hydro used for electricity generation.

Table 1 presents a set of criteria for assessing European countries in terms of sustainable share of renewable energy. It consists of indicators related to the share of energy from renewable sources included in the SHARES framework

Table 1. Evaluation criteria based on SHARES.

C_j	Criteria names
	Group G_1 - Electricity generation from RES (RES-E)
C_1	Annual electricity generation from Hydro
C_2	Annual electricity generation from Wind
C_3	Annual electricity generation from Solar
C_4	Annual electricity generation from Solid biofuels
C_5	Annual electricity generation from All other renewables
	Group G_2 - Electricity consumption in Transport from RES (RES-T)
C_6	Annual consumption of renewable electricity in road transport
C_7	Annual consumption of renewable electricity in rail transport
C_8	Annual consumption of renewable electricity in all other transport modes
C_9	Annual consumption of renewable electricity from compliant biofuels in transport
	Group G_3 - Electricity consumption in Heating and Cooling from RES (RES H&C)
C_{10}	Annual final energy consumption in heating and cooling
C_{11}	Annual derived RES based heat in heating and cooling
C_{12}	Annual derived RES based heat in heating and cooling for heat pumps
	Group G_4 - Gross final consumption of energy from RES (RES)
C_{13}	Gross final consumption of energy from renewable sources in electricity
C_{14}	Gross final consumption of energy from renewable sources in heating and cooling
C_{15}	Gross final consumption of energy from renewable sources in transport

focused on the harmonized calculation of the share of energy from renewable sources for EU countries. All criteria are profit type because the aim of each of them is maximization.

The SHARES dataset includes values in KTOE (Thousand Tonnes of Oil Equivalent). In assessing the energy economy of countries, it is important to consider the country's population. Therefore, SHARES values for the purposes of this study were converted to TOE (Tonnes of Oil Equivalent) per 1000 inhabitants. The population data for 2021 were collected from the Eurostat database from https://ec.europa.eu/eurostat/databrowser/view/TPS00001/default/table?lang=en (accessed on 8 June 2023). A decision matrix containing the performance of countries against the 15 SHARES criteria in TOE per 1000 inhabitants is presented in Table 2. The SSP-TOPSIS method was implemented in Python 3 and is available in open source repository provided on GitHub at https://github.com/energyinpython/SSP-TOPSIS with the processed research dataset and results.

4 Results

This section presents the research results conducted using the SSP-TOPSIS method to assess the sustainable share of renewable energy in the economy of

Table 2. Decision matrix including collected data in TOE per 1000 inhabitants unit for 2021.

Country	C_1	C_2	C_3	C_4	C_5	C_6	C_7	C_8	C_9	C_{10}	C_{11}	C_{12}	C_{13}	C_{14}	C_{15}
Belgium	2.40	98.30	41.81	20.56	14.35	0.40	2.39	0.06	62.79	128.41	7.39	16.34	174.57	152.14	65.64
Bulgaria	50.47	19.54	18.23	29.50	-27.71	0.12	1.20	0.03	24.11	138.92	2.00	18.92	88.68	160.22	25.46
Czechia	18.14	5.52	18.97	21.83	22.29	0.09	1.81	0.09	36.23	288.03	29.40	28.90	84.76	346.32	38.21
Denmark	0.24	255.47	19.27	53.70	21.06	2.28	4.40	0.00	46.18	213.13	273.28	59.61	343.06	547.67	52.86
Germany	20.76	131.58	51.02	11.28	40.71	0.59	4.88	0.00	35.61	164.81	25.02	18.45	249.89	208.41	37.94
Estonia	1.69	47.22	22.86	109.90	4.84	0.24	0.19	0.00	43.22	319.01	263.96	75.44	186.09	658.39	43.64
Greece	40.41	82.13	42.28	0.34	3.68	0.02	0.42	0.00	15.05	109.66	0.00	40.89	168.39	151.04	15.50
Spain	55.92	109.41	49.16	8.89	3.39	0.24	2.10	0.15	32.70	86.70	0.00	22.63	224.29	109.39	35.19
France	78.63	48.94	19.99	5.48	7.84	0.23	2.23	0.25	42.83	142.43	29.04	53.74	158.18	226.83	45.54
Croatia	144.36	44.35	3.17	14.05	11.29	0.08	2.54	0.41	22.59	291.46	27.96	3.95	214.21	323.49	25.61
Ireland	13.08	185.37	1.60	0.36	7.00	0.40	0.31	0.00	36.84	35.46	0.00	12.16	206.71	49.36	37.24
Italy	70.33	29.54	36.35	6.57	29.52	0.22	2.63	2.67	26.20	127.16	14.55	42.17	166.78	188.72	31.73
Cyprus	0.00	22.21	44.94	0.00	5.75	0.00	0.00	0.00	29.29	170.89	0.99	53.65	72.91	248.26	29.29
Latvia	131.59	6.81	0.31	25.88	13.26	0.70	1.59	0.08	24.17	510.26	223.03	0.33	175.48	737.28	26.54
Lithuania	13.35	47.47	5.87	11.89	9.19	0.27	0.07	0.12	45.39	233.02	214.17	17.08	87.32	464.27	45.85
Luxembourg	14.07	42.67	24.41	38.62	14.17	0.32	2.13	0.00	215.02	57.85	169.62	7.38	131.50	234.84	217.46
Hungary	2.16	5.98	33.54	14.62	2.69	0.07	1.01	0.01	29.15	177.62	17.86	2.24	57.90	198.16	30.25
Malta	0.00	0.01	42.64	0.00	1.21	0.06	0.00	0.00	20.90	14.40	0.00	31.54	43.80	61.51	20.97
Netherlands	0.44	93.78	56.56	38.67	11.45	0.97	1.47	0.00	36.06	66.18	28.54	22.14	198.46	119.49	38.51
Austria	407.34	69.01	26.78	33.92	9.34	0.10	13.53	9.63	51.49	400.45	127.33	47.55	523.13	578.17	74.76
Poland	5.28	37.58	8.94	14.54	3.77	0.02	1.05	0.04	29.59	198.82	14.83	9.78	69.01	223.47	30.70
Portugal	101.90	108.87	18.68	28.32	6.59	0.08	2.02	0.02	33.04	185.02	0.00	73.28	262.24	259.82	35.15
Romania	72.44	33.71	7.63	2.60	0.33	0.36	2.20	0.09	25.82	185.70	4.99	0.00	114.05	190.68	24.85
Slovenia	188.46	0.25	18.47	6.90	4.41	0.05	3.01	0.09	48.60	264.00	22.96	27.66	215.34	314.62	51.75
Slovakia	67.73	0.09	10.57	20.87	8.17	0.09	1.62	0.33	29.40	197.82	32.13	12.67	105.39	242.61	31.44
Finland	229.46	130.79	4.62	196.83	-11.57	1.39	4.08	0.00	119.97	897.28	366.17	103.60	544.66	1370.20	125.44
Sweden	558.05	237.82	12.64	92.57	15.20	8.45	15.10	0.00	139.87	534.16	327.50	152.26	892.73	1065.27	163.43
Iceland	3197.86	5.49	0.00	0.00	1352.66	9.72	0.00	0.00	84.84	190.80	2340.91	0.00	4546.30	2532.20	76.62
Norway	2339.95	191.55	2.78	0.47	2.86	24.30	9.47	0.00	68.03	156.86	36.74	109.71	2503.83	308.11	101.81

the 29 European countries examined. The investigation was conducted in two stages. The first stage involved applying a fixed compensation reduction for each criterion group. In contrast, in the second stage, the compensation reduction was gradually increased by incrementally increasing the sustainability coefficient s. The first stage involved evaluating countries using the TOPSIS method, which gives the same results as the SSP-TOPSIS method with no reduction in criteria compensation. The following steps included in the first stage were evaluating countries using the SSP-TOPSIS method with a reduced compensation of selected criteria using a sustainability coefficient s equal to the standard deviation calculated from a normalized matrix for each criterion. The SSP-TOPSIS column contains the ranking obtained after the described compensation reduction of all criteria. The other columns, on the other hand, include rankings obtained for compensation reductions within successive groups of criteria G_1, G_2, G_3, and G_4.

When analyzing the results obtained, one should be guided by the fact that if a reduction in compensation within a given criterion group results in a decrease in the ranking alternative, it means that the performance values in this criterion group compensated for worse performance within other criterion groups. It suggests limited sustainability within all criterion groups. If, on the other hand, a reduction in compensation within a given criterion group either does not cause a decrease in ranking or improves an alternative's score, it means that the alternative has performances within all criterion groups balanced enough to be invulnerable to a reduction in criterion compensation.

Table 3 includes SSP-TOSPIS preference values, which, when ordered in descending order, allow the construction of a ranking of evaluated alternatives. Rankings obtained for particular criteria groups are provided in Table 4. It can be noted that regardless of the group of criteria to which compensation reduction applies, the leader of the ranking is Sweden. It means that the country is highly resilient in terms of compensation reduction and has a high level of sustainability within all the criteria considered in the assessment in terms of the sustainable share of RES in considered branches of economy and industry.

The Netherlands, in the case of no reduction in criteria compensation (TOPSIS ranking), ranks second. The country dropped to third place when compensation reduction was applied to all criteria. However, the Netherlands remained in second place when compensation reductions were applied within the G_2, G_3, and G_4 criterion groups. When compensation reductions were performed within the G_1 group, it decreased to third place. This result suggests that performances within the criteria group G_1 (Electricity generation from RES) are at a good level in the Netherlands and compensate for worse performances within the other criteria groups. Improving performances from groups G_2, G_3, and G_4 would be advisable to enhance sustainability.

An analogous recommendation can be provided for Germany, as the situation observed for this country is corresponding. The difference is that the country descends from third place in the TOPSIS ranking to fourth place in the SSP-TOPSIS ranking after reducing the compensation of all criteria and within the G_1 group. For the remaining criterion groups, Germany maintains its third place.

A different, interesting situation, however, can be observed in Austria. The country ranks fourth in the TOPSIS ranking without considering compensation reduction. Regarding compensation reductions for all criteria and those in the G_1 group, Austria is promoted by two positions to second place in the ranking. In the case of compensation reductions of criteria from the remaining G_2, G_3, and G_4 groups, the country maintains its fourth place. The results show that Austria is well balanced in terms of the share of RES. An analogous analysis providing the discussed conclusions can be carried out for the other countries for which the results are collected in Table 4.

The second stage of the research involved incrementally increasing compensation reduction of criteria using the coefficient s modified incrementally from 0 to 1. The study was carried out by increasing compensation reduction for each group of criteria G_1, G_2, G_3, and G_4 and for all criteria simultaneously. Figure 1 displays charts representing shifts in the positions of examined countries in the ranking under the impact of gradually increasing compensation for each group of criteria.

It can be noticed that for Estonia there was a significant promotion during the increasing reduction of compensation of criteria of group G_1 from the 16th place

Table 3. Preference values of examined European countries obtained with SSP-TOPSIS setting compensation reduction in particular criteria groups.

Country	TOPSIS	SSP-TOPSIS (All criteria)	SSP-TOPSIS (G_1)	SSP-TOPSIS (G_2)	SSP-TOPSIS (G_3)	SSP-TOPSIS (G_4)
Belgium	0.3566	0.3780	0.3729	0.3590	0.3580	0.3580
Bulgaria	0.1724	0.2072	0.2041	0.1733	0.1748	0.1728
Czechia	0.1922	0.2291	0.2218	0.1941	0.1974	0.1935
Denmark	0.3520	0.3778	0.3691	0.3549	0.3558	0.3540
Germany	0.4108	0.4229	0.4182	0.4128	0.4125	0.4117
Estonia	0.2797	0.3167	0.3098	0.2806	0.2846	0.2814
Greece	0.3419	0.3601	0.3573	0.3421	0.3444	0.3421
Spain	0.3869	0.3997	0.3960	0.3884	0.3883	0.3875
France	0.2134	0.2543	0.2463	0.2161	0.2183	0.2148
Croatia	0.1063	0.1260	0.1164	0.1107	0.1116	0.1075
Ireland	0.2067	0.2150	0.2133	0.2074	0.2073	0.2074
Italy	0.3142	0.3385	0.3327	0.3164	0.3172	0.3148
Cyprus	0.3492	0.3622	0.3586	0.3495	0.3520	0.3498
Latvia	0.1339	0.1432	0.1359	0.1354	0.1382	0.1357
Lithuania	0.1159	0.1396	0.1295	0.1178	0.1221	0.1192
Luxembourg	0.3284	0.3531	0.3510	0.3295	0.3290	0.3286
Hungary	0.2733	0.2963	0.2941	0.2740	0.2744	0.2737
Malta	0.3221	0.3344	0.3327	0.3222	0.3236	0.3222
Netherlands	0.4178	0.4259	0.4225	0.4191	0.4190	0.4185
Austria	0.4029	0.4341	0.4267	0.4043	0.4066	0.4051
Poland	0.1114	0.1350	0.1297	0.1126	0.1154	0.1123
Portugal	0.2459	0.2835	0.2767	0.2477	0.2507	0.2469
Romania	0.1001	0.1215	0.1152	0.1037	0.1034	0.1007
Slovenia	0.1926	0.2303	0.2204	0.1968	0.1976	0.1948
Slovakia	0.1171	0.1413	0.1350	0.1193	0.1212	0.1181
Finland	0.3794	0.3921	0.3863	0.3820	0.3809	0.3811
Sweden	0.4665	0.4809	0.4775	0.4674	0.4676	0.4679
Iceland	0.3493	0.3524	0.3495	0.3503	0.3502	0.3502
Norway	0.3734	0.3832	0.3776	0.3755	0.3754	0.3751

to the 5th place. In addition, promotion was also observed in the case of reduction of compensation of criteria of groups G_3 and G_4. Only in the case of the reduction of compensation of group G_2 criteria was there a drop to 2 positions. Still, it was a minor one, and this only occurred with the full reduction of compensation. It indicates the sustainable participation of RES in the Estonian economy within all groups of criteria considered in the evaluation model under consideration.

It can be observed that the most shifts occurred for reductions in compensation of all criteria, which is obvious, and for reductions in compensation within the G_1 group criteria. It implies that the G_1 group of criteria is particularly important in the presented evaluation model. It is justified because G_1 includes the most significant number of criteria, and there are noticeable discrepancies in their performance values between alternatives. These facts make the evaluated countries the most sensitive to modifications in the model within this particular group of criteria.

Significant advancement during the reduction of compensation of group G_1 criteria is also observed for countries such as Luxembourg and Portugal. In the case of Luxembourg, it is a promotion of 10 positions from 13th place to 3rd place. On the other hand, Portugal, for the reduction of the compensation of the

Table 4. Rankings of examined European countries obtained with SSP-TOPSIS setting compensation reduction in particular criteria groups.

Country	TOPSIS	SSP-TOPSIS (All criteria)	SSP-TOPSIS (G_1)	SSP-TOPSIS (G_2)	SSP-TOPSIS (G_3)	SSP-TOPSIS (G_4)
Belgium	8	8	8	8	8	8
Bulgaria	23	23	23	23	23	23
Czechia	22	21	20	22	22	22
Denmark	9	9	9	9	9	9
Germany	3	4	4	3	3	3
Estonia	16	16	16	16	16	16
Greece	12	11	11	12	12	12
Spain	5	5	5	5	5	5
France	19	19	19	19	19	19
Croatia	28	28	28	28	28	28
Ireland	20	22	22	20	20	20
Italy	15	14	14	15	15	15
Cyprus	11	10	10	11	10	11
Latvia	24	24	24	24	24	24
Lithuania	26	26	27	26	25	25
Luxembourg	13	12	12	13	13	13
Hungary	17	17	17	17	17	17
Malta	14	15	15	14	14	14
Netherlands	2	3	3	2	2	2
Austria	4	2	2	4	4	4
Poland	27	27	26	27	27	27
Portugal	18	18	18	18	18	18
Romania	29	29	29	29	29	29
Slovenia	21	20	21	21	21	21
Slovakia	25	25	25	25	26	26
Finland	6	6	6	6	6	6
Sweden	1	1	1	1	1	1
Iceland	10	13	13	10	11	10
Norway	7	7	7	7	7	7

G_1 group criteria, is promoted by nine positions from 18th to 9th place. For the other groups of criteria, no significant changes are observed for these countries. Mentioned observations confirm the balanced use of RES by discussed countries.

On the other hand, a significant decrease in ranking from 10th to 22nd when reducing the compensation of G_1 criteria was observed for Iceland. A considerable dropping from rank 10 to rank 14 when reducing the compensation of G_3 criteria was also observed for this country. Iceland has an outstandingly high performance value for criteria C_1 and C_5 of G_1, C_{11} of G_3, C_{13} and C_{14} of G_4. As a result, when there is no reduction in compensation, the high values of these criteria are able to compensate for the poor performances of other criteria (C_3, C_4, C_7, C_8, C_{12}). On the other hand, when the possibilities for compensation are limited, as is required in the sustainability assessment, Iceland's rating worsens.

Similar to the first part of the research, for this section, analogous analyses can be carried out for the remaining countries, for which shifts in rankings under the impact of reductions in compensation of individual criterion groups are displayed in charts included in Fig. 1.

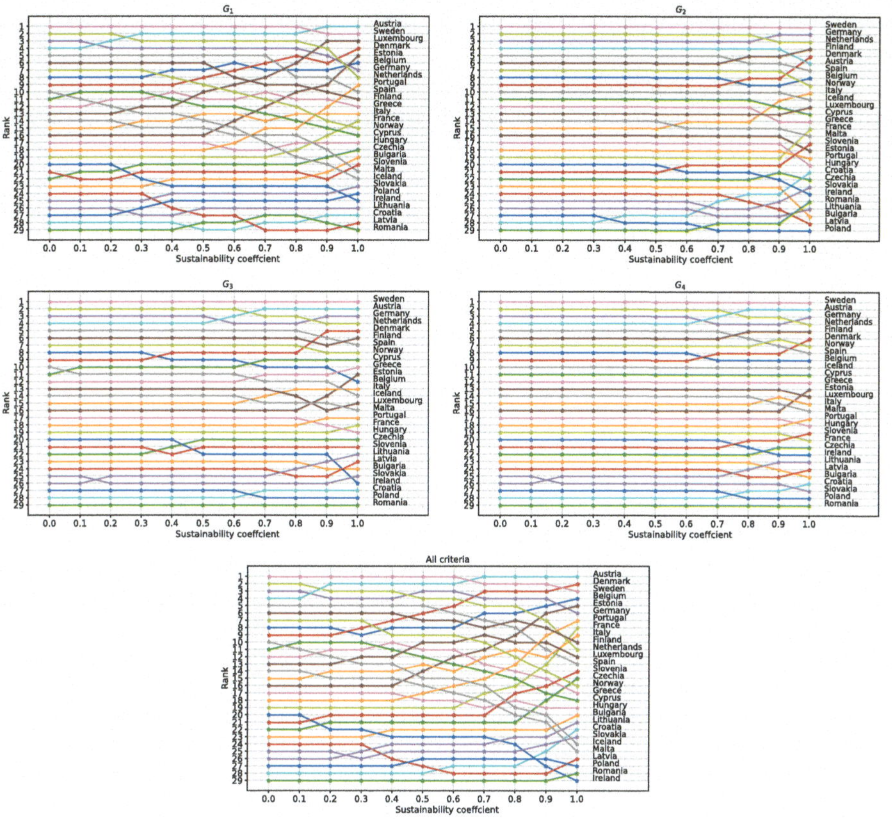

Fig. 1. Shifts in the positions of examined countries in the ranking caused by gradually increasing criteria compensation.

5 Conclusions

This paper presents a newly developed multi-criteria method SSP-TOPSIS based on TOPSIS assumptions providing the possibility to model the reduction of criteria compensation. The method was applied to the practical case of assessing the sustainable share of RES in the economies of European countries. Based on the results, it can be deduced that Sweden is the country leading the rankings in almost all cases of compensation reduction within different criteria groups. It confirms Sweden's balanced RES participation. Austria, Germany and Netherlands can also be mentioned among the well scored countries. SSP-TOPSIS provides stakeholders with a wide range of analytical options depending on their requirements for modeling the reduction of criteria compensation. The presented approach opens up new possibilities for adapting the well-known and widely used MCDA methods derived from the American school for which the phenomenon of criterion compensation has been a limitation. Companies can use the presented method to simulate rankings with the modified compensation of performances against criteria.

Since the research presented in this paper is an early attempt, directions for future work include further research to examine and confirm the reliability of SSP-TOPSIS in benchmarking procedures. Additional work also includes adapting other multi-criteria compensation methods to enable the reduction of criteria compensation and comparing their effects with the SSP-TOPSIS method demonstrated in this paper.

Acknowledgements. This research was funded in whole by the National Science Centre, Poland 2022/45/N/HS4/03050 (A.B., J.W.)

References

1. Bigerna, S., Polinori, P.: Convergence of KAYA components in the European Union toward the 2050 decarbonization target. J. Clean. Prod. **366**, 132950 (2022). https://doi.org/10.1016/j.jclepro.2022.132950
2. Candan, G., Cengiz Toklu, M.: A comparative analysis of the circular economy performances for European Union countries. Int. J. Sustain. Dev. World Ecol. **29**(7), 653–664 (2022). https://doi.org/10.1080/13504509.2022.2084794
3. Dogan, O., Deveci, M., Canıtez, F., Kahraman, C.: A corridor selection for locating autonomous vehicles using an interval-valued intuitionistic fuzzy AHP and TOPSIS method. Soft Comput. **24**, 8937–8953 (2020). https://doi.org/10.1007/s00500-019-04421-5
4. Kostetckaia, M., Hametner, M.: How sustainable development goals interlinkages influence European Union countries' progress towards the 2030 agenda. Sustain. Dev. **30**(5), 916–926 (2022). https://doi.org/10.1002/sd.2290
5. de Lima Silva, D.F., de Almeida Filho, A.T.: Sorting with TOPSIS through boundary and characteristic profiles. Comput. Ind. Eng. **141**, 106328 (2020). https://doi.org/10.1016/j.cie.2020.106328
6. Pamucar, D., Yazdani, M., Montero-Simo, M.J., Araque-Padilla, R.A., Mohammed, A.: Multi-criteria decision analysis towards robust service quality measurement. Expert Syst. Appl. **170**, 114508 (2021). https://doi.org/10.1016/j.eswa.2020.114508
7. Wątróbski, J.: Temporal PROMETHEE II-New multi-criteria approach to sustainable management of alternative fuels consumption. J. Clean. Prod. **413**, 137445 (2023). https://doi.org/10.1016/j.jclepro.2023.137445
8. Wątróbski, J., Bączkiewicz, A., Król, R., Sałabun, W.: Green electricity generation assessment using the CODAS-COMET method. Ecol. Ind. **143**, 109391 (2022). https://doi.org/10.1016/j.ecolind.2022.109391
9. Wątróbski, J., Bączkiewicz, A., Rudawska, I.: A strong sustainability paradigm based analytical hierarchy process (SSP-AHP) method to evaluate sustainable healthcare systems. Ecol. Ind. **154**, 110493 (2023). https://doi.org/10.1016/j.ecolind.2023.110493
10. Wątróbski, J., Bączkiewicz, A., Sałabun, W.: New multi-criteria method for evaluation of sustainable RES management. Appl. Energy **324**, 119695 (2022). https://doi.org/10.1016/j.apenergy.2022.119695

11. Wątróbski, J., Bączkiewicz, A., Sałabun, W.: pyrepo-mcda-Reference objects based MCDA software package. SoftwareX **19**, 101107 (2022). https://doi.org/10.1016/j.softx.2022.101107
12. Wątróbski, J., Bączkiewicz, A., Ziemba, E., Sałabun, W.: Sustainable cities and communities assessment using the DARIA-TOPSIS method. Sustain. Urban Areas **83**, 103926 (2022). https://doi.org/10.1016/j.scs.2022.103926
13. Włodarczyk, B., Firoiu, D., Ionescu, G.H., Ghiocel, F., Szturo, M., Markowski, L.: Assessing the sustainable development and renewable energy sources relationship in EU countries. Energies **14**(8), 2323 (2021). https://doi.org/10.3390/en14082323
14. Wong-Parodi, G., Mach, K.J., Jagannathan, K., Sjostrom, K.D.: Insights for developing effective decision support tools for environmental sustainability. Curr. Opin. Environ. Sustain. **42**, 52–59 (2020). https://doi.org/10.1016/j.cosust.2020.01.005

On Comparison of the Results of Selected Multicriteria Methods Applied in the Assessment Banking Websites in Poland in 2022

Witold Chmielarz and Marek Zborowski(✉)

University of Warsaw, Krakowskie Przedmiescie 26/28, 00-927 Warsaw, Poland
{witek,mzborowski}@wz.uw.edu.pl
http://wz.uw.edu.pl/en

Abstract. The primary objective of this article is to compare the customer ratings provided by Generation Z representatives regarding e-banking websites of the most popular banks in Poland and to examine the impact of the MCDA (Multiple Criteria Decision Analysis) methods on the objectivity of these ratings. The assessment involved a sample of over 200 clients (48% return rate) randomly selected from the academic community in Poland at the end of 2022 and over 150 people from the non academic circles. In both cases, the study was carried out using a survey questionnaire with the application of the CAWI technique. The survey questionnaire was obtained after verification of the pilot version created on the basis of previous research. The websites of 16 banks most familiar to respondents were tested using the MDCA methods including: TOPSIS, COMET, PROSA-C, PROMETHEE II, and two criterion weighting techniques (subjective and objective). Subsequently, the results obtained from each method were compared by considering their respective rankings. The study also investigated the sensitivity of the results in each bank to the development of 18 attributes, which served as criteria for evaluating the service. The convergence of the results was assessed using the Pearson correlation and the weighted Spearman correlation coefficient (r_w). To assess the discrepancies in the rankings for the selected methods, the Euclidean distances between the paired MCDA methods were calculated. The obtained results indicate the possibility of interchangeable use of the presented assessment methods, which may be helpful for business practitioners when analyzing and designing banking services.

Keywords: assessment of banking websites · multi-criteria methods · survey evaluation · internet banking

1 Introduction

The primary objective of this article is to compare the results of website evaluation methods used for the most popular online banks in Poland by the end of

M. Hernes and J. Wątróbski (Eds.): ECAI 2023, LNNS 1079, pp. 268–288, 2024.
https://doi.org/10.1007/978-3-031-66761-9_22

2022. The aim is to assess the similarities among the obtained scores and determine the specific positions of the websites in the rankings and the reasons behind them. The evaluation of banking websites is undoubtedly a complex decision-making problem. The number and variety of evaluation criteria (e.g. technical, financial attributes), often present challenges in making informed decisions. These criteria can be incomparable or even contradictory, further increasing the complexity involved in arriving at the most appropriate choice. In this situation, employing multi-criteria methods (MCDA) can prove beneficial in identifying compromise solutions [1–3]. In the course of the development of MCDA methods, a range of increasingly complex methods were devised. These methods varied in terms of the criteria employed, preference weights assigned to them, data aggregation techniques utilized, consideration of uncertainty phenomena, and diverse algorithms applied for calculating the final results [4,5]. The current research aims to identify MCDA methods that are appropriate for specific industries, phenomena, etc., acknowledging the absence of a universally applicable and reliable MCDA method [6]. It may also be that more sophisticated methods do not produce better results than simple methods, as shown in an earlier paper by the authors [7]. Therefore, more attention should be paid to the selection of the decision support method for the problem under consideration. The existence of alternative solutions obtained from multiple methods requires either additional preference scales, computational algorithms or practice-based approach to determine which method is best [8] and [9].

In order to accomplish the objective of this article and address the research problem, the authors used an approach based on comparing the results of the four MCDA methods. The set of methods applied included the scoring method and more sophisticated methods such as TOPSIS, VIKOR and COMET [10–12]. The authors utilized two techniques for determining the weighting criteria: the one applying equal weights and the one obtained by the entropy method.

The remaining sections of this article are structured as follows. The second section contains the introductory and main background of the TOPSIS, COMET, PROSA-C and PROMETHEE II methods described [13–16]. The third section presents the results of the study and a basic discussion of the topic. The final section provides conclusions, limitations of the study and directions for future research.

The article presents a case study, encompassing the evaluation of alternatives and their respective types. Section 4 shows the results of the assessment of alternatives and the influence of the methods used in the authors' approach on the scores. Section 5 contains the summary of the conducted survey and conclusions.

2 Preliminaries

2.1 Research Procedure

Years of extensive research on websites and banking applications [17–20] have guided the authors in adopting a well-established research procedure, whose stages are outlined below:

- a bibliographic analysis of recent developments in the field of website analysis and evaluation using MCDA methods, with a particular focus on banking websites and applications,
- construction of a pilot version of the survey questionnaire to verify the correctness and comprehensibility of the questions,
- development and adjustment of the first version of the survey questionnaire based on the evaluation, and then constructing the final version of the questionnaire, adopting an unambiguous attribute rating scale during data collection,
- random selection of groups of respondents and inviting them to complete the questionnaire using the CAWI (Computer Assisted Web Interview) technique,
- collection of completed questionnaires, data gathering and a preliminary verification of questionnaire completion accuracy,
- selection of a set of methods for the evaluation of banking services in order to make calculations and obtain results,
- comparative analysis and discussion of the results of calculations and their statistical analysis,
- drawing conclusions, discussing limitations and directions for further research.

2.2 Presentation of the Used Methods

Four of the most popular MCDA methods: TOPSIS, COMET, VIKOR, PROSA-C and PROMETHE II [21–23] were considered for the comparative study. A brief description of these methods is given below.

The TOPSIS Method. The TOPSIS method is made up of six stages [8]. Initially, the decision-maker (DM) describes the decision problem with n criteria and m alternatives. A decision matrix $D[x_ij]$ is then constructed, with rows representing the decision attributes of the alternatives, and columns representing the criteria (1):

$$D_[X_ij] = \begin{pmatrix} X_{11} \dots x_{1n} \\ \dots \dots \dots \\ x_{m1} \dots x_{mn} \end{pmatrix} \tag{1}$$

Subsequently, the decision matrix undergoes normalization in the second step, with the following formulae (2), (3):

$$r_{ij} = \frac{x_{ij} - min_i(x_{ij})}{max_i(x_{ij}) - min_i(x_{ij})} \tag{2}$$

$$r_{ij} = \frac{min_i(x_{ij} - x_{ij})}{max_i(x_{ij}) - min_i(x_{ij})} \tag{3}$$

for the benefit and cost criteria respectively.

In the next step, the weights are imposed on the normalized decision matrix, thus resulting in a weighted normalized decision matrix, whose each element is computed with the formula (4):

$$v_{ij} = w_j \cdot r_{ij} \tag{4}$$

In the fourth step, the positive (PIS) and negative (NIS) ideal solutions are obtained (V_j^+) (5) and (V_j^-) (6):

$$V_j^+ = v_1^+, v_2^+, ..., v_n^+ \tag{5}$$

$$V_j^+ = v_1^-, v_2^-, ..., v_n^- \tag{6}$$

The best alternative should be as close as possible to PIS and as far as possible from NIS. Therefore, the Euclidean distances between each alternative and PIS and NIS are computed in the fifth step. Finally, in the last step of the algorithm, relative closeness to the ideal solution is obtained (7):

$$CC_i = \frac{D_i^-}{D_i^- + D_i^+} \tag{7}$$

The obtained closeness coefficient CC_i is the score value produced by the TOPSIS method and is used to construct the ranking of alternatives [9].

The Characteristic of Objects METhod (COMET). The formal concept and notation of the COMET method can be presented by using the following five steps [24], where the fuzzy model is created in the result:

Step 1: Define the space of the problem as follows:
An expert determines the dimensionality of the problem by selecting number r of criteria, $C_1, C_2, ..., C_r$,. Subsequently, the set of triangular fuzzy numbers for each criterion C_i is selected, i.e., $\tilde{C}_{i1}, \tilde{C}_{i2}, ..., \tilde{C}_{ic_i}$. These numbers represent characteristic values in the considered domain. In this way, the following results are analyses (8):

$$\begin{array}{c} C_1 = \tilde{C}_{11}, \tilde{C}_{12}, ..., \tilde{C}_{1c_1} \\ C_2 = \tilde{C}_{21}, \tilde{C}_{22}, ..., \tilde{C}_{2c_2} \\ \cdots \\ C_r = \tilde{C}_{r1}, \tilde{C}_{r2}, ..., \tilde{C}_{rc_r} \end{array} \tag{8}$$

where $C_1, C_2, ..., C_r$ are numbers of the fuzzy numbers for all criteria.

Step 2: Generate the characteristic objects
The characteristic objects (CO) are obtained by using the Cartesian Product of triangular fuzzy numbers cores for each criterion as follows (9):

$$CO = C(C_1) \times C(C_2) \times ... \times C(C_r) \tag{9}$$

As the result of this, the ordered set of all (CO) is obtained (10):

$$\begin{array}{c} CO_1 = C(\tilde{C}_{11}), C(\tilde{C}_{21}), ..., C(\tilde{C}_{1c_1}) \\ CO_2 = C(\tilde{C}_{21}), C(\tilde{C}_{22}), ..., C(\tilde{C}_{2c_1}) \\ ... \\ CO_t = C(\tilde{C}_{1c_1}), C(\tilde{C}_{2c_2}), ..., C(\tilde{C}_{rc_r}) \end{array} \tag{10}$$

where t is the number of CO (11):

$$t = \Pi_{i=1}^{r} c_i \tag{11}$$

Step 3: Rank and evaluate the characteristic objects.
Determine the Matrix of Expert Judgment (MEJ). This is a result of the comparison of the characteristic objects by the knowledge of the expert. The MEJ structure is as follows (12):

$$MEJ = \begin{array}{c} \begin{pmatrix} \alpha_{11} & \alpha_{12} & ... & \alpha_{1t} \\ \alpha_{21} & \alpha_{22} & ... & \alpha_{2t} \\ ... & ... & ... & ... \\ \alpha_{t1} & \alpha_{t2} & ... & \alpha_{tt} \end{pmatrix} \begin{array}{c} CO_1 \\ CO_2 \\ ... \\ CO_t \end{array} \\ CO_1 \; CO_2 \; ... \; CO_t \end{array} \tag{12}$$

where α_{ij} is a result of comparing CO_i and CO_j by the expert. The more preferred characteristic object gets one point and the second object gets a null point. If the preferences are balanced, both objects get a half point. It depends solely on the knowledge and opinion of the expert and can be presented as (13):

$$\alpha_{ij} = f(CO_i, CO_j) = \begin{cases} 0.0, f_{exp}(CO_i) < f_{exp}(CO_j) \\ 0.5, f_{exp}(CO_i) = f_{exp}(CO_j) \\ 1.0, f_{exp}(CO_i) > f_{exp}(CO_j) \end{cases} \tag{13}$$

where F_{exp} is an expert judgment function. The most important properties are described by the formulas (14) and (15):

$$\alpha_{ii} = f(CO_i, CO_i) = 0,5 \tag{14}$$

$$\alpha_{ji} = 1 - \alpha_{ij} \tag{15}$$

On the basis of formulas (14) and (15), the number of comparisons is reduced from t^2 cases to p cases (16):

$$p = \binom{t}{2} = \frac{t(t-1)}{2} \tag{16}$$

Afterwards, we obtain a vertical vector of the summed Judgments (SJ) as follows (17):

$$SJ_i = \sum_{j=1}^{t} \alpha_{ij} \tag{17}$$

The last step assigns to each characteristic object the approximate value of the preference. In the result, we obtain a vertical vector P, where $i-th$ row contains the approximate value of preference for CO_i.

This algorithm is presented as a fragment of pseudo code:

```
1: k = length(unique(SJ));
2: P = zeros(t,1);
3: for i = 1:k
4:      ind = find(SJ == max(SJ));
5:      P(ind) = (k - i) / (k - 1);
6:      SJ(ind) = 0;
7: end
```

In line 1, we obtain number k as a number of unique value of the vector SJ. In line 2, we create vertical vector P of zeros (with identical size as vector SJ). In line 4, we obtain index with the maximum value of vector SJ. This index is used to assign the value of the preference to an adequate position in vector P (based on the principle of insufficient reason). In line 6, the maximum value of the vector SJ is reset.

Step 4: The rule base.
Each one characteristic object and value of preference is converted to a fuzzy rule as follows, general form (18) and detailed form (19):

$$IF\ CO_i\ THAN\ P_i \tag{18}$$

$$IF\ C(\tilde{C}1t)\ AND\ C(\tilde{C}_{2i})\ AND\ ...\ THAN\ P_i \tag{19}$$

In this way, the complete fuzzy rule base is obtained, which can be presented as (20):

$$\begin{array}{l} IF\ CO_1\ THAN\ P_1 \\ IF\ CO_2\ THAN\ P_2 \\ \dots\ \ \dots\ \ \dots\ \ \dots \\ IF\ CO_t\ THAN\ P_t \end{array} \tag{20}$$

Step 5: Inference in a fuzzy model and final ranking.
Each one alternative is a set of crisp number, which corresponding with criteria $C_1, C_2, ..., C_r$. It can be presented as follows (21):

$$A_i = \{\alpha_{1i}, \alpha_{2i}, ..., \alpha_{ri},\} \tag{21}$$

where condition (22) must be satisfied.

$$\begin{array}{l} \alpha_{1i} \in [C(\tilde{C}_{11}), C(\tilde{C}_{1c_1})] \\ \alpha_{2i} \in [C(\tilde{C}_{21}), C(\tilde{C}_{2c_2})] \\ \dots \\ \alpha_{ri} \in [C(\tilde{C}_{r1}), C(\tilde{C}_{rc_r})] \end{array} \tag{22}$$

Each one alternative activates the specified number of fuzzy rules, where for each one is determined the fulfilment degree of the conjunctive complex premise. Fulfilment degrees of all activated rules sum to one. The preference of alternative is computed as the sum of the product of all activated rules, as their fulfilment degrees, and their values of the preference. The final ranking of alternatives is obtained by sorting the preference of alternatives.

The VIKOR Method. The MCDM problem in VIKOR method is stated as follows: determine the best (compromise) solution in multicriteria sense from the set of j feasible alternatives $A1, A2, ..., AJ$, evaluated according to the set of n criterion functions. The input data are the elements f_{ij} of the performance (decision) matrix, where f_{ij} is the value of the $i-th$ criterion function for the alternative A_j. This method takes distance measurement into account, but in this approach, the goal is to identify the alternative closest to the ideal solution.

Therefore, the solution sought is a compromise solution [21]. The five steps of the VIKOR method are described below [22–24].

Step 1. Determinate the best f_j and the worst f_{-j} value for the function of a particular criterion. For profit criteria, the Equation is used (23).

$$F_J^* = \max_i f_{ij}, \quad F_J^- = \min_i f_{ij} \tag{23}$$

whereas in the case of the cost criteria, the following Equation is used (24).

$$F_J^* = \min_i f_{ij}, \quad F_J^- = \max_i f_{ij} \tag{24}$$

Step 2. Calculate Si and Ri with using Eqs. (25) and (26).

$$S_i = \sum_{j=1}^{n} w_j(f_j^* - f_{ij})/(f_j^* - f_j^-) \tag{25}$$

$$R_i = \max_j[w_j(f_j^* - f_{ij})/(f_j^* - f_j^-)] \tag{26}$$

Step 3. Calculate Qi with using Eq. (27).

$$Q_i = \frac{v(S_i - S^*)}{(S^- - S^*)} + (1 - v)\frac{(R_i - R^*)}{(R^- - R^*} \tag{27}$$

where:
$S^* = \min_i S_i$, $S^- = \max_i S_i$
$R^* = \min_i R_i$, $R^- = \max_i R_i$

And v - means the weight adopted for the strategy of "most criteria".
Step 4. Ranked alternatives S, R and Q are ordered in ascending order. Three ranked lists are the outcome.
Step 5. A compromise solution is proposed considering the conditions of good advantage and acceptable stability within the three vectors obtained in the previous step.

The best alternative is the one with the lowest value and the leading position in the ranking Q [25].

The PROSA Method. After determining the value of $\phi_{net}(a)$ and $\phi_j(a)$ for $j = 1...n$ the decision-maker can define the compensation of criteria for individual decision alternatives [26–28]:

1. $\phi_j(a) << \phi_{net}(a)$ means that for alternative a the performance of criterion j is compensated by other criteria (alternative a is not balanced with respect to criterion j),
2. $\phi_j(a) >> \phi_{net}(a)$ means that for alternative a the performance of criterion j compensates other criteria (alternative a is not balanced with respect to criterion j),
3. $\phi_j(a) \approx \phi_{net}(a)$ means that alternative a is balanced with respect to criterion j.

The operators $>>$, $<<$ denote the contractual relations "much greater than" and "much less than". These relations express the subjective view of the decision-maker as to whether the value on the right-hand side of the operator is much greater/much smaller than the value on the left-hand side, and therefore whether the alternative a is balanced in terms of criterion j or not.

In the next step, the value of the mean absolute deviation is determined in a weighted form, taking into account the balance (compensation) factor, according to the formula (28):

$$WMAD(a) = \sum_{j=1}^{n} \mid \phi_{net}(a) - \phi_j(a) \mid w_j s_j \tag{28}$$

where s_j is the equilibrium (compensation) factor for criterion j. It is easy to notice that $WMAD(a)$ is a specific weighted average distance of the solution $\phi_{net}(a)$ from the solutions $\phi_j(a)$ obtained for individual criteria.

The final assessment of the alternatives, i.e., PSV_{net} (PROSA Sustainable Value net), is calculated using the formula (29):

$$PSV_{net}(a) = \phi_{net}(a) - WMAD(a) \tag{29}$$

The PROMETHEE II Method. The second method applied in this study is Promethee II [29]. The Promethee II [30–32] method enables obtaining the full ranking of alternatives. In the case of the earlier version of this method, the resulting ranking is only partial [33,34]. Subsequently, after determining the compliance coefficients for each pair of variants, the dominance flows are determined for each of the variants:

- the leaving flow, describing how much the a_i variant outranks the other variants (30):

$$\phi^+(a_i) = \sum_{j=1}^{n} \pi(a_i, b_j) \tag{30}$$

- the entering flow, informing how much a_i variant is outranked by the other variants (31):

$$\phi^-(a_i) = \sum_{j=1}^{n} \pi(b_j, a_i) \tag{31}$$

Subsequently, the decision-maker can create a total ranking of the variants. In the Promethee II method [35,36] to create a full ranking of variants, the net dominance flow described by the formula (32) should be calculated:

$$\phi(a_i) = \phi^+(a_i) - \phi^-(a_i) \tag{32}$$

In the Promethee II method, the equivalence and preference relations in a broad sense are defined as follows [20]:

- a_i variant outranks b_j variant $(a_i L b_j)$, if $\phi(a_i) > \phi(b_j)$,
- a_i variant is equivalent to b_j variant $(a_i I b_j)$, if $\phi(a_i) = \phi(b_j)$.

Sample Characteristic. The website rankings used in this study were based on data collected in autumn 2022. A total of 738 people were surveyed, of which 356 people completed the survey fully and correctly, which amounted to a survey response rate of over 48%. A five-point, simplified, standardized Likert scale [37] was used to assess the individual criteria in the banks indicated by customers as those most frequently used by them:

- 1.00 - fully meets the criterion,
- 0.75 - almost fully meets the criterion,
- 0.50 - moderately meets the criterion,
- 0.25 - minimally meets the criterion,
- 0.00 - does not meet the criterion.

The original form of the questionnaire was validated on a pilot sample of 50 people as part of the study which was conducted in an academic environment. The individual criteria - attributes of the banking websites - were examined for their comprehensibility and relevance to the average website user. Following the verification, correction, and removal of the least significant criteria, the evaluation of each service incorporated 18 attributes (criteria) divided into three distinct groups: economic, technological, and anti-crisis criteria. A detailed list of attributes is provided in Table 1.

As a prerequisite for participation in the evaluation, respondents were requested to assess an e banking website they are familiar with and compare it to an e-banking service of another bank. This condition was established to ensure that responses were obtained from experienced participants who have interacted with various e-banking sites. Consequently, a total of 712 comprehensive evaluations of banking sites were obtained as a result.

Respondents rated sixteen (A1, A2, ..., A16) of the banking websites most frequently used by them. They were as follows: Alior Bank SA, Bank Handlowy w Warszawie SA, Bank Millenium SA, Bank Pocztowy SA, Bank Polska Kasa

Opieki (PKO SA), Bank Polskiej Spółdzielczości, BNP Paribas SA, Credit Agricole Bank Polska SA, Getin Noble Bank (now: VeloBank), ING - Bank Śląski SA, mBank SA, Nest Bank SA, PKO Bank Polski SA (PKO BP: Inteligo and IKE), Santander Bank Polska SA, Santander Consumer Bank SA and Toyota Bank Polska SA. All banking services with less than five ratings were excluded from this list, i.e. 16 banks' ratings were rejected.

The research sample was selected in a diverse manner: purposive - the research was conducted in an academic environment in randomly selected student groups and using a link to a survey posted on the Internet [38]. Therefore, the age range of the participants was between 19 and 35 years. Admittedly, this choice may have influenced the results of the survey (41 million people in Poland are potential customers of internet and mobile banking, more than 54% of the registered customers are active internet banking users and 44% active mobile banking users). The surveyed age group accounts for more than 65% of users) [39]. The sample included more than 70% female and nearly 30% male respondents. Secondary education was declared by 2% of respondents, bachelor's degree and incomplete higher education was indicated by 19% and secondary education was declared by 80%. The largest group of people came from large cities (over 200,000 inhabitants) and 19% came from rural areas. A quarter came from small medium-sized and large cities - up to 200,000 inhabitants. Among the respondents, there were 52% students, 31% working on a contract for specific work, a contract of mandate or being self employed and 17% working based on a contract of employment. The most common occupations were office workers (63%), service workers (16%), professionals (8%) and workers hired to perform simple technical tasks (7%). Most respondents describe their material situation as good (61%), very good (22%), average (16%) and adequate (2%).

In general, the evaluation data for banking services are relatively homogeneous and consistent. Once these were obtained, a reliability test in the form of Cronbach's alpha coefficient was applied. For all criteria (attributes), the Cronbach's alpha coefficient indicates internal consistency and the reliability of the sample was greater than 0.80. The internal consistency measure of the 16 dependent variables, based on Cronbach's alpha coefficient, was 0.85 (0.90 for Cronbach's alpha calculated from standardized items), for the 18 items in total.

The aim of this work is to investigate the impact of five different multi-criteria methods: TOPSIS, COMET, VIKOR, PROSA-C and PROMETHEE II on the evaluation results covering the websites of the sixteen most popular banks among Generation Z representatives in Poland at the end of 2022. The predetermined decision criteria were evaluated by users in a pilot sample in terms of preference weights. Establishing the significance of decision criteria is a crucial aspect in modelling decision-making problems. Various methods for obtaining the values of the criteria weights are presented in the literature [44]. Two objective criteria weighting methods were used in this study: equal weights and entropy weights (Table 1).

The Table 1 above presents the values of subjective weights (based on the calculated values of significance weights after normalization using the sum method

Table 1. Averaged value of relevance and significance weights of decision criteria: measured subjectively and randomly for individual attributes; Source: own work

Criterion	Attributes	Weight of significance	Subjective	Entropy
C1	Annual nominal interest rate on personal accounts	63.4242	0.0514	0.1320
C2	Account maintenance PLN/month	84.3820	0.0684	0.0712
C3	Fee for a transfer to a parent bank	77.2247	0.0626	0.0127
C4	Fee for a transfer to another bank	81.2219	0.0658	0.0199
C5	Direct debit	51.5702	0.0418	0.0074
C6	Fee for issuing a debit card	58.5674	0.0474	0.0146
C7	Fee for a card PLN/month	80.4916	0.0652	0.0249
C8	Additional services	53.7949	0.0436	0.0019
C9	Account access channels	75.6292	0.0613	0.0969
C10	Security	85.8146	0.0695	0.0441
C11	Visualisation	58.0028	0.0470	0.2078
C12	Navigation	65.7303	0.0532	0.0337
C13	Readability and ease of use	76.7978	0.0622	0.0803
C14	Range of functionalities	70.7107	0.0573	0.0538
C15	Interest rates on savings accounts	68.9860	0.0559	0.0563
C16	Interest rates on PLN 10,000 deposits	62.6517	0.0508	0.0371
C17	Interest rates on PLN 10,000 loans	58.3090	0.0472	0.0619
C18	Anti-crisis measures	61.1517	0.0495	0.0434

Table 2. Evaluation of bank websites - decision-making alternatives of choice; Source: own work

Criteria/Banks	A1	A2	A3	A4	A5	A6	A7	A8	A9	A10	A11	A12	A13	A14	A15	A16
C1	2.3232	2.3334	2.2926	2.3039	2.2977	2.2934	2.3007	2.3224	2.2895	2.2958	2.3011	2.3313	2.3012	2.2967	2.2804	2.2524
C2	2.3215	2.3273	2.3521	2.3095	2.3038	2.3046	2.3056	2.3183	2.2962	2.3078	2.3060	2.3352	2.3111	2.3077	2.2926	2.2837
C3	2.4327	2.4427	2.4258	2.4029	2.3958	2.3853	2.3954	2.4150	2.3851	2.3877	2.3987	2.4242	2.3972	2.3913	2.3740	2.3384
C4	2.7133	2.7056	2.7028	2.6589	2.6621	2.6485	2.6572	2.6703	2.6302	2.6457	2.6657	2.6991	2.6585	2.6604	2.6197	2.5863
C5	3.4635	3.4809	3.4405	3.4301	3.4316	3.4292	3.4325	3.4408	3.4207	3.4347	3.4344	3.4489	3.4392	3.4341	3.4254	3.4010
C6	3.9049	3.9138	3.9174	3.9199	3.9296	3.9204	3.9232	3.9295	3.9294	3.9253	3.9294	3.9231	3.9256	3.9226	3.9338	3.9318
C7	3.9948	3.9962	4.0080	4.0128	4.0124	4.0073	4.0121	4.0074	4.0262	4.0107	4.0152	4.0124	4.0122	4.0134	4.0279	4.0344
C8	4.1262	4.1244	4.0995	4.1845	4.1880	4.1928	4.1899	4.1808	4.1997	4.1941	4.1912	4.1686	4.1925	4.1952	4.2132	4.2311
C9	4.2043	4.2315	4.2136	4.2442	4.2410	4.2403	4.2414	4.2367	4.2573	4.2434	4.2427	4.2403	4.2423	4.2433	4.2634	4.2542
C10	4.2004	4.2084	4.2312	4.2437	4.2430	4.2496	4.2470	4.2304	4.2488	4.2511	4.2450	4.2268	4.2459	4.2482	4.2578	4.2859
C11	4.2165	4.2149	4.2805	4.3731	4.3707	4.3867	4.3815	4.3551	4.3817	4.3869	4.3754	4.3500	4.3793	4.3828	4.4045	4.4487
C12	4.3422	4.3609	4.3131	4.3766	4.3742	4.3783	4.3759	4.3895	4.3912	4.3771	4.3781	4.3780	4.3773	4.3792	4.4008	4.4029
C13	4.3865	4.3912	4.4224	4.4624	4.4613	4.4668	4.4653	4.4593	4.4710	4.4712	4.4626	4.4623	4.4668	4.4679	4.4850	4.4996
C14	4.4419	4.4431	4.4723	4.5131	4.5204	4.5331	4.5271	4.4996	4.5203	4.5328	4.5242	4.4850	4.5275	4.5316	4.5444	4.5822
C15	4.4600	4.4578	4.5271	4.5341	4.5460	4.5434	4.5430	4.5296	4.5411	4.5423	4.5477	4.5363	4.5439	4.5480	4.5592	4.5912
C16	4.5599	4.5679	4.5557	4.6092	4.6075	4.6119	4.6091	4.6107	4.6193	4.6102	4.6096	4.6048	4.6088	4.6136	4.6276	4.6222
C17	4.5331	4.5381	4.5666	4.6755	4.6817	4.6926	4.6842	4.6853	4.6819	4.6899	4.6840	4.6797	4.6877	4.6941	4.7104	4.7158
C18	4.9163	4.9136	4.8832	4.9483	4.9469	4.9479	4.9444	4.9576	4.9559	4.9466	4.9453	4.9525	4.9473	4.9520	4.9647	4.9627

- they should add up to 1, according to MCDA rules) and objective weights (determined by the Entropy method).

On the basis of the collected data, a decision matrix table was created for alternatives d A1, ..., A16, concerning the evaluation of the online services of individual banks (Table 2). It contains the values of each criterion for the evaluated alternatives.

Table 3. Standardized values of decision alternatives; Source: own work

Criteria/Banks	A1	A2	A3	A4	A5	A6	A7	A8	A9	A10	A11	A12	A13	A14	A15	A16
C1	0.4646	0.4667	0.4585	0.4608	0.4595	0.4587	0.4601	0.4645	0.4579	0.4592	0.4602	0.4663	0.4602	0.4593	0.4561	0.4505
C2	0.4643	0.4655	0.4704	0.4619	0.4608	0.4609	0.4611	0.4637	0.4592	0.4616	0.4612	0.4670	0.4622	0.4615	0.4585	0.4567
C3	0.4865	0.4885	0.4852	0.4806	0.4792	0.4771	0.4791	0.4830	0.4770	0.4775	0.4797	0.4848	0.4794	0.4783	0.4748	0.4677
C4	0.5427	0.5411	0.5406	0.5318	0.5324	0.5297	0.5314	0.5341	0.5260	0.5291	0.5331	0.5398	0.5317	0.5321	0.5239	0.5173
C5	0.6927	0.6962	0.6881	0.6860	0.6863	0.6858	0.6865	0.6882	0.6841	0.6869	0.6869	0.6898	0.6878	0.6868	0.6851	0.6802
C6	0.7810	0.7828	0.7835	0.7840	0.7859	0.7841	0.7846	0.7859	0.7859	0.7851	0.7859	0.7846	0.7851	0.7845	0.7868	0.7864
C7	0.7990	0.7992	0.8016	0.8026	0.8025	0.8015	0.8024	0.8015	0.8052	0.8021	0.8030	0.8025	0.8024	0.8027	0.8056	0.8069
C8	0.8252	0.8249	0.8199	0.8369	0.8376	0.8386	0.8380	0.8362	0.8399	0.8388	0.8382	0.8337	0.8385	0.8390	0.8426	0.8462
C9	0.8409	0.8463	0.8427	0.8488	0.8482	0.8481	0.8483	0.8473	0.8515	0.8487	0.8485	0.8481	0.8485	0.8487	0.8527	0.8508
C10	0.8401	0.8417	0.8462	0.8487	0.8486	0.8499	0.8494	0.8461	0.8498	0.8502	0.8490	0.8454	0.8492	0.8496	0.8516	0.8572
C11	0.8433	0.8430	0.8561	0.8746	0.8741	0.8773	0.8763	0.8710	0.8763	0.8774	0.8751	0.8700	0.8759	0.8766	0.8809	0.8897
C12	0.8684	0.8722	0.8626	0.8753	0.8748	0.8757	0.8752	0.8779	0.8782	0.8754	0.8756	0.8756	0.8755	0.8758	0.8802	0.8806
C13	0.8773	0.8782	0.8845	0.8925	0.8923	0.8934	0.8931	0.8919	0.8942	0.8942	0.8925	0.8925	0.8934	0.8936	0.8970	0.8999
C14	0.8884	0.8886	0.8945	0.9026	0.9041	0.9066	0.9054	0.8999	0.9041	0.9066	0.9048	0.8970	0.9055	0.9063	0.9089	0.9164
C15	0.8920	0.8916	0.9054	0.9068	0.9092	0.9087	0.9086	0.9059	0.9082	0.9085	0.9095	0.9073	0.9088	0.9096	0.9118	0.9182
C16	0.9120	0.9136	0.9111	0.9218	0.9215	0.9224	0.9218	0.9221	0.9239	0.9220	0.9219	0.9210	0.9218	0.9227	0.9255	0.9244
C17	0.9066	0.9076	0.9133	0.9351	0.9363	0.9385	0.9368	0.9371	0.9364	0.9380	0.9368	0.9359	0.9375	0.9388	0.9421	0.9432
C18	0.9833	0.9827	0.9766	0.9897	0.9894	0.9896	0.9889	0.9915	0.9912	0.9893	0.9891	0.9905	0.9895	0.9904	0.9929	0.9925

Subsequently, the table was normalized using the minimum-maximum normalization method for each weighting technique, adhering to the multi-criteria methods (Table 3). For the TOPSIS and COMET methods, the best alternative is the alternative with the highest preference value. Therefore, the lower the preference value, the lower the alternative is ranked. For the VIKOR method, the situation is reversed: the best alternative is the alternative for which the lowest preference value was calculated. Therefore, as the preference value increases, the alternative decreases in ranking. Correlations were then counted for the calculations with subjective and objective weights.

$$r_{xy} = \frac{\sum_{i=1}^{N}(x_i - \bar{x})(y_i - \bar{y})}{\sqrt{\sum_{i=1}^{N}(x_i - \bar{x})^2}\sqrt{\sum_{i=1}^{N}(y_i - \bar{y})^2}} \tag{33}$$

Weighted Spearman correlation coefficient (r_w). Formula for Weighted Spearman correlation coefficient:

$$r_w = 1 - \frac{\sigma \sum_{i=1}^{N}(x_i - y_i)^2(N - x_i + 1) + (N - y_i + 1)}{N^4 + N^2 - N^2 - N} \tag{34}$$

2.3 Calculation Results and Rankings for Subjective Weights

The calculations for subjective weights are presented as follows:

- the preference function values calculated in this way and the rankings resulting from these values are shown in Table 4,
- followed by a graphical interpretation of these results in Fig. 1 and
- the results of the calculations for Pearson's correlation coefficients and Spearman's weighted correlation coefficient (r_w) are presented in Fig. 2.

Table 4. Preference function values and rankings for calculations with subjective weights; Source: own work

Criteria/methods	Preference values					Rankings				
	TOPSIS	COMET	VIKOR	PROSA-C	PROMETHEE II	TOPSIS	COMET	VIKOR	PROSA-C	PROMETHEE II
A1	0.5414	0.606	0.3317	−0.3887	−0.2133	14	14	14	14	14
A2	0.4753	0.4259	0.6576	−0.5322	−0.4041	16	16	16	16	16
A3	0.7157	0.9307	0.0737	−0.0292	0.0989	2	4	2	8	8
A4	0.6888	0.9319	0.1756	−0.0012	0.1675	8	3	12	7	6
A5	0.6593	0.986	0.3266	0.244	0.4646	13	1	13	2	2
A6	0.5032	0.5084	0.5864	−0.5147	−0.3352	15	15	15	15	15
A7	0.7178	0.9264	0.0354	−0.0812	0.0383	1	5	1	10	10
A8	0.7059	0.9771	0.1282	0.27	0.4699	3	2	7	1	1
A9	0.7018	0.9155	0.1174	0.099	0.1494	4	7	4	5	7
A10	0.6824	0.891	0.1585	−0.1113	−0.0428	11	13	10	12	12
A11	0.6875	0.8988	0.1441	−0.1555	−0.0906	10	11	9	13	13
A12	0.7006	0.9233	0.1185	0.2366	0.3321	6	6	5	3	3
A13	0.7016	0.9154	0.0968	0.1201	0.17	5	8	3	4	5
A14	0.6893	0.9103	0.125	0.0686	0.1838	7	9	6	6	4
A15	0.6876	0.9002	0.1414	−0.0988	−0.0336	9	10	8	11	11
A16	0.6768	0.8981	0.1636	−0.0456	0.0934	12	12	11	9	9

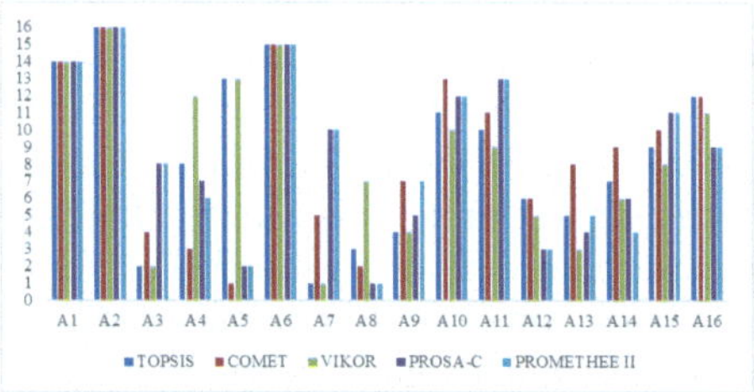

Fig. 1. Comparing rankings obtained for TOPSIS, COMET, VIKOR, PROSA-C and PROMETHEE II with equal weights (source: own work)

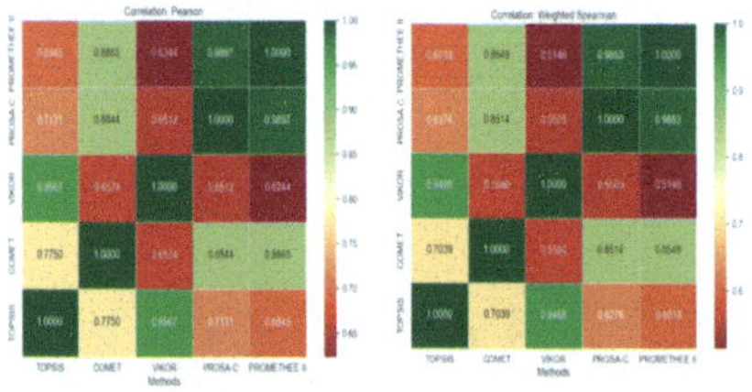

Fig. 2. WS and r_w correlation heat maps for TOPSIS, COMET, VIKOR, PROSA-C and PROMETHEE II with equal weights (source: own work)

The preference values for TOPSIS, COMET, VIKOR, PROSA-C and PROMETHEE II using equal weights are presented in Table 4. A comparison of the rankings is shown in Fig. 1). The same rankings for the five MCDA methods were obtained for only three alternatives A1 (Alior Bank SA), A6 (Bank Polskiej Spółdzielczości), A8 (Credit Agricole Bank Polska SA) when a weighted average was applied. All of them appear in the last positions of the ranking,

Table 5. Euclidean distances between pairs of positions in the ranking of the analyzed MCDA methods for equal weights; Source: own work

Banks/Euclidean distance	Topsis-Comet	Topsis-Vikor	Topsis-ProseC	Topsis-Promethee II	Comet-Vikor	Comet-ProseC	Comet-Promethee II	Vikor-ProseC	Vikor-Promethee II	ProseC-Promethee II	Total
A1	0	0	0	0	0	0	0	0	0	0	0
A2	0	0	0	0	0	0	0	0	0	0	0
A3	4	0	36	36	4	16	16	36	36	0	184
A4	25	16	1	4	81	16	9	25	36	1	214
A5	144	0	121	121	144	1	1	121	121	0	774
A6	0	0	0	0	0	0	0	0	0	0	0
A7	16	0	81	81	16	25	25	81	81	0	406
A8	1	16	4	4	25	1	1	36	36	0	124
A9	9	0	1	9	9	4	0	1	9	4	46
A10	4	1	1	1	9	1	1	4	4	0	26
A11	1	1	9	9	4	4	4	16	16	0	64
A12	0	1	9	9	1	9	9	4	4	0	46
A13	9	4	1	0	25	16	9	1	4	1	70
A14	4	1	1	9	9	9	25	0	4	4	66
A15	1	1	4	4	4	1	1	9	9	0	34
A16	0	1	9	9	1	9	9	4	4	0	46

in 16th, 15th and 14th place respectively. If we consider the average ranking from all results, the banks in first place are: Credit Agricole Bank Polska SA, Nest Bank SA and Bank Millenium SA. However, the highest differentiation (144) of results (calculated with the Euclidean distance) is found between the calculations obtained with the VIKOR and COMET as well as with TOPSIS and COMET method. The highest differentiations, calculated as the sum of the Euclidean distance of differences between the results of the analyzed methods, occur in the assessment of Bank Polska Kasa Opieki (774), BNP Paribas SA (406) and Bank Pocztowy SA (214).

The lowest - in the case of Bank Polskiej Spółdzielczości (0) and Alior Bank SA and Bank Handlowy w Warszawie SA (6 each). At the same time, these are the banks occupying the last positions in the ranking. On the other hand, the best banks, with the smallest differentiation, include Nest Bank SA and PKO Bank Polski SA. Two ranking similarity coefficients were used to examine the correlation: WS and rw. The highest value of the asymmetric WS coefficient (0.9858) was recorded for the PROMETHEE II - PROSA-C pair, followed by VIKOR-TOPSIS (0.9468). The lowest correlation was observed for VIKOR and PROMETHEE II (0.5146) and VIKOR and PROSA C (0.5505) and VIKOR and COMET (0.5580).

When examining the similarity of rankings using the rw coefficient, the highest correlation was found for PROSA-C and PROMETHEE II (0.9897) and VIKOR and TOPSIS (0.9567), and the lowest for VIKOR and PROMETHEE II (0.6244), VIKOR and PROSA-C (0.6512) and VIKOR-COMET (0.6574). Thus, the results of the ranking similarity test confirm the outliers in the ranking obtained by the VIKOR method, except for the relationship of this method with the TOPSIS method, where the highest value was obtained, for both coefficients.correlation was found for PROSA-C and PROMETHEE II (0.9897) and VIKOR and TOPSIS (0.9567), and the lowest for VIKOR and PROMETHEE II (0.6244), VIKOR and PROSA-C (0.6512) and VIKOR-COMET (0.6574). Thus, the results of the ranking similarity test confirm the outliers in the ranking obtained by the VIKOR method, except for the relationship of this method with the TOPSIS method, where the highest value was obtained, for both coefficients.

2.4 Calculation Results and Rankings for Objective Weights

The objective weights determined by the Entropy method were calculated using the following formulae:

- normalization of the decision matrix xij by the sum normalization method (35)

$$r_{ij} = \frac{x_{ij}}{\sum_{i=1}^{m} x_{ij}} \tag{35}$$

- calculation of the entropy value for each criterion (36):

$$E_j = -\frac{\sum_{i=1}^{m} p_{ij} ln p_{ij}}{ln m} \tag{36}$$

- hence the value of dj (37):

$$d_j = 1 - E_j \tag{37}$$

- determination of entropy weights wj for each criterion (38):

$$w_j = \frac{d_j}{\sum_{j=1}^{n} d_j} \tag{38}$$

The presentation of the results was carried out on a similar basis as for the case of subjective weights. Table 6 shows the values of the preference function for this case and the places in the ranking resulting from these values, Fig. 3 shows a graphical interpretation of these results, and Fig. 4 shows the results of calculations for Pearson's correlation coefficients and Spearman's weighted correlation coefficient (r_w). In addition, in order to determine the differentiation of the places in the ranking, the Euclidean distances shown in Table 7 were calculated for all combinations of the methods used.

Table 6. Preference function values and rankings for calculations with objective weights; Source: own work

Criteria/methods	Preference values					Rankings				
	TOPSIS	COMET	VIKOR	PROSA-C	PROMETHEE II	TOPSIS	COMET	VIKOR	PROSA-C	PROMETHEE II
A1	0.4934	0.4969	0.412	−0.3371	−0.1453	15	15	15	14	14
A2	0.5245	0.535	0.3418	−0.396	−0.2443	14	14	14	15	15
A3	0.6648	0.7081	0.079	−0.0914	0.0169	6	12	2	9	9
A4	0.632	0.6899	0.2048	−0.179	−0.0303	12	13	12	13	11
A5	0.6296	0.7797	0.2634	0.077	0.3382	13	1	13	6	2
A6	0.4362	0.4143	0.5184	−0.4953	−0.2834	16	16	16	16	16
A7	0.6919	0.7336	0	−0.1382	−0.0114	1	5	1	10	10
A8	0.6489	0.7441	0.1855	0.0862	0.3168	11	2	11	4	3
A9	0.6771	0.7312	0.09	0.0861	0.1237	3	6	3	5	7
A10	0.6605	0.7095	0.1242	−0.1544	−0.1045	10	11	8	11	12
A11	0.6631	0.7132	0.1163	−0.1614	−0.1087	9	10	7	12	13
A12	0.6806	0.744	0.0935	0.24	0.34	2	3	4	1	1
A13	0.6758	0.7342	0.1027	0.1183	0.1706	4	4	5	2	5
A14	0.6639	0.7294	0.142	0.0985	0.238	8	8	10	3	4
A15	0.6709	0.729	0.116	−0.0584	0.0171	5	9	6	8	8
A16	0.6645	0.73	0.1402	0.0205	0.1703	7	7	9	7	6

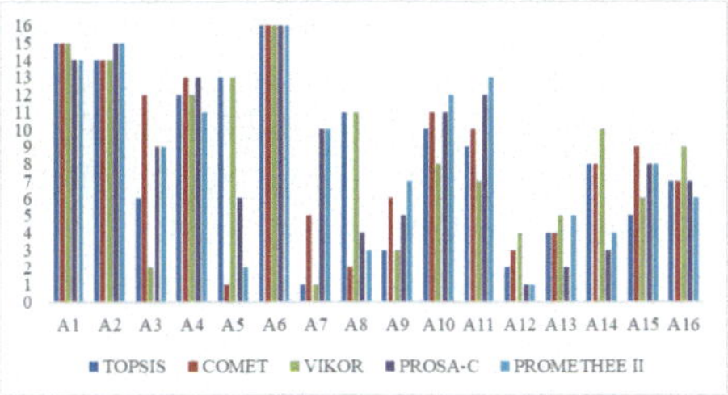

Fig. 3. Comparison of the rankings obtained for TOPSIS, COMET, VIKOR, PROSA-C and PROMETHEE II with entropy weights (source: own work)

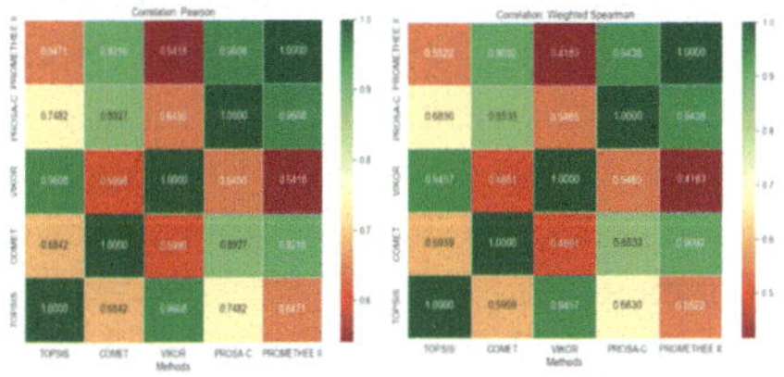

Fig. 4. WS and r_w correlation heat maps for TOPSIS, COMET, VIKOR, PROSA-C and PROMETHEE II with entropy weights (source: own work)

Table 7. Euclidean distances between pairs of places in the ranking of the MCDA methods analyzed for entropy weights; Source: own work

Banks/Euclidean distance	Topsis-Comet	Topsis-Vikor	Topsis-ProseC	Topsis-Promethee II	Comet-Vikor	Comet-ProseC	Comet-Promethee II	Vikor-ProseC	Vikor-Promethee II	ProseC-Promethee II	Total
A1	0	0	1	1	0	1	1	1	1	0	6
A2	0	0	1	1	0	1	1	1	1	0	6
A3	36	16	9	9	100	9	9	49	49	0	286
A4	1	0	1	1	1	0	4	1	1	4	14
A5	144	0	49	121	144	25	1	49	121	16	670
A6	0	0	0	0	0	0	0	0	0	0	0
A7	16	0	81	81	16	25	25	81	81	0	406
A8	81	0	49	64	81	4	1	49	64	1	394
A9	9	0	4	16	9	1	1	4	16	4	64
A10	1	4	1	4	9	0	1	9	16	1	46
A11	1	4	9	16	9	4	9	25	36	1	114
A12	1	4	1	1	1	4	4	9	9	0	34
A13	0	1	4	1	1	4	1	9	0	9	30
A14	0	4	25	16	4	25	16	49	36	1	176
A15	16	1	9	9	9	1	1	4	4	0	54
A16	0	4	0	1	4	0	1	4	9	1	24

Preference values for TOPSIS, COMET, VIKOR, PROSA-C and PROMETHEE II using objective weights are included in Table 5. A comparison of the rankings is shown in Fig. 4. The same ranking for the five MCDA methods was obtained for only one alternative, i.e. A6 (Bank Polskiej Spółdzielczości). It occupies the 16th position, indicating its placement at the bottom of the ranking. If we consider the average ranking obtained from all results, the banks in the first position were: PKO Bank Polski SA, Getin Noble Bank (now Velo Bank) and BNP Paribas SA.

However, the highest differentiation (144) of results (calculated with the Euclidean distance) occurs between the calculations obtained with the VIKOR method and COMET as well as with TOPSIS and COMET. The highest differentiations, calculated as the sum of Euclidean distances of differences between the results of the analyzed methods, occur in the assessment of Bank Polska Kasa Opieki (670), BNP Paribas SA (406) and Credit Agricole Bank Polska SA (394). The lowest - in the case of Nest Bank SA (2.2), PKO Bank Polski SA (4.0) and Getin Noble Bank (now Velo Bank - 4.8).

To test the correlation, two ranking similarity coefficients were used, as in the previous case: WS and rw. The highest value of the asymmetric WS coefficient (0.9457) was observed for the VIKOR TOPSIS pair. The lowest correlation was observed for VIKOR and COMET (0.4861) and VIKOR and PROSA-C (0.5465).

When examining the similarity of the rankings using the rw coefficient, the highest correlation was found for PROSA-C and PROMETHEE II (0.9608), VIKOR and TOPSIS (0.9608), and the lowest for VIKOR and PROMETHEE II (0.5418), VIKOR and PROSA-C (0.6450) and VIKOR COMET (0.5906). Thus, as in the previous case, the results of the ranking similarity test confirm the outliers in the ranking obtained by the VIKOR method, except for the relationship of this method with the TOPSIS method, where the highest value was obtained for both coefficients.

Analysis of Table 5 and Table 7 (for subjective and objective weights) allows us to conclude that the greatest differences between the obtained rankings were observed for the VIKOR method (similar to the study: [24]). For the method applied with subjective weights, this is particularly true for A5, A3 and A4. For the method utilized with objective weights: for A5, A7 and A8.

The results of the research carried out indicate the high complexity of multi-criteria decision making problems. The complexity involved in evaluating e-banking services and m-banking applications poses a significant challenge as regards identifying a universally applicable method for determining the best choice. The authors' first study related to this research problem concerned the possibility of replacing the results of complex multi-criteria methods with simple methods [45]. The set of five methods currently analyzed showed a high convergence of the results obtained by the four methods, apart from the results received in the case of the VIKOR method. The authors' own method, the Conversion method, initially appeared to be an optimal solution, particularly in terms of data acquisition. However, the authors faced a similar situation where the results diverged, which prompted them to reevaluate its effectiveness and investigate the research problem further [46]. However, later analyses showed that, as for the VIKOR method, the obtained results of the Conversion method deviated from the other MCDAs analyzed [47]. Nevertheless, the hybrid approach used in this work, where data and first analyses were collected for the scoring method and then verified with the Conversion method, seemed reasonable. The hybrid approach used is worthy of attention and consideration in further research. Moreover, the significant convergence observed among the results of the four analyzed methods suggests their interchangeability.

3 Conclusion

This article aims to assess the websites of the most popular banks among Generation Z student representatives. Additionally, it seeks to examine how the application of various MCDA methods and weighting techniques influences the objectivity of the resulting rankings. The case study in this article was related to the research problem of selecting the best e-banking service for a potential customer.

The results obtained strongly support the conclusion that specific conditions must be satisfied to ensure an effective evaluation of a multi-criteria problem. They are as follows:

- firstly, it is necessary to select evaluation criteria (attributes) that characterize the most relevant features of the analyzed issue from the user's point of view,
- secondly, the selection of a method that will guarantee that the collected data will be properly used,
- thirdly, the method of comparison of the obtained results,
- it is also recommended to choose appropriate weights for the criteria reflecting the preferences of the decision-maker.

The present study demonstrates that the TOPSIS and COMET methods yielded the most similar results in the rankings, whereas the VIKOR method exhibited the largest disparities compared to the results obtained from other methods. Therefore, making definitive assessments about the optimality or objectivity of the employed methods based solely on this premise proves challenging. However, in light of the obtained results and the insights from previous research, the authors hold the belief that by strategically selecting and integrating multiple MCDA methods, and conducting a thorough comparative analysis, it is possible to determine a set of methods producing convergent and optimal results for a particular industry.

This work encountered two fundamental barriers: firstly, the data collection was limited to academic representatives of Generation Z, who are perceived as the main future users of banking services, and secondly, only a restricted number of the five most popular MCDA methods were employed.

The obtained results motivate further research into additional MCDA methods to expand the range of methods that facilitate the convergence of results and enhance the objectivity of assessments in the addressed problem. This direction is expected to dominate the authors' future research in this field.

References

1. Kizielewicz, B., Wątróbski, J., Sałabun, W.: Identification of Relevant Criteria Set in the MCDA Process-Wind Farm
2. Greco, S.: A new PCCA method: IDRA. Eur. J. Oper. Res. **98**, 587–601 (1997). https://doi.org/10.1016/S0377-2217(96)00022-7
3. Voogd, H.: Multicriteria evaluation with mixed qualitative and quantitative data. Environ. Plan. B. **9**, 221–236 (1982)

4. Beaudrie, C., Corbett, J.C., Lewandowski, T.A., Malloy, T., Zhou, X.: Evaluating the application of decision analysis methods in simulated alternatives assessment case studies: potential benefits and challenges of using MCDA. Integr. Environ. Assess. Manag. **17**, 27–41 (2020). https://doi.org/10.1002/ieam.4316
5. Fortemps, P., Greco, S., Słowiński, R.: Multicriteria choice and ranking using decision rules induced from rough approximation of graded preference relations. In: Tsumoto, S., Słowiński, R., Komorowski, J., Grzymała-Busse, J.W. (eds.) RSCTC 2004. LNCS, pp. 510–522. Springer, Berlin, Heidelberg (2004). https://doi.org/10.1007/978-3-540-25929-9_62
6. Tsotsolas, N., Alexopoulos, S.: MCDA approaches for efficient strategic decision making. Presented at the May 30 (2018)
7. Chmielarz, W., Zborowski, M.: Analysis of e-banking websites' quality with the application of the TOPSIS method - a practical study. Procedia Comput. Sci. **126**, 1964–1976 (2018). https://doi.org/10.1016/j.procs.2018.07.256
8. Wątróbski, J., Jankowski, J., Ziemba, P., Karczmarczyk, A., Zioło, M.: Generalised framework for multi-criteria method selection. Omega **86**, 107–124 (2019). https://doi.org/10.1016/j.omega.2018.07.004
9. Papapostolou, A., Mexis, F.D., Sarmas, E., Karakosta, C., Psarras, J.: Web-based application for screening energy efficiency investments: a MCDA approach. In: 2020 11th International Conference on Information, Intelligence, Systems and Applications IISA, pp. 1–7 (2020)
10. Salabun, W., Watrobski, J., Piegat, A.: Identification of a multi-criteria model of location assessment for renewable energy sources. In: Rutkowski, L., Korytkowski, M., Scherer, R., Tadeusiewicz, R., Zadeh, L., Zurada, J. (eds.) ICAISC 2016. LNCS, vol. 9692, pp. 321–332. Springer, Cham (2016). https://doi.org/10.1007/978-3-319-39378-0_28
11. Cinelli, M., Kadziński, M., Gonzalez, M., Słowiński, R.: How to support the application of multiple criteria decision analysis? Let us start with a comprehensive taxonomy. Omega **96**, 240–261 (2020). https://doi.org/10.1016/j.omega.2020.102261
12. Duckstein, L., Opricovic, S.: Multiobjective optimization in river basin development. Water Resour. Res. **16**, 14–20 (1980). https://doi.org/10.1029/WR016i001p00014
13. Sałabun, W., Piegat, A.: Comparative analysis of MCDM methods for the assessment of mortality in patients with acute coronary syndrome. Artif Intell. Rev. **48**, 557–571 (2017). https://doi.org/10.1007/s10462-016-9511-9
14. Sałabun, W.: The characteristic objects method: a new distance-based approach to multicriteria decision-making problems. J. Multi-Criteria Decis. Anal. **22**, 37–50 (2015). https://doi.org/10.1002/mcda.1525
15. Jankowski, J., Sałabun, W., Wątróbski, J.: Identification of a multi-criteria assessment model of relation between editorial and commercial content in web systems. Presented at the January 1 (2017)
16. Hwang, C.-L., Yoon, K.: Methods for multiple attribute decision making. In: Hwang, C.-L., Yoon, K. (eds.) Multiple Attribute Decision Making: Methods and Applications A State-of-the-Art Survey, pp. 58–191. Springer, Berlin, Heidelberg (1981). https://doi.org/10.1007/978-3-642-48318-9_3
17. Chmielarz, W., Zborowski, M.: Comparative analysis of electronic banking websites in selected banks in Poland in 2014. Presented at the October 11 (2015)
18. Chmielarz, W., Zborowski, M.: On a comparative analysis of internet banking services by the conversion method and the scoring method. J. Econ. Manag. **38**, 26–45 (2019). https://doi.org/10.22367/jem.2019.38.02

19. Chmielarz, W., Zborowski, M.: A hybrid method of assessing individual electronic banking services in 20190. The Case of Poland. Procedia Comput. Sci. **176**, 3881–3889 (2020)
20. Chmielarz, W., Zborowski, M.: Towards VES function for creating a sustainable method for evaluating e-banking websites quality. Procedia Comput. Sci. **192**, 5139–5155 (2021). https://doi.org/10.1016/j.procs.2021.09.292
21. Piegat, A., Sałabun, W.: Comparative Analysis of MCDM Methods for Assessing the Severity of Chronic Liver Disease (2015)
22. Dhurkari, R.K.: MCGL: a new reference dependent MCDM method. Int. J. Oper. Res. **36**, 477–495 (2019). https://doi.org/10.1504/IJOR.2019.104053
23. Martel, J.-M., Matarazzo, B.: Other outranking approaches. In: Figueira, J., Greco, S., Ehrogott, M. (eds.) Multiple Criteria Decision Analysis: State of the Art Surveys, pp. 197–259. Springer, New York, NY (2005). https://doi.org/10.1007/0-387-23081-5_6
24. Sałabun, W.: The characteristic objects method: a new approach to identify a multi-criteria group decision-making model. Int. J. Comput. Technol. Appl. 2229-6093 **5**, 1597–1602 (2014)
25. Sałabun, W., Wątróbski, J., Shekhovtsov, A.: Are MCDA methods benchmarkable? A comparative study of TOPSIS, VIKOR, COPRAS, and PROMETHEE II methods. Symmetry **12**, 1–55 (2020). https://doi.org/10.3390/sym12091549
26. Imandasari, T., Sadewo, M., Windarto, A., Wanto, A., Wijaya, H., Kurniawan, R.: Analysis of the selection factor of online transportation in the VIKOR method in Pematangsiantar city. In: Journal of Physics: Conference Series, vol. 1255, p. 012008 (2019). https://doi.org/10.1088/1742-6596/1255/1/012008
27. Mardani, A., Zavadskas, E., Govindan, K., Amat Senin, A., Jusoh, A.: VIKOR technique: a systematic review of the state of the art literature on methodologies and applications. Sustainability **8**, 37 (2016). https://doi.org/10.3390/su8010037
28. Baczkiewicz, A., Watróbski, J., Kizielewicz, B., Sałabun, W.: Towards objectification of multi-criteria assessments: a comparative study on MCDA methods. Presented at the September 26 (2021)
29. Kumar, M., Samuel, C.: Selection of best renewable energy source by using VIKOR method. Technol. Econ. Smart Grids Sustain. Energy **2**, 8 (2017). https://doi.org/10.1007/s40866-017-0024-7
30. Ziemba, P., Watróbski, J., Ziolo, M., Karczmarczyk, A.: Using the PROSA method in offshore wind farm location problems. Energies **10**, 1755 (2017). https://doi.org/10.3390/en10111755
31. Download citation of Using the PROSA Method in Offshore Wind Farm Location Problems. https://www.researchgate.net/publication/320789134_Using_the_PROSA_Method_in_Offshore_Wind_Farm_Location_Problems
32. Ziemba, P.: Towards strong sustainability management—a generalized PROSA method. Sustainability **11**, 1555 (2019). https://doi.org/10.3390/su11061555
33. Qiao, D., Shen, K., Wang, J., Wang, T.: Multi-criteria PROMETHEE method based on possibility degree with Z-numbers under uncertain linguistic environment. J. Ambient Intell. Hum. Comput. **11**, 2187 (2020). https://doi.org/10.1007/s12652-019-01251-z
34. Ostovare, M., Shahraki, M.R.: Evaluation of hotel websites using the multicriteria analysis of PROMETHEE and GAIA: evidence from the five-star hotels of Mashhad. Tourism Manag. Perspect. **30**, 107–116 (2019). https://doi.org/10.1016/j.tmp.2019.02.013

35. Andreopoulou, Z., Koliouska, C., Galariotis, E., Zopounidis, C.: Renewable energy sources: using PROMETHEE II for ranking websites to support market opportunities. Technol. Forecast. Soc. Chang. **131**, 31–37 (2018). https://doi.org/10.1016/j.techfore.2017.06.007
36. Kuang, H., Kilgour, D.M., Hipel, K.W.: Grey-based PROMETHEE II with application to evaluation of source water protection strategies. Inf. Sci. **294**, 376–389 (2015). https://doi.org/10.1016/j.ins.2014.09.035
37. Starfield, T.: Simple multi-attribute ranking technique smart (2005). http://www.uvm.edu/~tdonovan/modeling/Module12/12_SMART_transcript.pdf
38. Shekhovtsov, A., Kołodziejczyk, J., Sałabun, W.: Fuzzy model identification using monolithic and structured approaches in decision problems with partially incomplete data. Symmetry **12**, 1541 (2020). https://doi.org/10.3390/sym12091541
39. Brans, J.P., Vincke, P., Mareschal, B.: How to select and how to rank projects: the PROMETHEE method. Eur. J. Oper. Res. **24**, 228–238 (1986). https://doi.org/10.1016/0377-2217(86)90044-5
40. Brans, J.-P.: L'ingénierie de la décision: l'élaboration d'instruments d'aide a la décision. Université Laval, Faculté des sciences de l'administration (1982)
41. Likert, R.: A Technique for the Measurement of Attitudes. New York University, New York (1932)
42. Respondenci do ankiet online. https://ankieteo.pl/program-do-ankiet/respondenci
43. Barbrich, P., Nocoń, B.: NetB@nk, bankowość internetowa i mobilna, płat-ności bezgotówkowe, 1 kwartał 2022. Zwiazek Banków Polskich, Warszawa
44. Shekhovtsov, A., Kozlov, V., Nosov, V., Sałabun, W.: Efficiency of methods for determining the relevance of criteria in sustainable transport problems: a comparative case study. Sustainability **12**, 7915 (2020). https://doi.org/10.3390/su12197915
45. Chmielarz, W., Zborowski, M.: Scoring method versus TOPSIS method in the evaluation of e-banking services. In: 2018 Federated Conference on Computer Science and Information Systems (FedCSIS), pp. 683–689 (2018)
46. Chmielarz, W., Zborowski, M.: Conversion method in comparative analysis of e-banking services in Poland. In: Kobyliński, A., Sobczak, A. (eds.) Perspectives in Business Informatics Research, pp. 227–240. Springer, Berlin, Heidelberg (2013). https://doi.org/10.1007/978-3-642-40823-6_18
47. Chmielarz, W., Zborowski, M.: On the assessment of e-banking websites supporting sustainable development goals. Energies **15**, 378 (2022). https://doi.org/10.3390/en15010378

Application of the Multi-criteria VMCM Method to Study the Dynamics of Changes in the Level of Early Childhood Education in Poland Related to EU Countries

Małgorzata Nermend, Mateusz Piwowarski(✉), and Kesra Nermend

University of Szczecin, Cukrowa 8, 71-004 Szczecin, Poland
mateusz.piwowarski@usz.edu.pl

Abstract. Early school education has a very important role in the growth of young children. They get fundamental skills at this level of education, basic reading, and writing as the basis for their further development. Therefore, countries launch various types of programs supporting this process. This process is very complicated because it is influenced by many factors which exceed the educational obligations at the school level, including the very important role of state action. The success or failure of this process will depend on the State's actions in this regard. In the article, the authors attempt to use the opportunities offered by multi-criteria methods to analyze the level of development of Early School Education in Poland related to selected countries of the European Union.

Keywords: early childhood education · multi-criteria method · VMCM method

1 Introduction

Modern education is a process that consists of many elements that are complementary components. Without them, this process is disrupted causing the level of education to differ in European countries, which significantly affects the educational opportunities of young people. A child starting school depends on the capabilities of educational units and the teachers working in them. Among the factors that affect the level of education are not only those related to didactics but also those that affect the financing of educational units, which is directly related to the educational structure. Education responds to social needs but also causes them.

When thinking about education, we usually associate it with the teaching effect of the didactic process. The authors wanted to show that other factors also have a significant role in the entire educational process, showing early childhood education as the one that is most often overlooked but which is nevertheless the basis for further educational stages. The quality of edutainment is influenced, among other things, by the funding of the system, the conditions for hiring teachers, the number of teachers, and the demographics. Nowadays, we can see that it is a force that causes change and undergoes change, playing an important role in modern globalization processes. The financial cost of education

M. Hernes and J. Wątróbski (Eds.): ECAI 2023, LNNS 1079, pp. 289–302, 2024.
https://doi.org/10.1007/978-3-031-66761-9_23

should, in the long run, give benefits in the form of good preparation of a person for professional activity, through the acquisition of key competencies and entrepreneurship, and should be a driving force for development and economic progress [1].

Education in Europe is diverse and includes many different systems and criteria, where funding is important as a fundamental right of every citizen. Accordingly, countries are investing in the development of education systems. There are many different educational systems in Europe, which differ in terms of min. Curricula, examinations, and the number of educational stages. For example, some countries, such as France and Italy, have centralized education systems in which curricula and exams are defined at the national level. Other countries, such as Germany and Sweden, have more decentralized systems where responsibility is essentially determined at the regional level.

Nowadays, education is widely discussed by politicians, parents, educators, psychologists, and many other social groups. This has been caused, among other things, by socio-economic changes in European countries giving a diverse picture of education. In the literature we can find numerous studies on education, however, they refer to the educational process and only its results.

The first years of a child's life affect his cognitive, social, and emotional development. This is one of the most important developmental periods for a child. That is why it is so important to take proper care of them and provide appropriate conditions not only at home but also in educational institutions. Children should receive education in schools, this is their fundamental right in the modern world. This has a significant impact not only on individuals, but on society as a whole contributing to its development, having a significant impact on the fight against global poverty. The more children graduate from elementary school, the more they continue their education in later years, which translates significantly into further functioning. Access to education should be the same for both boys and girls and should eliminate any discrimination. Gender equality is one of the UN sustainable development goals [2], where countries in the community are obligated to take steps to end discrimination and violence, ensure women's full participation in decision-making processes, and enact effective legislation that promotes gender equality and empowers women and girls at all levels [3]. The purpose of this paper is to use the selected method of multi-criteria/multidimensional analysis, to solve this problem in the context of research on the level of early childhood education in EU countries in European countries. The methodological contribution is the use of MCDA (Multi-Criteria Decision Analysis) method called VMCM (Vector Measure Construction Method) [4]. Allowing us to consider the dynamics of changes over time of decision alternatives in the solution of the decision problem.

1.1 Multi-criteria Decision Analysis in the Context of Research on the Level of Early Childhood Education in EU Countries

Various studies on education are conducted around the world, however, they usually refer to the educational process and its results. One of the most popular is the international survey PISA (Programme for International Student Assessment), coordinated by the Organization for Economic Cooperation and Development [5]. It is the largest international survey of students' skills in the world. It is realized every 3 years in all OECD member countries, as well as in several dozen other countries. Poland has been participating in this survey since 2000 [6]. Its main goal is to obtain comparable data on the skills of students who are 15 years of age or older. It uses research tools that are geared toward examining the ability to practically associate and apply knowledge and skills in different areas. It covers three areas: reading literacy, mathematical literacy, and scientific literacy. In PISA 2018, in all three areas covered by the survey, the performance of Polish students is above the average for OECD countries and places them among the world's top performers [7].

Another one of the world's largest research projects on educational achievement is the PIRLS (Progress in International Reading Literacy Study) [8]. The main purpose of the survey is to measure reading comprehension skills in the form of literary text and applied text. The survey is conducted every five years by the (International Association for the Measurement of School Achievement - IEA). Dozens of countries from around the world take part in the survey, and students who attend school and are completing their fourth year at the ISCED 1 level, and whose average age at the time of the survey is no less than 9.5 years old, participate [9].

Another major survey conducted every four years is TIMSS (Trends in International Mathematics and Science Study) to measure students' school achievement. It is one of the world's largest educational achievement research projects organized by the IEA. Dozens of countries from around the world participate in the study, measuring students' knowledge and reasoning in mathematics and science [10]. This study consists of measuring students' skills in the fourth and eighth years of schooling.

The study that covers the ranking of education systems is the Pearson Ranking. It is based on the results of tests conducted regularly every few years, testing knowledge of the sciences and humanities, in which Poland also ranked a high 10th. Ranking in addition to the tests, additional measures, creating a more comprehensive index than, for example, PISA. The index created for The Learning Curve includes 60 different indicators, divided into three themes: the contribution a country makes to education i.e. min. Expenditures, student-teacher ratios, staff salaries, life expectancy of students, etc., education outcomes taking into account PISA, TIMSS, and PIRLS results, graduation rates, unemployment rates by educational attainment, literacy rates, etc.) and indicators of the socio-economic environment such as crime rates, GDP per capita, unemployment, social inequality, etc. [11].

The studies described above are the largest conducted globally in the world. However, the author's contribution to this article is, the authors focus not only on test scores but also on important factors that also directly affect the quality of education in a dynamic perspective. The authors in this publication use the VMCM method, which is widely used in various social areas. It is used, for example, to analyze a complex socio-economic process described by many factors examining the assessment of the quality of life of consumers in European countries [12]. It also found its application in the study of the dynamics of change in the development of the information society in EU countries [13], Assessing the impact of renewable energy in EU countries [14], Study the level of poverty in the EU countries [15], analysis of the dynamics of change in the investment attractiveness of the region [16], assessments of the level of housing attractiveness of municipalities [17], make complex decisions in management business processes [18], multi-criteria assessment of bureaucracy [19], study on the impact of healthcare capacity in European countries on Covid-19 infections and mortality [20], measurement of economic growth factors [21]. Multi-criteria methods were also applied to the study of edutainment in the broadest sense. They are used for analysis in e-learning training design and decision support [22] or in optimizing the selection of e-learning courses [23]. In most studies where MCDA methods were used to measure the level of education, they were limited to only one year. Authors in the study use a new approach, using the possibilities which given the VMCM method, to study the changes in the level of early childhood education in a dynamic approach.

2 Materials and Methods

In the research used VMCM (Vector Measure Construction Method) method was applied as a methodical apparatus, to assess the problem related to, the dynamics of changes in the level of early childhood education development described with selected indicators in the selected European Union countries. The indicators value was collected from the Eurostat database in the time horizon (2013–2021) years. At the beginning of the study the authors (based on the adopted variable), the pattern, and the anti-pattern objects were calculated based on the data from the base year (2013) as a reference point (so-called artificial pattern) to assess the level of early childhood education development from 2013 as a reference year do 2021. The purpose of this research concept was to enable the general ranking of EU countries, due to, the level of early childhood education development and the possibility of making comparison with respect to the result of the ranking for each country separate in years (2013–2021), which was the pattern or anti-pattern to estimate. The indicators used in this research were selected from the early childhood education area:

x_1 – Pupils enrolled in early childhood education by sex, type of institution, and intensity of participation total.
x_2 – Pupils enrolled in early childhood education by sex and age total.
x_3 – Pupils in early childhood education - as a percentage of the population at the appropriate school age (in %).
x_4 – Public expenditure on education by education level and program orientation - as % of public expenditure or % of GNI (in %).
x_5 – Level (0–2) of education per 1,000 inhabitants.
x_6– Expenditure on public educational institutions - as % of total or current expenditure.
x_7 – Public expenditure on Early childhood education per pupil based on Full-Time Equivalent.
x_8 –Households without access to internet at home, because the access costs are too high (in %).

Analyzed selected countries (decision variants) were: Austria (AT), Belgium(BE), Bulgaria (BG), Croatia (HR), Cyprus (CY), Czechia (CZ), Denmark (DK), France (FR), Germany (DE),Hungary (HU), Ireland (IE), Italy (IT), Latvia (LV), Lithuania (LT), Luxembourg (LU), Malta (MT), Netherlands (NL), Poland (PL), Portugal (PT), Romania(RO), Slovakia (SK), Slovenia (SI), Spain (ES), Sweden (SE), Data referred to the year 2013–2021 and were obtained from the Eurostat database. Estonia (EE), Finland

(FI), Greece (EL), and Norway (NO), were not included in the analysis due to a lack of data for most of the adopted criteria.

2.1 Description of Vector Measure Construction Method (VMCM)

The vector Measure Construction Method - VMCM is a relatively new multi-criteria method (methodological bases have been developed in 2009 [4]. The VMCM method uses the properties of vector calculus to construct vector aggregate measures (based on the definition of an inner product). This approach allows for ranking, classification of objects, and the study of the dynamics of changes. The procedure of VMCM comprises 8 stages:

Stage 1–2. Selection of Variables and Elimination of Variables.
The criteria in the study were selected, using the approach related to substantive and logical selection and the approach related to the elimination of variables such as those characterized by a high degree of collinearity [24, 25]. Elimination of variables is carried out by using significance coefficient of features [26]:

$$V_{x_i} = \frac{\sigma_i}{\overline{x}_i} \tag{1}$$

where: x_i – i-th variable, σ_i – standard deviation of the i-th variable, $\overline{x}_i$ – mean value of the i-th variable,

Variables, for which significance coefficient values are within the range 0; 0, 1, are quasi-constant and such variables should be eliminated from the set of variables under consideration [26, 27].

Stage 3. Defining the Diagnostic Variables Character.
The diagnostic variables character can as stimulants (benefits), destimulants (no benefit), and nominatns (natural, interval). Stimulants (benefits) are such variables, in which greater values mean a higher level of benefits. Destimulants are such variables which smaller values mean a higher level of benefits, while nominatns are such variables, in which desired values are within a specific range. in the study, the criteria ($x_1 - x_7$) had character stimulants (benefits) and (x_8) was destimulants (no benefit).

Stage 4. Assigning Weights to Diagnostic Variables.

In the article, ten experts from the research area (primary school teachers) were asked to determine the importance of criteria on a scale [4, 24–32]. The average rating of the experts' evaluation showed that the criteria (x_1 – [10], x_2 – [7], x_3 – [6], x_4 – [10], x_5 – [9], x_6– [10], x_7 – [10], x_8 – [8]). In the next step, the weights were normalized according to the formula:

$$w_i = \frac{w_i}{\sum_{j=1}^{m} w_j} \tag{2}$$

The criteria will have a greater impact on the value of the measure, the higher the w_i. However, it should be remembered that the weights are to be positive and meet the following condition:

$$\sum_{i=1}^{m} w_i = 1 \tag{3}$$

In practice, researchers use the statistical weights approach, which is based on statistical resources. Criteria weighting can be carried out, for example, using a relative value meter [31], an example of such an approach is:

$$w_i = \frac{V_{x_i}}{\sum_{k=1}^{m} V_{x_k}} \tag{4}$$

where w_i means variable's weight.

Nevertheless, regardless of the system adopted the weights should be normalized.

Stage 5. Normalization of Variables.

The criteria used in studies are heterogeneous because they can occur in different units of measure, which can hinder any arithmetic operations. Therefore, the criteria must be normalizing. This process leads not only to the elimination of units of measurement, but also to equalize the values of criteria. Standardization is the most used normalization techniques:

$$x'_{ij} = \frac{a_i}{\sigma_i} \tag{5}$$

where nominator a_i can be defined in any way, e.g.:

$$a_i = x_{ij} - \overline{x}_i \tag{6}$$

where x'_{ij} is normalized value of the i-th variable for the j-th object.

Stage 6. Determination of Pattern and Anti-pattern.

Patterns and anti-patterns can be selected as real-life objects. In the study Pattern and anti-pattern are automatically determined based on the first and third quartiles [32]. Where stimulants (benefit) values of the third quartile and for destimulants (no benefit) values of the first quartile are taken as coordinates of the pattern accordingly:

$$x'_{i\,w} = \begin{cases} x'_{i_{qIII}} & \textit{for stimulants} \\ x'_{i_{qI}} & \textit{for destimulants} \end{cases} \tag{7}$$

where:

$x'_{i\,w}$ is the value of the i-th normalized variable for the pattern, $x'_{i_{qI}}$ - the value of the i-the normalized variable for the first quartile, and $x'_{i_{qIII}}$ - the value of the i-th nomalized variable for the third quartile. In case of anti-pattern the procedure is inversed, values in the first quartile are anti-pattern coordinates for stimulants and values in the third quartile for destimulants:

$$x'_{i\,aw} = \begin{cases} x'_{i_{qI}} & \textit{for stimulants} \\ x'_{i_{qIII}} & \textit{for destimulants} \end{cases} \tag{8}$$

where $x_{iaw'}$ means the value of the i-th normalized variable for anti-pattern.

Stage 7. Construction of the Synthetic Measure.

The values of the variables of the examined objects in the vector space are interpreted as vector coordinates. Each object therefore determines a specific direction in space. The pattern and anti-pattern difference is also a vector which determines a certain direction in space. Along this direction, the aggregate measure value for each object is calculated. This difference can be treated as a monodimensional coordinate system, in which the coordinates are calculated based on the formula [8, 11]:

$$c = \frac{\left(\vec{A}, \vec{B}\right)}{\left(\vec{B}, \vec{B}\right)} \tag{9}$$

In turn, $\vec{A}$ and $\vec{B}$ are vectors, and $\left(\vec{A}, \vec{B}\right)$ is the scalar product, which can be defined as follows:

$$\left(\vec{A}, \vec{B}\right) = \sum_{k=1}^{n} a_k b_k \tag{10}$$

where: a_k, b_k – coordinates of the appropriate vector $\vec{A}$ and $\vec{B}$.

We consider the $\vec{B}$ vector as the monodimensional coordinates system, thus it represents a difference between the pattern and anti-pattern. By entering the coordinates of the pattern and anti-pattern as well as the object into the formula (11) the result is as follows:

$$m_{aj} = \frac{\sum_{i=1}^{m}\left(x_{ij'} - x_{iaw'}\right)\left(x_{iw'} - x_{iaw'}\right)}{\sum_{i=1}^{m}\left(x_{iw'} - x_{iaw'}\right)^2} \tag{11}$$

Stage 8. Classification of Objects.

The last step of the VMCM method is decision variant rankings depending on the obtained aggregate measure value values, thus it is possible to determine which of them are "better" and which are "worse". They also allow determining which are like each other in terms of adopted criteria. To better visualize the results of calculations, decision variants can be divided into classes with similar measurement values. In the simplest case objects can be classified based on mean value $\overline{m}_{m_{a0}}$ and standard deviation of the aggregate measure $\sigma_{m_{a0}}$. Objects are classified into four classes [4]:

$$\begin{cases} 1 \text{ for } m_{aj} \in \langle \overline{m}_{m_{a0}} + \sigma_{m_{a0}}; \infty), \\ 2 \text{ for } m_{aj} \in \langle \overline{m}_{m_{a0}}; \overline{m}_{m_{a0}} + \sigma_{m_{a0}}), \\ 3 \text{ for } m_{aj} \in \langle \overline{m}_{m_{a0}} - \sigma_{m_{a0}}; \overline{m}_{m_{a0}}), \\ 4 \text{ for } m_{aj} \in \left(-\infty; \overline{m}_{m_{a0}} - \sigma_{m_{a0}}\right). \end{cases} \tag{12}$$

3 Research Results

The level of early childhood education in EU countries was conducted using the VMCM method. The obtained results of the aggregate measure and the determined affiliation to classes are presented in the tables. The research was carried out from 2013 to 2021 the results are illustrated in three tables due to limitations in the number of pages by the conference organizers were chosen every other year in Tables 1, 2 and 3 the results of the last two years. To study the dynamics of change in the level of early childhood education in EU countries. All values of the aggregate measure are calculated in each year separately and as the pattern referred to (the pattern variant built based on data from the base year 2013) as a reference point.

Table 1. Ranking of EU countries in 2013 and 2015 in respect of the level of early childhood education VMCM method.

Country	Measure value	Class	Country	Measure value	Class
Pattern form 2013 and Data (2013)			Pattern form 2013 and Data (2015)		
Denmark	0,181	1	Germany	0,159	1
Sweden	0,154	1	Sweden	0,152	1
Germany	0,150	1	France	0,148	1
France	0,148	1	Denmark	0,137	1
Luxembourg	0,137	1	Luxembourg	0,118	2
Lithuania	0,135	1	Italy	0,111	2
Italy	0,116	2	Spain	0,109	2
Netherlands	0,104	2	Netherlands	0,097	2
Spain	0,104	2	Latvia	0,082	2
Malta	0,087	2	Malta	0,080	2
Latvia	0,082	2	Slovenia	0,069	3
Slovenia	0,060	3	Austria	0,064	3
Austria	0,052	3	Poland	0,058	3
Czechia	0,046	3	Czechia	0,056	3
Poland	0,043	3	Lithuania	0,055	3
Bulgaria	0,042	3	Slovakia	0,038	3
Slovakia	0,030	3	Ireland	0,031	3
Hungary	0,013	3	Bulgaria	0,030	3
Romania	0,012	3	Hungary	0,030	3
Ireland	−0,001	4	Romania	0,019	3
Croatia	−0,010	4	Portugal	0,006	4
Portugal	−0,023	4	Cyprus	−0,015	4
Cyprus	−0,050	4	Croatia	−0,016	4

Table 1 shows the results of the classification of EU countries due to the level of early childhood education obtained. Based on data from 2013 and 2015, Two differences in-class assignments can be observed. Denmark changed its position in 2013 from first to fourth in 2015, while Germany changed its position in 2013 from third to first in 2015. These changes result from the fact that in 2015 in Denmark the value of the criteria x1 – Pupils enrolled in early childhood education by sex, type of institution, and intensity of participation total, x2 – Pupils enrolled in early childhood education by sex and age total and, x5- Level (0–2) of education per 1,000 inhabitants, decreased and amounts to 4.87%, 4.87%, and 4.49%. This results from the reduction in the number of pupils in early childhood education, generally approx. 5%. On the other hand, the x8- Households without access to internet at home, because the access costs are too high (in %) criterion increased by 8.6%, which means that the number of households without access to the Internet has increased due to the high cost. This process may be of concern, as access to the Internet can be very important in the education process, which happened, during the pandemic. The situation in Germany is quite the opposite, as criteria x1, x2, and x5 increased by 4.3%, 4.3%, and 14%. However, criterion x8 in Germany has decreased by 42% compared to 2013, which shows that household access to the Internet in Germany is developing rapidly. This process may be since 2015 was the year in which the emigration wave to Germany was very large. An interesting point about Germany is that the x6- Expenditure on public educational institutions - as % of total or current expenditure, criterion decreased by 7.35% for education spending.

Table 2. Ranking of EU countries in 2017 and 2019 in respect of the level of early childhood education VMCM method.

Country	Measure value	Class	Country	Measure value	Class
Pattern form 2013 and Data (2017)			Pattern form 2013 and Data (2019)		
Germany	0,173	1	Germany	0,192	1
Sweden	0,151	1	France	0,153	1
France	0,150	1	Sweden	0,150	1
Denmark	0,116	2	Luxembourg	0,123	2
Spain	0,114	2	Spain	0,120	2
Luxembourg	0,113	2	Denmark	0,118	2
Italy	0,111	2	Italy	0,116	2
Netherlands	0,094	2	Latvia	0,102	2
Latvia	0,088	2	Netherlands	0,097	2
Malta	0,080	2	Malta	0,087	2
Slovenia	0,074	3	Poland	0,086	2
Poland	0,072	3	Czechia	0,082	3
Austria	0,072	3	Slovenia	0,075	3
Czechia	0,068	3	Austria	0,074	3
Ireland	0,054	3	Lithuania	0,068	3

(continued)

Table 2. (*continued*)

Country	Measure value	Class	Country	Measure value	Class
Pattern form 2013 and Data (2017)			Pattern form 2013 and Data (2019)		
Hungary	0,048	3	Ireland	0,068	3
Slovakia	0,040	3	Hungary	0,067	3
Lithuania	0,033	3	Slovakia	0,038	3
Bulgaria	0,026	4	Bulgaria	0,033	4
Romania	0,026	4	Romania	0,033	4
Portugal	0,019	4	Croatia	0,017	4
Cyprus	0,005	4	Portugal	0,014	4
Croatia	−0,006	4	Cyprus	0,008	4

Table 2 shows the results of the classification of EU countries due to the level of early childhood education obtained. Based on data from 2017 and 2019. The created ranking in shows that in total 3 counties were classified into Class 1: Germany, France, and Sweden. In this respect, France changed their ranking from place 3 to 2, and Sweden from place 2 to 3. Germany is the leading country in this ranking. These are countries providing potentially the best environment for early childhood education. The oposite pole (Class 4) includes 4 countries, of which Croatia and Cyprus had the lowest ranking. Denmark in relation to 2013 which was the leading country in this ranking, moved down from Class 1 to 2 and changed their ranking from place 1 to 7, in 2019. These changes result from the fact that in 2019 in Denmark the value of the criteria x1 – Pupils enrolled in early childhood education by sex, type of institution, and intensity of participation total, x2 – Pupils enrolled in early childhood education by sex and age total and, x4- Public expenditure on education by education level and program orientation - as % of public expenditure or % of GNI (in %), decreased and amounts to 9.30%, 9.30%, and 22.22%. This results from the reduction in the number of pupils in early childhood education, generally approx. 9%. However, the x8- Households without access to the Internet at home, because the access costs are too high (in %) criterion down up by 55,88%, in relation to 2013 which means that the number of households without access to the Internet has down up due to the high cost. This process impacted very positive prospects in the education process.

Table 3 shows the results of the classification of EU countries due to the level of early childhood education obtained. Based on data from 2020 and 2021. Most countries retained their position due to 2019, except at the end of the ranking there is a minor change. Croatia was the next to last, which at the same time changed its position to 18. Hungary moved up from position 17 to 14. While Austria moved down from position 13 to 17 in relation to 2019.

Table 3. Ranking of EU countries in 2020 and 2021 in respect of the level of early childhood education VMCM method.

Country	Measure value	Class	Country	Measure value	Class
Pattern form 2013 and Data (2020)			Pattern form 2013 and Data (2021)		
Germany	0,205	1	Germany	0,213	1
France	0,155	1	France	0,156	1
Sweden	0,150	1	Sweden	0,151	1
Luxembourg	0,134	2	Luxembourg	0,140	2
Denmark	0,128	2	Denmark	0,133	2
Spain	0,124	2	Lithuania	0,129	2
Italy	0,121	2	Spain	0,127	2
Latvia	0,112	2	Italy	0,123	2
Lithuania	0,104	2	Latvia	0,118	2
Netherlands	0,101	2	Netherlands	0,104	2
Malta	0,092	2	Poland	0,098	2
Poland	0,092	2	Malta	0,096	2
Czechia	0,085	3	Czechia	0,090	3
Slovenia	0,077	3	Hungary	0,081	3
Austria	0,074	3	Slovenia	0,079	3
Hungary	0,074	3	Ireland	0,075	3
Ireland	0,072	3	Austria	0,074	3
Slovakia	0,037	4	Croatia	0,041	4
Romania	0,037	4	Romania	0,041	4
Bulgaria	0,037	4	Bulgaria	0,041	4
Croatia	0,029	4	Slovakia	0,036	4
Portugal	0,011	4	Portugal	0,007	4
Cyprus	0,003	4	Cyprus	−0,001	4

4 Conclusion

The article presents an original approach to assessing the dynamics of changes in the level of early childhood education in EU countries by using an author's VMCM method. The presented method is dedicated to the study of complex socio-economic phenomena described by many factors over time was applied. is a relatively new approach to studying the level of changes in the level of early childhood education. The research shows that the educational situation in the European countries studied varied. For each year, the resulting aggregate measures value was divided into four classes. In 2013–2014 the educational situation in Denmark was the best, but from 2015–2021 Germany was the best in the ranking. This situation was due to a drastic decrease in the number of children participating in the education process in Denmark while this number in Germany increased to 9.32% related to 2013. In this respect, Poland changed their ranking from place 15 in 2013 to 11 in 2021, and Latvia from place 11 to 9. Poland moved up from Class 3 to 2. This shows that situation in Poland in the context of the level of early childhood education has improved compared to 2013 in 2021. In this respect, Poland changed their ranking from place 15 in 2013 to 11 in 2021, and Latvia from place 11 to 9. Poland moved up from Class 3 to 2. This shows that situation in Poland in the context of the level of early childhood education has improved compared to 2013 in 2021. in Poland, the value of the criteria x_1 – Pupils enrolled in early childhood education by sex, type of institution, and intensity of participation total, x_2 – Pupils enrolled in early childhood education by sex and age total, x_3 – Pupils in early childhood education - as a percentage of the population at the appropriate school age (in %) and, x_5- Level (0–2) of education per 1,000 inhabitants, decreased and amounts to 17.10%, 17,10%, 44.32%, and 26,25%. This results from the reduction in the number of pupils in early childhood education, while the x_8- Households without access to the internet at home, because the access costs are too high (in %) criterion increased by 68,25%, which means that the number of households without access to the Internet has increased due to the high cost. This process may be of concern, as access to the Internet can be very important in the education process. In addition, Croatia has become the leading country in the ranking in class four in 2021 compared to 2013. Moreover, the situation of Austria moved down in Class 3. In addition, countries that have centralized education systems, such as France and Italy, in 2013–2017 ranked worse than Germany and Sweden, which have more decentralized education systems. Such studies are needed for state authorities to monitor the educational process in each country and relate it to the rest of the European countries. Making it possible to identify the factors inhibiting this process and intervene quickly to improve it.

References

1. Skibska, J., Wojciechowska, J. (eds.): Współczesna edukacja: wielopłaszczyznowość zadań. Wydawnictwo Libron - Filip Lohner, Kraków (2016)
2. Cele zrównoważonego rozwoju ONZ: https://www.un.org/sustainabledevelopment/
3. Koniewski Maciej, M.R.: Teoretyczne uzasadnienia edukacji zróżnicowanej ze względu na płeć. Edukacja. 5–18 (2018). https://doi.org/10.24131/3724.180101
4. Nermend, K.: Vector calculus in regional development analysis: comparative regional analysis using the example of Poland. Physica-Verlag, Berlin (2009)
5. OECD: https://www.oecd.org
6. Badanie PISA: https://pisa.ibe.edu.pl
7. Wyniki PISA: https://pisa.ibe.edu.pl/wp-content/uploads/2019/12/raport-wyniki-badan-pisa-2018.pdf
8. PIRLS 2021 INTERNATIONAL RESULTS IN READING. https://pirls2021.org
9. Lista krajów biorących udział w poszczególnych edycjach badania. https://www.iea.nl/studies/iea/pirls
10. TIMSS Trends in International Mathematics and Science Study. https://www.iea.nl/studies/iea/timss
11. Ranking Pearson. https://www.pearson.com/

12. Piwowarski, M., Maison, D., Wątróbski, J.: Application of VMCM method (Vector Measure Construction Methods) to estimate consumer's quality of life in EU countries – dynamic perspective. Procedia Comp. Sci. **159**, 2404–2413 (2019). https://doi.org/10.1016/j.procs.2019.09.415
13. Ogonowski, P.: Application of VMCM, to investigate, the dynamics of changes information society development. Procedia Comp. Sci. **176**, 3182–3190 (2020). https://doi.org/10.1016/j.procs.2020.09.131
14. Ogonowski, P.: Application of VMCM, to assess of renewable energy impact in european union countries. Procedia Comp. Sci. **192**, 4762–4769 (2021). https://doi.org/10.1016/j.procs.2021.09.254
15. Miłaszewicz, D.: Change of the paradigm and method of modern economics – the role of behavioral economics. Studia i Prace WNEiZ. **47**, 65–82 (2017). https://doi.org/10.18276/sip.2017.47/3-06
16. Marchewka, A.: Applying Vector Measure Construction Method (VMCM) to analyze change dynamics of investment attractiveness of the region. Procedia Comp. Sci. **192**, 3252–3261 (2021). https://doi.org/10.1016/j.procs.2021.09.098
17. Szaja, M.: Applying the VMCM method to assess the level of residential attractiveness of municipalities in the West Pomeranian Voivodeship. Procedia Comp. Sci. **192**, 4701–4710 (2021). https://doi.org/10.1016/j.procs.2021.09.248
18. Piwowarski, M., Nermend, K.: Issues of multi-criteria methods applicability supporting complex business process decision-making in management. Procedia Comp. Sci. **207**, 4161–4170 (2022). https://doi.org/10.1016/j.procs.2022.09.479
19. Klimek, J.A.: Multi-Criteria Assessment of the Bureaucracy in Poland - the Individual Decision-Making Aspect. Procedia Comp. Sci. **176**, 2434–2444 (2020). https://doi.org/10.1016/j.procs.2020.09.321
20. Borawski, M., Duda, J., Biercewicz, K.: Impact of the healthcare potential in the European countries on infections and mortality caused by Covid-19. Procedia Comp. Sci. **192**, 4037–4046 (2021). https://doi.org/10.1016/j.procs.2021.09.178
21. Miłaszewicz, D., Piwowarski, M., Nermend, K.: Application of Vector Measure Construction Methods to estimate growth factors of fundamental importance for the economy on the example of nations in transition. Procedia Comp. Sci. **176**, 2913–2922 (2020). https://doi.org/10.1016/j.procs.2020.09.262
22. Korniejenko, K.: Możliwości wykorzystania metod wielokryterialnych w projektowaniu e-learningu. Zeszyty Naukowe Wydziału Elektrotechniki i Automatyki Politechniki Gdańskiej. 51–54
23. Zastosowanie metody PROMETHEE II do wyboru kursu e-learningowego w przedsiębiorstwie. e-mentor. 39–45. https://doi.org/10.15219/em80.1398
24. Nermend, K.: Metody analizy wielokryterialnej i wielowymiarowej we wspomaganiu decyzji. Wydawnictwo Naukowe PWN, Warszawa (2017)
25. Grabiński, T.: Metody statystyki międzynarodowej. PWE, Warszawa
26. Kukuła, K.: Metoda unitaryzacji zerowanej. Wydaw. Naukowe PWN, Warszawa (2000)
27. Nermend, K.: Taxonomic Vector Measure of Region Development (TWMRR). Polish Journal of Environmental Studies (2007)
28. Nermend, K.: Badanie własności metod normowania wykorzystywanych w rangowaniu obiektów społeczno-gospodarczych. Zeszyty Naukowe/Uniwersytet Szczeciński; Ekonomiczne Problemy Usług. 171–186
29. Nermend, K.: Properties of normalization methods used in the construction of aggregate measures. Folia Oeconomica Stetinensia. **12**, 31–45 (2012). https://doi.org/10.2478/v10031-012-0030-9
30. Nermend, K., Borawski, M.: Modeling user's preferences in the Decision Support System. Indian J. Fundam. App. Life Sci. **4**, 1480–1491 (2014)

31. Panek, T., Zwierzchowski, J.: Statystyczne metody wielowymiarowej analizy porównawczej: teoria i zastosowania. Szkoła Główna Handlowa. Oficyna Wydawnicza, Warszawa (2013)
32. Kolenda, M.: Taksonomia numeryczna: klasyfikacja, porządkowanie i analiza obiektów wielocechowych. Wydawn. Akademii Ekonomicznej im. Oskara Langego we Wrocławiu, Wrocław (2006)

A Fuzzy Multi-criteria Approach for the Quality Assessment of Online Shopping Platforms

Paula Bajdor(✉)

Czestochowa University of Technology, Czestochowa, Poland
paula.bajdor@pcz.pl

Abstract. The article presents the fuzzy TOPSIS method used to prioritize the criteria contributing to the overall quality of Internet platforms' functioning. The article's primary purpose is to present the fuzzy TOPSIS method applying as an appropriate tool when the respondents' answers were given using an ordinal scale. Thus, the subject of consideration was the presentation of the fuzzy TOPSIS method on the example of primary data obtained during research on assessing the quality of Internet platforms' functioning. Twenty-four qualitative criteria were tested, contributing to the overall quality of Internet platforms' functioning. The method used made it possible to create a ranking of these criteria, indicating which, according to the respondents, are the most important and which play a minor role in determining the level of quality of the online platforms' functioning. In addition, the created ranking of criteria contributing to the overall quality of the functioning of online platforms can be a guideline for managers responsible for online platforms' managing. Proper management of an online platform should take into account the emphasis on increasing the level of quality of its functioning, and the ranking of criteria, which is the effect of the fuzzy TOPSIS method used in this case, is a valuable reference point for managers responsible for online platforms' managing.

Keywords: Fuzzy method · F-TOPSIS · online shopping · criteria · management

1 Introduction

The very concept of online shopping is not new because, for the first time, the framework of this concept was presented in 1967 by A. Doody and W. Davidson, who stated that "*in the not too distant future, a substantial share of [staple items] may be marketed through ... large central distribution facilities ... Consumers will never set foot inside these centers ... Instead, retail transactions will be made by electronic telecommunications and push-button devices installed in private homes and hooked online to data processing networks*" [1]. In just 30 years, this prediction has become a reality – online shopping is now just another form of shopping. In 2022, the percentage of Internet users in Poland declared making such purchases was 77% [2]; worldwide, the number of people making such purchases exceeded 2 billion [3]. The COVID-19 pandemic, which lasted over two years, certainly contributed to their great popularity. Many factors contribute to their

M. Hernes and J. Wątróbski (Eds.): ECAI 2023, LNNS 1079, pp. 303–315, 2024.
https://doi.org/10.1007/978-3-031-66761-9_24

popularity, including convenience, speed, transparency, simplicity, price, availability or security, and many others. One should also not forget about the universal access to the Internet using various devices and the growing number of shops offering online shopping, which also contributed to the popularization of this form of shopping [4].

The universality of online shopping has resulted in the fact that currently, there are many Internet platforms on the market offering online shopping. It means that these platforms, in order to attract new customers, must place great emphasis on quality, which in their case is a whole consisting of a large number of criteria. However, these criteria are perceived in different ways by users of online platforms. Taking into account both the large number of these criteria as well as their diversity, the main goal of this article is to prioritize these criteria using the fuzzy TOPSIS method. As a result, it will be possible to determine which of the identified criteria are the most important and which are the least important in perceiving the overall quality of the functioning of online platforms.

2 The Methodology

2.1 TOPSIS Theoretical Approach

Multi-Criteria Decision Making (MCDM) methods are used to make a choice or decision based on selecting the best solution from a set of available solutions in terms of many, usually conflicting, criteria. These methods are characterized by a limited and small number of established decision variants, the selection of which is made by pre-determined criteria. These valuable tools enable decision-makers to make multi-criteria decisions or select a solution for discrete problems. They enable individual and group decisions, such as choosing the best location, making decisions related to supply chain management, identifying people in the risk group, choosing the best product in a given category, or choosing the best factors affecting the quality of specific products or services.

Currently, thanks to the use of computers that significantly facilitate the implementation of individual decision-making stages, MCDMs are valuable tools that enable easier decision-making in various decision-making areas, both concerning economic decisions and decisions related to management.

Among these methods, several individual methods can be distinguished:

- based on outranking relationships that represent the decision maker's preferences, as well as dependencies between given decision variants, e.g. PROMETHEE (Preference Ranking Organization Method for Enrichment Evaluations) method, characterized by analyzing the differentiation of assessments of individual variants for all criteria. The more varied the assessments are in the light of a given criterion, the better one of the variants is, while when the difference is not significant, variants are equivalent, or one of the variants slightly prevails over the others [5];

- based on minimizing the error of the utility function using a linear programming model, e.g., the SMART (Simple Multi-Attribute Rating Technique) method, which evaluates the performance of alternatives by selecting values from the range between the worst alternative and the best. The advantage of this method is that the values of the alternatives are mutually independent [6];

- based on the assumption of direct and proportional dependence of the significance and degree of utility of the analyzed decision variants on the system of criteria describing the alternatives as well as on the values and weights of the criteria, e.g., the COPRAS (Complex Proportional Assessment) method, which is used in the maximizing and minimizing the criteria values [7].

The TOPSIS method (Technique for Order Preference by Similarity to Ideal Solution) is currently one of the most popular methods for solving multi-criteria discrete tasks [8]. Its main idea is to determine the distance of the considered objects from the ideal and anti-ideal solutions. On the other hand, the final result is a synthetic indicator that creates a ranking of the examined criteria. The first position of a given criterion means that this criterion is characterized by the smallest distance from the ideal solution and, at the same time, the most significant distance from the anti-ideal solution [9].

Although explicit data may be insufficient to model real situations properly in some conditions, C-T Chen proposed an extension of this method - fuzzy Topsis. It was based on the assumption that personal views and judgments may be insufficiently explicit, making it impossible to estimate them using an exact numerical value. Therefore, he proposed linguistic evaluations instead, assuming that the evaluations and weightings of criteria in the selection process would be evaluated using linguistic variables. These variables are used to evaluate the shortcomings of all analyzed criteria and each possible solution concerning each criterion. Thus, the fuzzy F-TOPSIS (Fuzzy Technique for Order Preference by Similarity to Ideal Solution) method differs from the TOPSIS method in that the values of the criteria characterizing the decision variants are given as fuzzy triangular numbers [10]. This method was created to develop a methodology for solving multi-person or multi-criteria decision-making problems undertaken in a fuzzy environment. F-Topsis allows for the analysis of real decision-making problems in which there are difficulties with precise evaluations or preferred evaluations expressed using linguistic expressions represented by fuzzy numbers [11].

A wide range of applications characterizes this method because it is used to analyze issues concerning, among others, sustainable development, supply chain and logistics management, business and marketing management, engineering and production systems or even health care. The main advantages of this method include its simplicity, comprehensibility, and the guarantee of straightforward interpretation of the results [12]. Therefore, F-Topsis was used to prioritize the criteria that make up the overall quality of the functioning of online platforms.

2.2 TOPSIS Phases/stages

The F-TOPSIS application procedure consists of the following stages [13]:

1. Collecting subjective assessments of decision-makers on the importance of scales.
2. Calculation of fuzzy significance factors or weights based on the subjective assessments of decision makers using the following table and equations:

$$w_{j1} = min\left\{w_{j1}^k\right\}$$

$$w_{j2} = \frac{\sum_{k=1}^{K} w_{j2}^k}{K}$$

$$w_{j3} = max\left\{w_{j3}^{k}\right\}$$

3. Calculation of aggregated values of fuzzy weights based on the following formulas:

$$\mu_A = \left\{ \begin{array}{c} 0x < l \\ \frac{(x-l)}{(m-l)} l \leq x \leq m \\ \frac{(u-x)}{u-m} l \leq x \leq u \\ 0x > u \end{array} \right\}$$

4. Determining the coordinates for the fuzzy positive ideal solution (FPIS) and fuzzy negative ideal solution (FNIS):

$$\overline{A}_j^{+} = \left\{ \begin{array}{l} Max\tilde{N}_{ij} | j \in B \\ Min\tilde{N}_{ij} | j \in C \end{array}, wherei = 1, \ldots, m \right\}$$

$$\overline{A}_j^{-} = \left\{ \begin{array}{l} Min\tilde{N}_{ij} | j \in B \\ Max\tilde{N}_{ij} | j \in C \end{array}, wherei = 1, \ldots, m \right\}$$

5. Calculate the Euclidean distance of each alternative from the fuzzy positive and negative ideal using the following formula:

$$S_i^{+} = \sum_{j=1}^{n} d\left(\tilde{N}_{ij}, \ddot{A}_j^{+}\right), i = 1, \ldots, m; j = 1, \ldots, n$$

$$S_i^{-} = \sum_{j=1}^{n} d\left(\tilde{N}_{ij}, \ddot{A}_j^{-}\right), i = 1, \ldots, m; j = 1, \ldots, n$$

where d represents the distance between two fuzzy numbers calculated by:

$$d\left(\ddot{A}, \ddot{B}\right) = \sqrt{\frac{1}{3}[(a_l - b_l)^2 + [(a_m - b_m)^2 + [(a_u - b_u)^2]}$$

6. Determination of the ranking factor to determine the ranking for selected criteria.

3 The Results of Applying the Fuzzy TOPSIS Method to Evaluate the Quality Criteria of the Functioning of Purchasing Platforms

3.1 The Research Framework

The conducted research aimed to evaluate the criteria determining the quality of perception of shopping platforms by their users. The study itself included the following steps (Fig. 1):

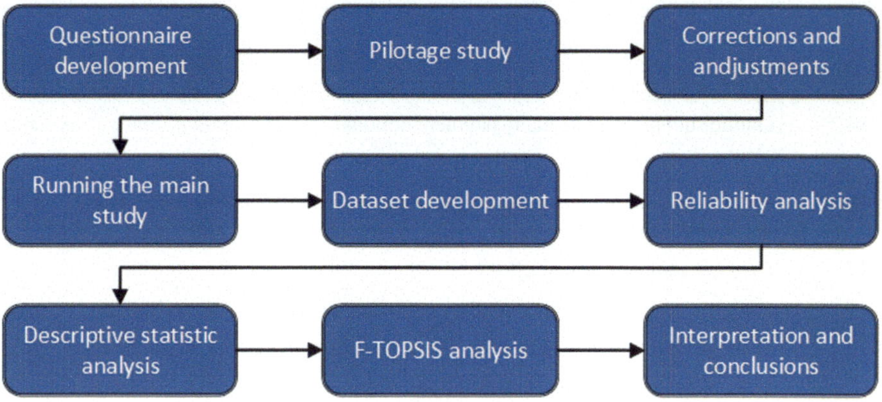

Fig. 1. A research framework.

Based on the literature analysis, a questionnaire was created regarding the perception of the quality of functioning of Internet platforms offering online shopping. Because the study was aimed at users of these platforms, therefore, in the first question, respondents declared that they would make such purchases. If the answer is no, the study has ended. The questionnaire contained research questions and metric questions. Based on research questions, 24 criteria were distinguished, which make up the general perception of the quality of functioning of online platforms (Table 1).

Table 1. Quality criteria for the functioning of purchasing platforms

Symbols	Criteria
V1	Online shopping is quick
V2	Online shopping is transparent
V3	Online shopping is easy
V4	Online shopping offers additional options
V5	Online shopping offers access to products unavailable in traditional shopping
V6	Online shopping offers additional offers or promotions
V7	Online shopping is highly interactive
V8	Each stage of online shopping runs smoothly
V9	The latest IT solutions are used in online shopping
V10	The online platform's interface is very intuitive
V11	Online shopping is characterized by a transparent pricing policy
V12	Users have access to all information needed
V13	Online shopping can be done at convenient times for users

(*continued*)

Table 1. (*continued*)

Symbols	Criteria
V14	Online shopping offers many payment options
V15	Online shopping offers many delivery options
V16	When shopping online, contacting customer service is easier compared to a traditional store
V17	Payment transactions are safe
V18	Online shopping is accompanied by an understandable privacy policy
V19	While shopping online, users have more control over the purchase process
V20	The online shopping platform is trustworthy
V21	Online shopping ensures anonymity for buyers
V22	Online shopping responds to the individual needs of users
V23	The purpose of online shopping is to meet the needs and expectations of users
V24	The shopping platform displays additional product recommendations in line with users' preferences

On the other hand, the metric questions concern such issues as gender, age, time of making online purchases, and frequency. The final structure of respondents was as follows (Table 2):

Table 2. Descriptive statistical results.

Classification	Frequency	Percent
Gender		
Male	134	33%
Female	266	67%
Age		
18–24	194	49%
25–34	17	4%
35–44	129	32%
45–54	53	13%
Above 54	7	2%
Online shopping duration		
Less than one month	2	1%
Less than one year	18	5%

(*continued*)

Table 2. (*continued*)

Classification	Frequency	Percent
1–2 years	54	14%
2–5 years	172	43%
More than five years	154	39%
Online shopping frequency		
Few times a week	8	2%
Few times a month	90	23%
At least once a month	143	36%
Only for a special occasion	34	9%
Few times a year	125	31%

Before conducting the main survey, a pilot study was conducted to identify any inaccuracies and possible ambiguities that could make it difficult for the respondents to understand the survey questions. After introducing the necessary corrections, the main study was carried out. The minimum sample size for such research, with an acceptable confidence level of 1–2 = 0.90 and the maximum estimation error e = 5%, should be 384 questionnaires. Because 400 questionnaires were qualified for the research, this condition has been met, and the sample obtained meets one of the assumptions of the representative method. The main study was conducted from December 2019 – April 2020 using the CAWI method. Next, an integrated response database was developed containing all the answers to the issues contained in the surveys. To develop the database and encode raw data, the Excel 2019 spreadsheet was used. In the next step, reliability analysis and frequency tests were carried out using Statistica 13 software. Then, the following stages of F-TOPSIS were performed, and the results were interpreted.

3.2 The Reliability Analysis

The nature of the answers to individual questions was a single-choice and a 5-fold answer choice, so the 5-step Likert scale was adopted, a bipolar interval scale. The scale used in the study consists of 5 categories of answers: "very high" (VH), "high" (H), "medium" (M), "low" (L), "very low" (VL) ordered in the correct order. Therefore, before analyzing the obtained research results, the reliability of the measurement scale was analyzed using Cronbach's alpha statistics (Table 3).

For all the variables, Cronbach's alpha values are equal to or higher than 0,90, which means that the scale's reliability is highly acceptable.

3.3 The Research Results

As mentioned earlier, using the TOPSIS method to evaluate the quality criteria of purchasing platforms' functioning involves several stages arranged correctly. The first step

Table 3. The Reliability Test results.

Criteria	The average of the scale after item removal	The variance of the scale after item removal	Item correlation	Cronbach's alpha after item removal
V1	89,97	140,51	0,65	0,90
V2	90,27	141,12	0,64	0,90
V3	90,44	144,34	0,28	0,91
V4	90,10	142,58	0,55	0,91
V5	90,28	139,29	0,60	0,90
V6	90,42	138,10	0,67	0,90
V7	90,51	140,18	0,63	0,90
V8	90,46	143,05	0,52	0,91
V9	90,58	140,51	0,61	0,90
V10	90,36	142,75	0,53	0,91
V11	90,53	141,46	0,51	0,91
V12	90,41	139,05	0,64	0,90
V13	89,84	141,10	0,61	0,90
V14	89,99	142,12	0,56	0,91
V15	90,06	141,54	0,57	0,91
V16	90,84	147,28	0,21	0,91
V17	90,43	143,49	0,53	0,91
V18	90,61	140,87	0,60	0,90
V19	90,61	144,31	0,41	0,91
V20	90,50	140,72	0,67	0,90
V21	90,99	145,36	0,29	0,91
V22	90,30	146,04	0,43	0,91
V23	90,18	142,90	0,55	0,91
V24	90,33	141,42	0,53	0,91

was to obtain the respondents' ratings for each quality criterion. In the study, respondents rated the importance of each quality criterion using an ordinal scale that included the following categories: "very high" (VH), "high" (H), "medium" (M), "low" (L), "very low" (VL) ordered in the correct order. Therefore, in the second step, fuzzy weight coefficients were calculated based on the respondents' ratings, which resulted in the above categories being transformed into fuzzy triangular numbers (Table 4).

Then, a fuzzy data matrix was built, a fragment of which is presented in the Table 5 below:

Table 4. Parameters of fuzzy triangular numbers assigned to particular categories.

Categories	Symbol	Pessimistic value (parameter l)	The best-fit value (parameter m)	Optimistic value (parameter u)
Very high	VH	0,7	0,9	1
High	H	0,5	0,7	0,9
Medium	M	0,3	0,5	0,7
Low	L	0,1	0,3	0,5
Very low	VL	0	0,1	0,3

Table 5. Fragment of a fuzzy data matrix.

Criteria	Respondents			
	$Respondent_1$	$Respondent_2$	...	$Respondent_{400}$
V1	(0,7; 0,9; 1)	(0,7; 0,9; 1)	...	(0,5; 0,7; 0,9)
V2	(0,5; 0,7; 0,9)	(0,7; 0,9; 1)	...	(0,5; 0,7; 0,9)
V3	(0,7; 0,9; 1)	(0,7; 0,9; 1)	...	(0,5; 0,7; 0,9)
...	...	...	...	
V24	(0,3; 0,5; 0,7)	(0,5; 0,7; 0,9)	...	(0,5; 0,7; 0,9)

At this point, the stage of creating a normalization matrix was omitted because the values assigned to successive quality criteria were of the same category. Therefore, there was no need to standardize them. The fuzzy matrix presented above was the basis for all subsequent calculations and successive stages implemented in the fuzzy TOPSIS method.

In the next step, the aggregated values of the fuzzy weights were calculated (Table 6).

Table 6. The aggregated values of the fuzzy weights.

Criteria	L-FW	M-FW	U-FW	Criteria	L-FW	M-FW	U-FW
V1	0,58	0,76	0,91	**V13**	0,61	0,79	0,92
V2	0,51	0,71	0,87	**V14**	0,57	0,76	0,91
V3	0,49	0,67	0,83	**V15**	0,56	0,74	0,90
V4	0,55	0,74	0,90	**V16**	0,39	0,59	0,77
V5	0,50	0,71	0,87	**V17**	0,48	0,68	0,86
V6	0,47	0,68	0,85	**V18**	0,43	0,64	0,83
V7	0,45	0,66	0,84	**V19**	0,44	0,64	0,82

(*continued*)

Table 6. (*continued*)

Criteria	L-FW	M-FW	U-FW	Criteria	L-FW	M-FW	U-FW
V8	0,47	0,67	0,85	**V20**	0,46	0,66	0,85
V9	0,44	0,65	0,83	**V21**	0,36	0,57	0,76
V10	0,49	0,69	0,87	**V22**	0,50	0,70	0,88
V11	0,44	0,66	0,84	**V23**	0,53	0,72	0,89
V12	0,48	0,68	0,85	**V24**	0,48	0,70	0,87

In the following steps:

- coordinates for the fuzzy positive ideal solution (FPIS) and fuzzy negative ideal solution (FNIS) were estimated;
- the Euclidean distance for each fuzzy alternative of the positive and negative ideal values was calculated;
- the closeness coefficient CCi of the alternatives are calculated;
- ordering the obtained values in descending order - ranking

The final results are presented in the Table 7 below.

The obtained research results indicated that the most important thing for the respondents is that online shopping can be done at any time (V13), that it is a fast process (V1), and the fact that this type of shopping offers various forms of payment (V14). On the other hand, the last places were taken by the anonymity of buyers (V21), the fact that in the case of online shopping, contact with customer service is more accessible than in the case of traditional purchases (V16), and the fact that when shopping, respondents have more control over the purchase process (V19).

Similar research was carried out by a broad group of scientists who used this method:

- concerning problems related to the choice of network services, the effect of the research was the development of an approach to modeling software and hardware as services in order to meet dynamic customer requirements [14];
- to assess the competitive advantages of shopping sites operating in Taiwan, it was indicated that security and trust are the most critical factors in increasing the competitive advantage of a shopping site [15].
- to rank websites by proposing a decision-making approach and model based on the AHP and F-TOPSIS methods [16];
- to select IT system providers using an interactive decision-making methodology to select/rank IS providers based on multiple criteria [17].

None of these studies, however, concerned the analysis of criteria relating to the quality of shopping websites, which means that the research results are a very good supplement to the existing knowledge.

Table 7. Final results

Criteria	Distance from the ideal solution	Distance from the non-ideal solution	CC_i	Rank position
V1	67,55	285,35	0,81	2
V2	110,54	242,60	0,69	7
V3	136,28	217,63	0,61	14
V4	86,86	266,02	0,75	5
V5	113,20	239,94	0,68	8
V6	129,66	223,63	0,63	13
V7	143,83	209,09	0,59	17
V8	137,02	216,20	0,61	16
V9	153,68	199,28	0,56	20
V10	125,52	227,78	0,64	11
V11	147,84	205,59	0,58	19
V12	128,78	225,04	0,64	12
V13	50,72	302,37	0,86	1
V14	72,03	280,85	0,80	3
V15	83,20	270,01	0,76	4
V16	192,81	160,24	0,45	23
V17	136,62	216,66	0,61	15
V18	158,52	194,94	0,55	21
V19	162,09	191,58	0,54	22
V20	144,46	208,37	0,59	18
V21	212,50	140,91	0,40	24
V22	117,65	235,46	0,67	9
V23	99,57	253,51	0,72	6
V24	122,61	230,87	0,65	10

4 Conclusions

The use of the F-TOPSIS method made it possible to identify which of many criteria for the quality of the functioning of online platforms are the most important for people making online purchases. The final effect of the analysis in the form of a ranking made it possible to rank these criteria from the best to the worst from the point of view of the perception of the quality of the functioning of online platforms.

The results of the conducted analysis may constitute guidelines for managers in effectively managing Internet platforms. Bearing in mind that the freedom of time to make purchases, many payment options, and the simplicity of the entire purchasing process is the most important criteria for the quality of functioning of online platforms, in the process of managing these platforms, care should be taken to ensure that these criteria are not lowered. On the other hand, lower scores for other criteria indicate to managers areas that require further improvement to achieve the highest possible level of quality.

In addition, the results obtained in the form of a ranking of quality criteria can be a guideline for managers managing Internet platforms, thanks to which they can plan the sequence of individual actions aimed at improving the overall quality level of Internet platforms. However, it should be borne in mind that the subject matter discussed in this article covers an area subjected to constant changes - the functioning of online platforms is changing, and the attitudes and expectations of online shopping customers are changing. This means that this subject should be constantly explored - further research can be carried out covering a more comprehensive range of criteria grouped into cultural, social, or technological groups.

References

1. Ignatius, A.: HBR Predicts the Future. Harv. Bus. Rev. **45**, 4–16 (1967)
2. Majchrzyk, Ł.: e-Commerce w Polsce 2021: 77% internautów kupuje online, https://mobirank.pl/2021/09/28/e-commerce-w-polsce-2021-77-proc-internautow-kupuje-online/. Last accessed 27 June 2023
3. Mohsin, M.: 10 Online Shopping Statistics You Need to Know in 2023, https://www.oberlo.com/blog/online-shopping-statistics. Last accessed 27 June 2023
4. Bucko, J., Kakalejcik, L., Ferencova, M.: Online shopping: factors that affect consumer purchasing behavior. Cogent Business & Management **5**(1), 1–15 (2018)
5. Broniewicz, E., Dziurdzikowska, E.: Multi-criteria methods in balancing socio-economic processes. Res. Papers of Wrocław Univ. Econ. **491**, 53–62 (2017)
6. Patel, M., Bhatt, B., Vashi, M.: SMART-Multi-criteria decision-making technique for use in planning activities (2017)
7. Organ, A., Yalcin, E.: Performance evaluation of research assistants by copras method. Eur. Sci. J. **8**, 102–109 (2016)
8. Trzaskalik, T.: Wielokryterialne wspomaganie decyzji. Przegląd Metod i zastosowań. Zeszyty Naukowe Politechniki Śląskiej **74**, 239–263 (2014)
9. Hwang, C.-L., Yoon, K.: Multiple Attribute Decision Making. Lecture Notes in Economics and Mathematical Systems, p. 186, Springer (1981)
10. Jahanshahloo, G.R., Hosseinzadeh, L.F., Izadikhah, M.: Extension of the TOPSIS method for decision-making problems with fuzzy data. Appl. Math. Comput. **185**, 1544–1551 (2006)
11. Kacprzak, D.: Podwójna rozmyta metoda TOPSIS wspomagająca podejmowanie decyzji grupowych. Przegląd Statystyczny **66**, 27–50 (2019)
12. Roszkowska, E., Wachowicz, T.: Application of fuzzy TOPSIS to scoring the negotiation offers in ill-structured negotiation problems. Eur. J. Oper. Res. **242**(3), 920–932 (2015)
13. Chen, C.-T.: Extensions of the TOPSIS for group decision-making under fuzzy environment. Fuzzy Sets Syst. **114**(1), 1–9 (2000)
14. Cheng, D.Y., Chao, K.M., Lo, C.C., Tsai, C.F.: A user centric service-oriented modeling approach. World Wide Web **14**, 431–459 (2011)

15. Sun, C.C., Lin, G.T.R.: Using fuzzy TOPSIS method for evaluating the competitive advantages of shopping websites. Expert Syst. Appl. **36**, 11764–11771 (2009)
16. Yu, X., Guo, S., Guo, J., Huang, X.: Rank B2C e-commerce websites in e-alliance based on AHP and fuzzy TOPSIS. Expert Syst. Appl. **38**, 3550–3557 (2011)
17. Kahraman, C., Engin, O., Kabak, Ö., Kaya, İ.: Information systems outsourcing decisions using a group decision-making approach. Eng. Appl. Artif. Intell. **22**, 832–841 (2009)

Use of the Technique for Order Preference by Similarity Ideal Solution Method for the Assessment of Education Level in Poland Related to EU Countries

Małgorzata Nermend, Kinga Flaga-Gieruszyńska(✉), and Mariusz Borawski

University of Szczecin, Cukrowa 8, 71-004 Szczecin, Poland
kinga.flaga-gieruszynska@usz.edu.pl

Abstract. The problem of education can be categorized as a multi-criteria issue because it can be described by multiple factors. The authors of the study attempt to utilize the multi-criteria TOPSIS method to assess the level of Pre-primary to tertiary education in the countries of the European Union using dynamic approaches. Pre-primary to tertiary education plays a crucial role in the growth and development of the younger generation. At this level of education, they acquire fundamental skills that are essential for their further development. Therefore, this process necessitates continuous monitoring to improve the quality and adaptation of educational programs to the dynamic of ever-changing world. The Education process is very complicated because it is influenced by many factors which exceed the educational obligations at the school level, including the very important role of state action. The success or failure of this process will depend on the State's actions in this regard. The use of the multi-criteria TOPSIS method for assessing the level of pre-primary to tertiary education in EU countries will allow each country to monitor its development in education in relation to the rest of the EU member countries, which can be helpful in rapid intervention in improving this process.

Keywords: Pre-primary to tertiary education · multi-criteria method · TOPSIS method

1 Introduction

Elementary school is the foundation of education and at the same time a very important period in a child's life as during this time their educational, social and emotional development takes place. In Europe, all children are covered by compulsory schooling, however, they start school at different ages [1]. In Poland, a child covered under compulsory education must be at least 7 years old. In other European countries, this obligation usually varies between 6 and 7 years of age, for example, in Austria, Denmark, France, Spain, and Luxembourg children start school at the age of 6, in Greece, the Netherlands, and Germany at the age of 5, in Lithuania and Estonia at the age of 7 [2]. Educational institutions should be well-equipped to provide students with adequate care and learning

M. Hernes and J. Wątróbski (Eds.): ECAI 2023, LNNS 1079, pp. 316–330, 2024.
https://doi.org/10.1007/978-3-031-66761-9_25

conditions. Education is a fundamental right in the modern world. It has a significant impact not only on the individual, but also on society as a whole, contributing to the development of the country [3]. There are many studies on educational outcomes. A very well-known one is the PISA (Programme for International Student Assessment) survey coordinated by the Organization for Economic Cooperation and Development [4], which is also conducted in Poland among students who are 15 years of age or older [5, 6]. Another one of the world's largest research projects on educational achievement is the PIRLS (Progress in International Reading Literacy Study) [7], implemented by the International Association for the Measurement of School Achievement - IEA, which aims to measure reading comprehension skills including students aged 9.5 or less [8]. Similarly, another important global study organized by the IEA is TIMSS (Trends in International Mathematics and Science Study) to measure school achievement in mathematics and science [9]. It includes the measurement of students' skills in the fourth and eighth years of schooling. Another notable study is the Pearson Ranking, which includes a ranking of education systems that test proficiency in the sciences and humanities [10].

The above-described surveys are the largest conducted in the world that address the school skills of students. It should be noted that currently highly developed societies are characterized by having fewer children, which gives rise to certain consequences such as demographic decline and an aging population. The differences in this context have significant consequences for society or the labor market. However, as mentioned earlier, the quality of education is also a matter of economic factors such as the financing of educational units, social and economic factors such as adequate infrastructure, computerization, provision of school meals, space for play and rest, outdoor playgrounds, extra-curricular and day-care activities [11–14].

European countries also differ in their levels of access to education and school systems as a result of economic, political, and social changes. Access to education should be the same for both boys and girls and should eliminate any discrimination. Gender equality is one of the UN Sustainable Development Goals and the right to education is now one of the fundamental rights of an individual. It is traditionally classified as a cultural right, but some look at it – and rightly so – as a certain economic aspect [15]. The doctrine of constitutional law indicates that this right must be seen both in terms of the good and value of an individual and an important fundamental social good. The realization of the right to science is not so much a special privilege of modern man, but is an indispensable condition for the development of society and at the same time the full participation of the individual in social life. Universal education has become the most important driver of economic and civilizational development in the modern world [16].

The beneficiary of the right to education is everyone. On the other hand, its content is the possibility of acquiring knowledge (education), conducted in organized forms, on a regular and continuous basis, covering first a certain basic canon of general knowledge, and then enabling the acquisition of in-depth specialist knowledge, completed by obtaining documents, uniformly on a national scale giving the opportunity to continue to study or pursue a particular profession. There are disputes in the doctrine as to whether the right to education should be considered just in terms of classical social labor or also as a freedom right [17].

The starting point for the proper formation of the right to education of children is the provisions of Article 28 para. 1 of the Convention on the Rights of the Child on 20.11.1989 [18]. According to which, they recognize the child's right to education and to progressively exercise this right on an equal footing. In particular make basic education compulsory and free for all, and take steps to ensure regular school attendance and to reduce school leaving rates [19].

In turn, the purpose of the Convention on Combating Discrimination in the Field of Education drawn up in Paris 15.12.1960 [20] is not only the elimination of all discrimination in the field of education, but also the recognition as a principle of equal opportunities and equal treatment of all people in this field [21, 22].

The right of everyone to education was also reflected in the International Covenant on Economic, Social and Cultural Rights on 19.12.1966 [23]. Article 13 of the Covenant stipulates that teaching should aim at the full development of personality and a sense of human dignity and promote respect for human rights and fundamental freedoms. In order to establish the standards of the right to education in the Member States of the European Union, the provisions of the Charter of Fundamental Rights of the European Union (hereinafter referred to as CFREU), but also the provisions of the European Convention on Human Rights and Fundamental Freedoms on 4.11.1950 [24] (ECHR) including Protocol No 1 of 20.3.1952 [25] are of paramount importance because of their membership in the Council of Europe. In addition to the Convention, the Council of Europe also presents a wide range of recommendations by the Committee of Ministers focusing on the right to education, including recommendations on quality assurance of education [26]. In accordance with Art. 14 para. 1 CFREU everyone has the right to education and access to vocational and continuing training. This right includes the possibility of receiving compulsory education free of charge (Article 14 para. 2 CFREU). Moreover, that provision corresponds with the principle concerning the impact of the European Union on the functioning of education in the Member States [27]. In accordance with Article 2 of Protocol No 1 to the ECHR, no one may be deprived of the right to education. In carrying out its functions in the field of education and teaching, the State recognizes the right of parents to provide education and teaching in accordance with their own religious and philosophical beliefs [28–30].

In the framework of the Council of Europe recommendations, it is crucial to ensure quality education for all and to define the role of public authorities in achieving this goal. It should be emphasized that it is necessary to adapt the quality of education and the responsibility of public authorities to ensure it to the demands of modern, complex societies. High-quality education should be recognized by public authorities as an important element of their policies. Public authorities are responsible for ensuring the quality of the entire educational offer within the national education system, whether in public or private establishments. They decide which educational institutions and programmes, whether public or private, belong to the national system under the responsibility of the state authorities. Decisions should be made public and should be based on transparent requirements, criteria, and procedures, considering the quality of institutions and programmes. All school-age children should have the right and duty to participate in quality education provided by public or private institutions. Public education should be free of charge and all parents or legal guardians should have the right and duty to include their children in the education system. If parents or legal guardians do not enroll their children in school, public authorities should have appropriate powers and measures to ensure the right of children to quality education.

To sum up, ensuring the quality of education is one of the manifestations of "equal access", which has been recognized as a duty of public authority. The implementation of this obligation must not be limited to eliminating legal differences, but must also focus on removing factual inequalities, especially those resulting from the material situation, and in Polish conditions still emerging, for example, between the city and the countryside.

In this article, the authors examine these relationships using the TOPSIS method. The TOPSIS method is an example of a multi-criteria decision analysis (MCDA) tool useful for making decisions to rank or select alternatives based on multiple criteria [31–33]. Studies using this method have applications in the most diverse areas and fields of science. Their range is so large that we can find studies in the field of environmental protection [34–37], medical waste processing [38], In engineering and technology, economic management and treatment [39] stress testing of doctors during COVID-19 [40], and even in determining the best escape route [41].

2 Materials and Methods

The research used Technique for Order Preference by Similarity Ideal Solution (TOPSIS) method was applied as a methodical apparatus, to assess of education level in Poland related to EU countries, in the dynamics approach. The problem in the research is a multi-criterion described with many criteria in the selected European Union countries. The criteria value was collected from the Eurostat database in the time horizon (2013–2021) years. At the beginning of the study based on the adopted criteria, has create one matrix with the data from all analysis periods (2013–2021) duo to create a common ideal and anti-ideal variant as a reference point. Moreover, was calculated based on the created matrix the mean and standard deviations of which will be used at the stage of normalization of the data matrix for each year separately when using the TOPSIS method. In this way, we eliminate the possibility of better-than-ideal and anti-ideal variants appearing in the research sample and the possibility of creating a ranking over time. The criteria used in this research were selected from the Pre-primary to tertiary education area:

x_1 – Number of classroom teachers in primary education per 1,000 inhabitants.
x_2 – Pupils enrolled in primary education by sex, type of institution and intensity of participation per 1,000 inhabitants.
x_3 – Public expenditure on pre-primary to tertiary education - as % of GDP.
x_4 – Expenditure on public educational institutions - as % of total or current expenditure - c.
x_5 – Financial aid to students by Pre-primary to tertiary education - as % of total public expenditure.
x_6 –Households without access to internet at home because the access costs are too high (in %).

Analyzed selected countries (decision variants) were: Austria (AT), Belgium(BE), Bulgaria (BG), Cyprus (CY), Czechia (CZ), Denmark (DK), Finland (FI), France (FR),

Germany (DE), Greece (EL), Hungary (HU), Ireland (IE), Italy (IT), Latvia (LV), Lithuania (LT), Luxembourg (LU), Malta (MT), Netherlands (NL), Poland (PL), Portugal (PT), Romania(RO), Slovakia (SK), Slovenia (SI), Spain (ES), Sweden (SE), Data referred to the year 2013–2021 and were obtained from the Eurostat database. Croatia (HR), Estonia (EE), and Norway (NO) were not included in the analysis due to a lack of data for most of the adopted criteria.

2.1 Description of Technique for Order Preference by Similarity Ideal Solution (TOPSIS)

The research used Technique for Order Preference by Similarity Ideal Solution (TOPSIS) method is used to order decision variants in terms of their similarity to an ideal variant, i.e. the most desirable variant. This is accomplished by minimizing the distance to the benchmark, the so-called ideal reference solution, and maximizing this distance to the anti-ideal, reference solution. Distances are calculated between each variant and the ideal and anti-ideal benchmarks, from which the value of the measure is determined. It allows the formation of the final order [42, 43]. At the beginning of the study, criteria selection and criterion elimination were performed as additional step in the TOPSIS method. The criteria in the study were selected using a substantive and logical selection approach and a variable elimination approach such as those with a high degree of collinearity [43, 44]. Elimination of variables is carried out by using significance coefficient of features [45]:

$$V_{x_i} = \frac{\sigma_i}{\overline{x}_i} \tag{1}$$

where: x_i – i-th variable, σ_i – standard deviation of the i-th variable, $\overline{x}_i$ – mean value of the i-th variable,

Variables, for which significance coefficient values are within the range 0; 0, 1, are quasi-constant and such variables should be eliminated from the set of variables under consideration [45, 46]. The first step of the TOPSIS method is the normalization of variables, which can be done using various methodological approaches. The approach proposed here is expressed by the formula [47]

$$x''_{ij} = \frac{x'_{ij}}{\sqrt{\sum_{k=1}^{n} x'^2_{ik}}}, \tag{2}$$

where $A_i = 0$ (shift of normalized values with respect to zero), and B_i (normalization base of the i-th criteria) determined from the formula:

$$B_i = \sqrt{\sum_{j=1}^{n} x'^2_{ij}}, \tag{3}$$

where: x''_{ij}- value of the *i-th* criteria for the *j-th* variant after normalization, x'_{ij} - value of the *i-th* criteria for the *j-th* variant

of the variant after moving and rescaling the set of points, but before coordinate correction. The normalized values are multiplied by the weights:

$$v_{ij} = x''_{ij}.w_i, \tag{4}$$

where: w_i - weight value for the *i-th* variant.

In the article, ten experts from the research area (Pre-primary to tertiary education teachers) were asked to determine the importance of criteria on a scale [43–46, 48–53]. The average rating of the experts' evaluation showed that the criteria (x_1 – [10], x_2 – [7], x_3 – [8], x_4 – [8], x_5 – [9], x_6– [6]). In the next step, the weights were normalized according to the formula:

$$w_i = \frac{w_i}{\sum_{j=1}^{m} w_j} \tag{5}$$

The criteria will have a greater impact on the value of the measure, the higher the w_i. However, it should be remembered that the weights are to be positive and meet the following condition:

$$\sum_{i=1}^{m} w_i = 1 \tag{6}$$

In practice, researchers use the statistical weights approach, which is based on statistical resources. Criteria weighting can be carried out, for example, using a relative value meter [52], an example of such an approach is:

$$w_i = \frac{V_{x_i}}{\sum_{k=1}^{m} V_{x_k}} \tag{7}$$

Determining the character criteria is very important for calculating the value of the ideal and ani-ideal variants. The criteria character can as benefits. The criteria with benefits character are such variables, in which greater values mean a higher level of benefits and no benefit are such variables which smaller values mean a higher level of benefits. in the study, the criteria ($x_1 - x_5$) had character benefits and (x_6) was no benefit. The ideal reference solution is determined based on the relationship:

$$v_i^+ = \begin{cases} max_j\, v_{ij} \text{ for benefit} \\ min_j\, v_{ij} \text{ for no benefit} \end{cases} \tag{8}$$

And the anti-ideal reference solution based on the record:

$$v_i^- = \begin{cases} min_j\, v_{ij} \text{ for benefit} \\ max_j\, v_{ij} \text{ for no benefit} \end{cases} \tag{9}$$

The distance from the reference object (ideal reference solution) is calculated according to the formula:

$$d_j^+ = \sqrt[p]{\sum_{i=1}^{m} \left|v_{ij} + v_i^+\right|^p} \tag{10}$$

and the distance from the anti-ideal reference solution by the formula:

$$d_j^- = \sqrt[p]{\sum_{i=1}^{m} \left|v_{ij} + v_i^-\right|^p} \tag{11}$$

where: p - A constant that specifies the type of metric.

Finally, the value of the measure is determined from the formula:

$$S_j = \frac{d_j^-}{d_j^+ + d_j^-}, \tag{12}$$

where: S_j is a value in the range from 0 to 1, with 0 being the worst variant, and 1 variant of the best. The last step of the study is rankings of the variant depending on the obtained aggregate measure value values; thus, it is possible to determine which of them are "better" and which are "worse". They also allow determining which are like each other in terms of adopted criteria. To better visualize the results of calculations, decision variants can be divided into classes with similar measurement values. To better visualize the results of the EU country classification, a map was generated. In the simplest case objects can be classified based on mean value $\overline{m}_{S_j}$ and standard deviation of the aggregate measure. σ_{S_j} Objects are classified into four classes [48]:

$$\begin{cases} 1 \text{ for } S_j \in \langle \overline{m}_{S_j} + \sigma_{S_j}; \infty), \\ 2 \text{ for } S_j \in \langle \overline{m}_{S_j}; \overline{m}_{S_j} + \sigma_{S_j}), \\ 3 \text{ for } S_j \in \langle \overline{m}_{S_j} - \sigma_{S_j}; \overline{m}_{S_j}), \\ 4 \text{ for } S_j \in \left(-\infty; \overline{m}_{S_j} - \sigma_{S_j}\right). \end{cases} \tag{13}$$

3 Research Results

The level of Pre-primary to tertiary education in EU countries was carried out using the TOPSIS method. The obtained results of the aggregate measure and the determined class membership are presented in maps. The research was conducted from 2013 to 2021, the results are illustrated in three figures due to limitations in the number of pages by the organizers of the conference, every other year was selected in Figs. 1 and 2, and in Fig. 3 the results of the last two years. To study the dynamics of changes in the level of Pre-primary to tertiary education in EU countries, each grade in a separate figure was illustrated in four figures. Regarding the ideal and anti-ideal variants, a matrix with data from all periods of analysis (2013–2021) was created as a reference point.

The created ranking Fig. 1. a) and b) shows that in total four counties were classified as Class 1: Denmark, Sweden, Netherlands, and Ireland. The situation in class one between 2013 and 2015 is stable. These are countries providing potentially the best environment for Pre-primary to tertiary education. Denmark is the leading country in this ranking. The opposite pole (Class 4) Also includes four countries, of which Romania had the lowest ranking. In this respect, Slovakia changed its ranking from place 23 to 15 and moved up from Class 4 to 3. These changes result from the fact that in 2015 in Slovakia the value of the criteria, x_1 – Number of classroom teachers in primary education per 1,000 inhabitants, x_2 – Pupils enrolled in primary education by sex, type of institution, and x_3- Public expenditure on pre-primary to tertiary education - as % of GDP, decreased and amounts to 21,64%, 30,31%, and 13,58%. Lithuania from place 18 to 22 moved down from Class 3 to 4. These changes result from the fact that in 2015 in Lithuania

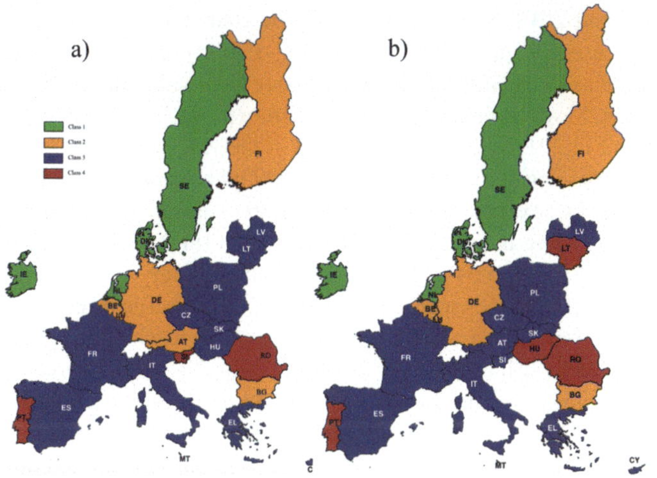

Fig. 1. Ranking of EU countries in 2013 and 2015 in respect of the level of Pre-primary to tertiary education Topsis method.

the value of the criteria x_6- Households without access to the internet at home because the access costs are too high (in %) criterion increased by 6.9%, which means that the number of households without access to the Internet has increased due to the high cost. This process may be of concern, as access to the Internet can be very important in the education process this shows that in the context of friendliness for Pre-primary to tertiary education, these countries are closer to Denmark.

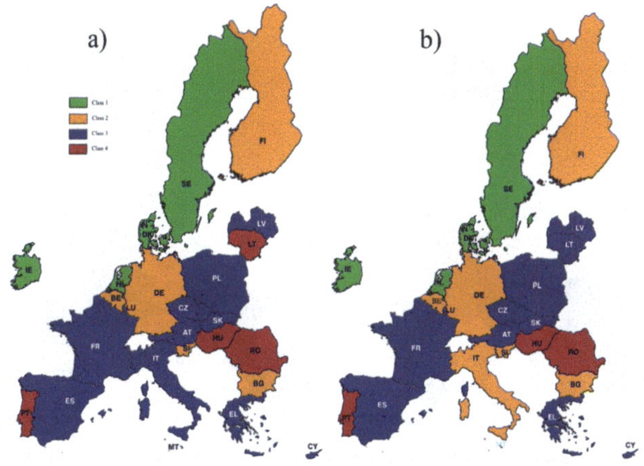

Fig. 2. Ranking of EU countries in 2017 and 2019 in respect of the level of Pre-primary to tertiary education Topsis method.

The created ranking Fig. 2. a) and b) shows the results of the classification of EU countries due to the level of Pre-primary to tertiary education obtained based on the data from 2017 and 2019. Slovenia shows a huge change in reference to 2013. It is a very interesting case, even though education (x3 – Public expenditure on pre-primary to tertiary education as % of GDP) declined by about 11% with respect to 2013.

In this respect, Slovenia changed its ranking from place 23 to 5. These changes result from the fact that in 2019 in Slovenia increase in the value of critters x1 – Number of classroom teachers in primary education per 1.000 inhabitants. x2 – Pupils enrolled in primary education by sex, type of institution and x6 –Households without access to the internet at home because the access costs are too high (in %)., decreased and amounts to 64.39%, 16.21%, and 496.64%. This process impacted very positive prospects in the education process. Malta is down up in reference to 2013.It is a next very interesting case, even though education criteria x1 – Number of classroom teachers in primary education per 1,000 inhabitants. is increased by about 5.41%, x2 – Pupils enrolled in primary education by sex, type of institution, and intensity of participation per 1,000 inhabitants down up about 3.25%, x3 – Public expenditure on pre-primary to tertiary education - as % of GDP down up about 22%, x4 – Expenditure on public educational institutions - as % of total or current expenditure - c down up about 16.66%, x5 – Financial aid to students by Pre-primary to tertiary education - as % of total public expenditure down up about 30.91%, and x6 –Households without access to the internet at home because the access costs are too high (in %) down up about 84.95%. In this respect, Malta changed their ranking from place 7 to 16 and moved down from Class 2 to 3 This process impacted very negative prospects in the education process in Malta. In this respect, the Lithuania changed their ranking from place 18 to 22, and Hungary from place 22 to 24. Both countries moved down from Class 3 to 4. While Poland showed stability and did not show much change between classes. There were small changes within the third class.

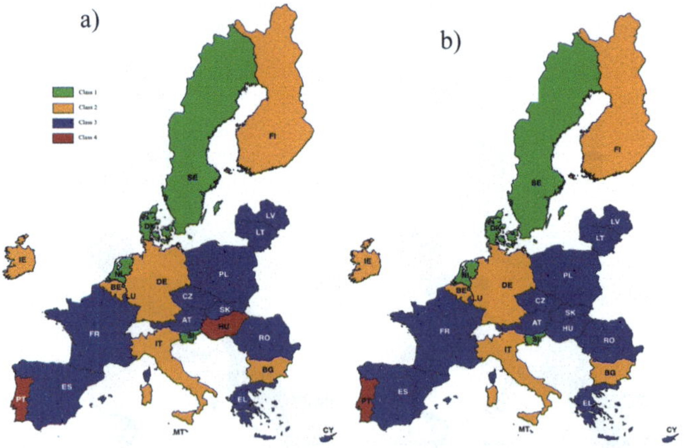

Fig. 3. Ranking of EU countries in 2020 and 2021 in respect of the level of Pre-primary to tertiary education Topsis method.

The created ranking Fig. 3. a) and b) shows the results of the classification of EU countries by the level of education obtained from Pre-primary to tertiary education based on the data from 2020 and 2021. A huge change shows that the fourth class, in relation to 2013, contains only one variant (Portugal). With to the other variants, Hungary and Romania changed their ranking position and moved up from class 4 to 3, and Slovenia changed its ranking position from place 24 to 4 and moved up from class 4 to 1. This shows that in the context of Slovenia, friendliness due to the level of education from pre-school to higher is closer to Denmark.

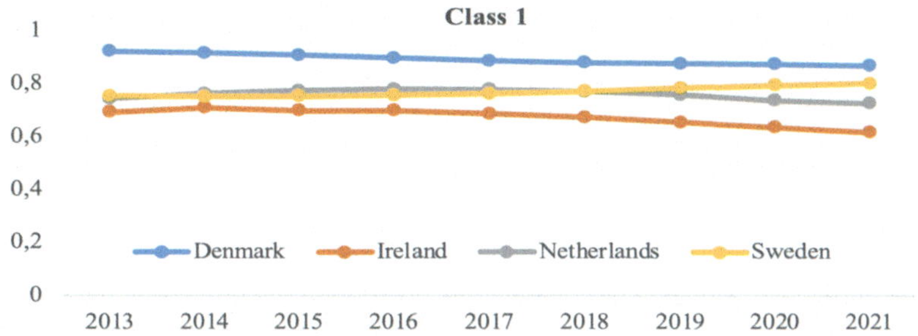

Fig. 4. Dynamics of changes in EU countries in the first class (2013–2021) in respect of the level of Pre-primary to tertiary education by the TOPSIS method.

The Fig. 4. Shows dynamics of changes in the level of early childhood education. In the first class, Ireland has changed its position in the ranking. The value of Ireland's aggregate measure falling from 0.693 to 0.617, in 2021, thus moving from the first class to the second class. While Denmark is the leading country in this ranking (2013–2021) (Fig. 5).

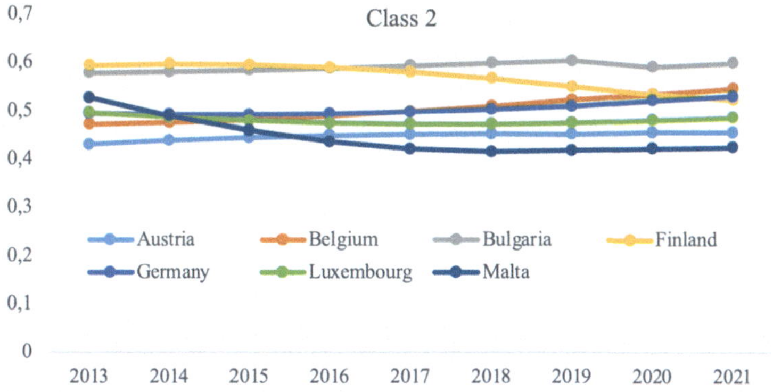

Fig. 5. Dynamics of changes in EU countries in the second class (2013–2021) in respect the level of Pre-primary to tertiary education by the TOPSIS method.

In the second class, Luxembourg has changed its position in the ranking. The value of Luxembourg's aggregate measure falling from 0.50 to 0.48, in 2021, thus moving from the second class to the third class. While Malta showed a lot of variability dropping from eighth position in class two to position 18 in class three. The value of the aggregate measure dropped from 0.53 to 0.42. (2013–2021) (Fig. 6).

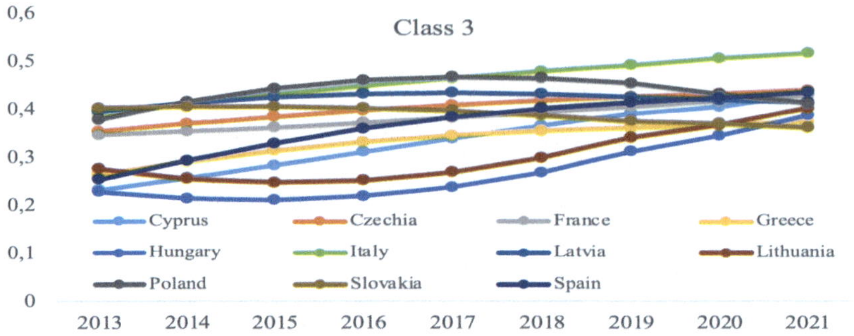

Fig. 6. Dynamics of changes in EU countries in the third class (2013–2021) in respect of the level of Pre-primary to tertiary education by the TOPSIS method.

In the third class, Slovakia has changed its position in the ranking in the third class. The value of Slovakia's aggregate measure fell from 0.40 to 0.36, in 2021. While Spain and Cyprus showed a lot of variabilities increases from positions 21 and 22 to 15 and 16 positions in class three. The value of the aggregate measure increases from 0.25 to 0.44 and Cyprus from 0.23 to 0.43 in the time between (2013–2021) (Fig. 7).

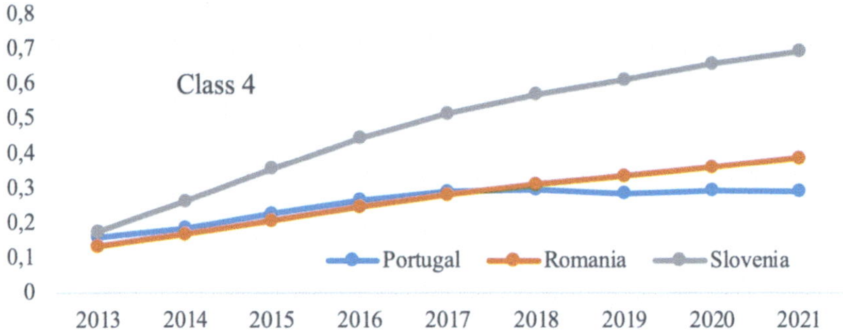

Fig. 7. Dynamics of changes in EU countries in the fourth class (2013–2021) in respect of the level of Pre-primary to tertiary education by the TOPSIS method.

The most variabilities in the third class show Slovenia have changed its position in the ranking from the fourth class to the first class. The value of Slovenia's aggregate measure increases from 0.18 to 0.69, in 2021. While Romania and Hungary showed a lot of variabilities increases from positions 26 and 23 to 23 and 22 positions and moved to class three. The value of the aggregate measure increases from 0.13 to 0.39 and Hungary from 0.23 to 0.39 in the time between (2013–2021).

4 Conclusion

The article presents an original approach to assessing the dynamics of changes in the level of Pre-primary to tertiary education in EU countries by using the TOPSIS method. The presented method is dedicated to the study of complex socio-economic phenomena described by many factors. The criteria value was collected from the Eurostat database in the time horizon (2013–2021) years. The authors in the study used the original approaches using the TOPSIS method in a dynamic approach. At the beginning of the study based on the adopted criteria, the author created one matrix with the data from all analysis periods (2013–2021) duo to create a common ideal and anti-ideal variant as a reference point. Moreover, was calculated based on the created matrix the mean and standard deviations of which will be used at the stage of normalization of the data matrix for each year separately when using the TOPSIS method. In this way, we eliminate the possibility of better-than-ideal and anti-ideal variants appearing in the research sample and the possibility of creating a ranking over time. The research shows that the educational situation in the European countries studied varied. For each year, the resulting aggregate measures value was divided into four classes. From 2013–2020 the educational situation in Denmark was the best.

A huge change shows Slovenia in reference to the year 2013. It is a very interesting case, even though education (x_3 – Public expenditure on pre-primary to tertiary education - as % of GDP) declined by about 1,86% with respect to 2013. In this respect, Slovenia changed its ranking from place 23 to 4. These changes result from the fact that in 2020 in Slovenia increase in the value of critters x_1 – The number of classroom teachers in primary education per 1.000 inhabitants. x_2 – Pupils enrolled in primary education by sex, type of institution and x_6 –Households without access to the internet at home because the access costs are too high (in %)., decreased and amounts to 65.83%, 16.60%, and 496.64%. This process impacted very positive prospects in the education process. In this respect, Poland changed their ranking from place 16 in 2013 to 20 in 2021, and Malta from place 8 to 18. Malta moved up from Class 2 to 3. This shows that situation in Poland in the context of the level of Pre-primary to tertiary education have gotten worse compared to 2013 in 2021. In Poland, the value of the criteria x1 – The number of classroom teachers in primary education per 1.000 inhabitants., x2 – Pupils enrolled in primary education by sex, type of institution, x3 – Public expenditure on pre-primary to tertiary education - as % of GDP and, decreased and amounts to 55.63%, 66,99%, and 1,86%. This results from the reduction in the number of pupils in pre-primary to tertiary education, while the x6- Households without access to the internet at home, because the access costs are too high (in %) criterion increased by 214,95%, which means that the number of households without access to the Internet has increased due to the high cost. This process may be of concern, as access to the Internet can be very important in the education process. Such studies are needed for state authorities to monitor the educational process in each country and relate it to the rest of the European countries. Making it possible to identify the factors inhibiting this process and intervene quickly to improve it.

References

1. Koniewski Maciej, M.R.: Teoretyczne uzasadnienia edukacji zróżnicowanej ze względu na płeć. Edukacja. 5–18 (2018). https://doi.org/10.24131/3724.180101
2. Kształcenie obowiązkowe w wybranych krajach europejskich, https://www.senat.gov.pl/gfx/senat/pl/senatopracowania/4/plik/ot-604.pdf

3. Nermend, M., Singh, U.S., Singh, S.: Educational background as decision criteria in selection of private sector healthcare service providers. Procedia Comp. Sci. **207**, 4651–4659 (2022). https://doi.org/10.1016/j.procs.2022.09.529
4. OECD: https://www.oecd.org
5. Badanie PISA – Badanie PISA (Programme for International Student Assessment, czyli Program Międzynarodowej Oceny Umiejętności Uczniów) realizowane jest przez międzynarodowe konsorcjum nadzorowane przez OECD (Organizację Współpracy Gospodarczej i Rozwoju) i przedstawicieli krajów członkowskich, https://pisa.ibe.edu.pl/. Last accessed 29 June 2023
6. raport-wyniki-badan-pisa-2018.pdf, https://pisa.ibe.edu.pl/wp-content/uploads/2019/12/raport-wyniki-badan-pisa-2018.pdf
7. PIRLS 2021 INTERNATIONAL RESULTS IN READING, https://pirls2021.org
8. Lista krajów biorących udział w poszczególnych edycjach badania, https://www.iea.nl/studies/iea/pirls
9. TIMSS Trends in International Mathematics and Science Study, https://www.iea.nl/studies/iea/timss
10. Ranking Pearson, https://www.pearson.com/
11. Skibska, J., Wojciechowska, J. (eds.): Współczesna edukacja: wielopłaszczyznowość zadań. Wydawnictwo Libron - Filip Lohner, Kraków (2016)
12. Bałachowicz, J., Korwin-Szymanowska, A., Lewandowska, E., Witkowska-Tomaszewska, A. (eds.): Zrozumieć uczenie się: zmienić wczesną edukację. Wydawnictwo Akademii Pedagogiki Specjalnej. Warszawa (2017)
13. Nermend, M., Singh, S., Singh, U.S.: An evaluation of decision on paradigm shift in higher education by digital transformation. Procedia Comp. Sci. **207**, 1959–1969 (2022). https://doi.org/10.1016/j.procs.2022.09.255
14. Singh, S., Singh, U.S., Nermend, M.: Decision analysis of e-learning in bridging digital divide for education dissemination. Procedia Comp. Sci. **207**, 1970–1980 (2022). https://doi.org/10.1016/j.procs.2022.09.256
15. Jabłoński, M., Jarosz-Żukowska, S.: Prawo konstytucyjne w formie pytań i odpowiedzi. Wydawn. Uniwersytetu Wrocławskiego, Wrocław (2003)
16. D20001091.pdf, https://isap.sejm.gov.pl/isap.nsf/download.xsp/WDU20001011091/O/D20001091.pdf
17. Banaszak, B., Preisner, A. (eds.): Prawa i wolności obywatelskie w Konstytucji RP. C.H. Beck, Warszawa (2002)
18. Konwencja o prawach dziecka, przyjęta przez Zgromadzenie Ogólne Narodów Zjednoczonych dnia 20 listopada 1989 r, https://isap.sejm.gov.pl/isap.nsf/DocDetails.xsp?id=wdu19911200526. Last accessed 29 June 2023
19. D20140821.pdf, https://isap.sejm.gov.pl/isap.nsf/download.xsp/WDU20140000821/O/D20140821.pdf
20. Konwencja w sprawie zwalczania dyskryminacji w dziedzinie oświaty, sporządzona w Paryżu dnia 15 grudnia 1960 r., https://isap.sejm.gov.pl/isap.nsf/DocDetails.xsp?id=WDU19640400268. Last accessed 29 June 2023
21. Konwencja przeciw dyskryminacji w edukacji | UNESCO, https://www.unesco.org/en/right-education/convention-against-discrimination. Last accessed 29 June 2023
22. Daudet, Y., Eisemann, P.M.: COMMENTARY ON THE CONVENTION AGAINST DISCRIMINATION IN EDUCATION
23. Międzynarodowy Pakt Praw Gospodarczych, Społecznych i Kulturalnych otwarty do podpisu w Nowym Jorku dnia 19 grudnia 1966 r., https://isap.sejm.gov.pl/isap.nsf/DocDetails.xsp?id=WDU19770380169. Last accessed 29 June 2023

24. Konwencja o Ochronie Praw Człowieka i Podstawowych Wolności sporządzona w Rzymie dnia 4 listopada 1950 r., zmieniona następnie Protokołami nr 3, 5 i 8 oraz uzupełniona Protokołem nr 2, https://isap.sejm.gov.pl/isap.nsf/DocDetails.xsp?id=wdu19930610284. Last accessed 29 June 2023
25. Protokół nr 1 do Konwencji o ochronie praw człowieka i podstawowych wolności, sporządzony w Paryżu dnia 20 marca 1952 r. oraz Protokół nr 4 do powyższej konwencji, sporządzony w Strasburgu dnia 16 września 1963 r, https://isap.sejm.gov.pl/isap.nsf/DocDetails.xsp?id=WDU19950360175. Last accessed 29 June 2023
26. Dokumenty Rady Europy - Biuro Pełnomocnika Rządu do Spraw Osób Niepełnosprawnych, https://niepelnosprawni.gov.pl/index.php?c=article&id=50&print=1. Last accessed 29 June 2023
27. Perkowski, M., Zoń, W. (eds.): Umiędzynarodowienie krajowego obrotu prawnego. T. 1. Wydawnictwo Prawo i Partnerstwo, Białystok (2016)
28. Balcerzak, M., et al. (eds.): Konstytucja RP. T. 1: Komentarz art, pp. 1–86. Wydawnictwo C. H. Beck, Warszawa (2016)
29. HUDOC - European Court of Human Rights. Accessed 29 June 2023
30. Degener, R., Garlicki, L. (eds.) Konwencja o Ochronie Praw Człowieka i Podstawowych Wolności. T. 2: Komentarz do artykułów 19–59 oraz do protokołów dodatkowych. Wydawnictwo C. H. Beck, Warszawa (2011)
31. Susmaga, R., Szczęch, I., Zielniewicz, P., Brzezinski, D.: MSD-space: Visualizing the inner-workings of TOPSIS aggregations. Eur. J. Oper. Res. **308**, 229–242 (2023). https://doi.org/10.1016/j.ejor.2022.12.003
32. Xie, S., Zhang, J.: TOPSIS-based comprehensive measure of variable importance in predictive modelling. Expert Syst. Appl. **232**, 120682 (2023). https://doi.org/10.1016/j.eswa.2023.120682
33. Corrente, S., Tasiou, M.: A robust TOPSIS method for decision making problems with hierarchical and non-monotonic criteria. Expert Syst. Appl. **214**, 119045 (2023). https://doi.org/10.1016/j.eswa.2022.119045
34. Kumar Deb, B., Chakraborti, P.: Multi-response optimization of the performance and emission characteristics of a diesel engine fueled with palm kernel oil methyl ester using TOPSIS-RSM hybrid approach. Materials Today: Proceedings. S2214785323029784 (2023). https://doi.org/10.1016/j.matpr.2023.05.345
35. Hasanzadeh, R., Mojaver, P., Azdast, T., Khalilarya, S., Chitsaz, A., Rosen, M.A.: Decision analysis for plastic waste gasification considering energy, exergy, and environmental criteria using TOPSIS and grey relational analysis. Process Saf. Environ. Prot. **174**, 414–423 (2023). https://doi.org/10.1016/j.psep.2023.04.028
36. Alhassan, H., Peleato, N., Sadiq, R.: Mercury risk reduction in artisanal and small-scale gold mining: A fuzzy AHP-Fuzzy TOPSIS hybrid analysis. Resour. Policy **83**, 103744 (2023). https://doi.org/10.1016/j.resourpol.2023.103744
37. Lv, B., et al.: Evaluation of the water resource carrying capacity in Heilongjiang, eastern China, based on the improved TOPSIS model. Ecol. Ind. **150**, 110208 (2023). https://doi.org/10.1016/j.ecolind.2023.110208
38. Patel, A., Jana, S., Mahanta, J.: Intuitionistic fuzzy EM-SWARA-TOPSIS approach based on new distance measure to assess the medical waste treatment techniques. Appl. Soft Comput. **144**, 110521 (2023). https://doi.org/10.1016/j.asoc.2023.110521
39. Jin, J., Garg, H.: Intuitionistic fuzzy three-way ranking-based TOPSIS approach with a novel entropy measure and its application to medical treatment selection. Adv. Eng. Softw. **180**, 103459 (2023). https://doi.org/10.1016/j.advengsoft.2023.103459
40. Fatima, M., Sherwani, N.U.K., Singh, V.: Comparative analysis among doctors working in private and government hospitals in identifying and prioritizing essential stress factors during

COVID-19- an AHP-TOPSIS approach. Intelligent Pharmacy **1**, 17–25 (2023). https://doi.org/10.1016/j.ipha.2023.04.005
41. Kacprzyk, J., Bozhenyuk, A., Gerasimenko, E.: Lexicographic maximum dynamic evacuation modeling with partial lane reversal based on hesitant fuzzy TOPSIS. Appl. Soft Comput. **144**, 110435 (2023). https://doi.org/10.1016/j.asoc.2023.110435
42. Hwang, C., Yoon, K.: Multiple attribute decision making: methods and applications; a state-of-the-art-survey. Springer, Berlin Heidelberg (1981)
43. Nermend, K.: Metody analizy wielokryterialnej i wielowymiarowej we wspomaganiu decyzji. Wydawnictwo Naukowe PWN, Warszawa (2017)
44. Grabiński, T.: Metody statystyki międzynarodowej. PWE, Warszawa
45. Kukuła, K.: Metoda unitaryzacji zerowanej. Wydaw. Naukowe PWN, Warszawa (2000)
46. Nermend, K.: Taxonomic Vector Measure of Region Development (TWMRR). Polish J. Environ. Stud. (2007)
47. Trzaskalik, T.: Wprowadzenie do badań operacyjnych z komputerem. Polskie Wydawnictwo Ekonomiczne, Warszawa (2008)
48. Nermend, K.: Vector calculus in regional development analysis: comparative regional analysis using the example of Poland. Physica-Verlag, Berlin (2009)
49. Nermend, K.: Badanie własności metod normowania wykorzystywanych w rangowaniu obiektów społeczno-gospodarczych. Zeszyty Naukowe/Uniwersytet Szczeciński; Ekonomiczne Problemy Usług. 171–186
50. Nermend, K.: Properties of normalization methods used in the construction of aggregate measures. Folia Oeconomica Stetinensia. **12**, 31–45 (2012). https://doi.org/10.2478/v10031-012-0030-9
51. Nermend, K., Borawski, M.: Modeling user's preferences in the decision support system. Indian J. Fundam. App. Life Sci. **4**, 1480–1491 (2014)
52. Panek, T., Zwierzchowski, J.: Statystyczne metody wielowymiarowej analizy porównawczej: teoria i zastosowania. Szkoła Główna Handlowa. Oficyna Wydawnicza, Warszawa (2013)
53. Kolenda, M.: Taksonomia numeryczna: klasyfikacja, porządkowanie i analiza obiektów wielocechowych. Wydawn. Akademii Ekonomicznej im. Oskara Langego we Wrocławiu, Wrocław (2006)

Application of Machine Learning Algorithms for Identification of Key Criteria Groups in Public Tendering Proceedings in Poland

Robert Król[1], Aleksandra Bączkiewicz[1(✉)], and Jarosław Wątróbski[1,2]

[1] Institute of Management, University of Szczecin, ul. Cukrowa 8, 71-004 Szczecin, Poland
robert.krol@phd.usz.edu.pl,
{aleksandra.baczkiewicz,jaroslaw.watrobski}@usz.edu.pl
[2] National Institute of Telecommunications, ul. Szachowa 1, 04-894 Warsaw, Poland

Abstract. Identifying a consistent set of criteria is essential to structuralize decision support models. It is especially relevant for the transparency of such models in public proceedings, where identifying a consistent family of criteria is a key factor in their objectification. It is worth pointing out that the criteria in public proceedings in Poland have arbitrary content and weighting determined each time by the contracting authority. There are no dictionaries that would make it easier for experts to analyze the criteria used, their categories, their dependence on the type of purchase, or their relevance to the subject of the contract. Following these shortcomings, this research uses natural language processing (NLP) to classify the criteria involved. The dataset was obtained from the Supplement to Tenders Electronic Daily (TED). The classification tasks used lemmatization, embeddings, and cosine similarity between embeddings where distances of 0.6 and 0.8 were checked. For comparison, the Lingo clustering algorithm was used. The work analyzed 113 373 proceedings in which 25 535 unique criterion names were used. Using NLP made it possible to group the criteria into 473 to 1 414 groups with a precision coefficient of 0.74 obtained.

Keywords: Choosing criteria for tendering proceedings · Natural Language Processing · Machine learning · Artificial intelligence

1 Assessment Criteria in Public Proceedings in Poland

Public proceedings in Poland are conducted according to the rules established by Directive 2014/24 of 26 February 2014 [10] and further detailed by the Act of 11 September 2019. - Public Procurement Law (UZP) [21]. Every entity implementing public tasks in Poland is obliged to carry out purchases in accordance with the rules of the UZP. The law defines the obligation to use criteria for

M. Hernes and J. Wątróbski (Eds.): ECAI 2023, LNNS 1079, pp. 331–341, 2024.
https://doi.org/10.1007/978-3-031-66761-9_26

selecting contractors, including the requirement that quality criteria be used in addition to the price criterion [9,35].

The selection of criteria and their relevance in a particular proceeding is up to the decision of the contracting authorities. Relevant criteria constitute the basis of frameworks for evaluating offers in tender proceedings [32]. Multi-criteria evaluation requires consideration of multiple aspects, which implies the involvement of multiple criteria that are often in conflict with each other [30]. The selected criteria are indicators in the framework used for evaluation in procedures for choosing the most advantageous alternative [33]. Such an approach gives many possibilities, like assigning significance values (weights) to criteria [4,7]. As the Guide published by the Public Procurement Office points out - "this is the right approach. Procurements are very diverse and are organized under different circumstances [27]. The legislator is unable to know the legitimate preferences of the procuring entities or the specific factual situation in which the procuring entity will have to proceed, and therefore cannot a priori impose tender evaluation criteria" [15]. However, the decision to keep the choice of criteria free does not make the task of analyzing the range of criteria already in use any easier [8]. It is caused by the lack of dictionaries from which procurers select or are inspired by criteria that can be applied under the circumstances [11]. The fact that public entities are looking for a source of inspiration for their own applications is evidenced by the handbook issued by the Public Procurement Office, indicated above [18].

2 The Challenge of Analyzing Tender Data in Poland

Selecting criteria for a framework for evaluating alternatives is not a trivial procedure and usually requires an extensive literature review or the involvement of experts in the field [19,31]. The guide for non-price criteria mentioned in the introduction was created with the participation of experts and analysts as a result of a hand-made review of the database of proceedings in the Public Procurement Bulletin [23]. There are 49 505 procurement notices in the Bulletin from 1 January 2023 to 28 May 2023 only. From 2021, when the new version of the Bulletin is available - there are 352,804 announcements. Selection criteria, due to the free nature of their selection, may address a similar issue but be formulated differently, for example as presented in Table 1.

Analyzing such a large data set is an organizational challenge if it is based on human effort. Therefore, in this paper, we attempt to use natural language processing algorithms for criteria analysis in terms of the possibility of creating groups of criteria.

3 Natural Language Processing Techniques for Clustering in Language Polish

A language can be defined as a set of rules and symbols where symbols are combined and used to transfer information [28,34]. Natural language processing

Table 1. Examples of criteria with different content but the same meaning.

Refers to	Example 1	Example 2
Consideration of disadvantaged people	Employment of unemployed persons on a contract of employment	Employment of person(s) with disabilities
Delivery date	Delivery date	Delivery time of the purchased items in days
Positive impact on the environment	Exhaust gas emissions	Total pollutant emissions

(NLP, Natural Language Processing) is a field of artificial intelligence (technical sciences) and linguistics (social sciences) devoted to techniques that enable computers to "understand" and execute commands conveyed in human language [13,16].

The development of NLP techniques makes it possible to extract text and analyze enormous data sets, and find hidden relationships between these data [6,26]. NLP techniques are used to generate text summarization, similarity identification, and clustering, discourse analysis, machine translation, optical character recognition, part-of-speech tagging, etc [2,17]. This paper focuses on analyzing the text of public proceedings to identify groups of criteria used by procurers.

Two techniques were used to analyze the content of the criteria. The first is based on clustering using cosine distance in steps:

- Lemmatization - reducing the inflectional form of a word to its dictionary form using the National Corpus of the Polish Language, Spacy library with the Morpheus 2 [14] inflectional analyzer;
- Tokenization - removal of non-distinctive words and punctuation using the Sentence-BERT model [24];
- Embedding - the creation of a matrix with numbers representing a given word in a multidimensional space using the Sentence-BERT model [24];
- To calculate the similarity, the cosine similarity algorithm between embeddings was chosen [12].

The criteria are compared with each other. The name of the cluster becomes the name of the criterion that was identified as the first similar pair. The result is a flat structure consisting of clusters and the criteria assigned to them in their original form.

The second technique is based on the Lingo algorithm created for clustering search results [20], but it also works well in other applications [25]. The operation of the algorithm is based on the assumption that the most important thing is to provide precise and grammatically, and logically correct labels for the clusters. The person evaluating the clustering results should immediately understand them. It is acceptable to assign a document to several groups or not to assign it to any.

The Lingo algorithm implements this scheme in the following steps:

- Text preprocessing - the text undergoes tokenization and removal of non-letter characters. Individual words are subjected to lemmatization, which is analogous to the first approach.
- The basic finding of frequent phrases - using suffix arrays, phrases that satisfy the following conditions are searched for:
  1. Appear in a sufficiently large number of documents.
  2. Originate from a single sentence (strings of words that form the end of one sentence and the beginning of the next are not considered).
  3. Consist of as many words as possible.
  4. Do not begin or end with a function word (such may appear in the middle of a phrase).
- Creating labels for groups - first, dominant topics (represented by a set of terms) describing the input collection of criteria are detected. These are obtained using a decomposition of the term matrix into particular values. Then the frequent phrases obtained in the previous step are assigned to these topics (by calculating and comparing cosine distance similarity), and these phrases become human-understandable labels for groups.
- Assigning criteria to clusters - the criteria and labels obtained in the previous step are converted to a vector representation. If the cosine similarity between a particular criterion and label reaches a sufficiently large value, this document is assigned to a cluster with this label.
- Sorting the groups - the last phase is to sort the groups so that those most numerous and with the most descriptive descriptions appear first.

4 Character of Data to Analyze

The data were collected from the supplement to Tenders Electronic Daily on European public procurement [1]. This database is hosted on the official website of the European Union, which provides data on European countries in many different domains [29]. Poland is obliged to send information on every public proceeding whose anticipated cost exceeds the so-called "EU threshold", i.e., in the current state of the law, €130 000. This source was used because it publishes a complete set of data that cannot be analogously retrieved from the sources of the Polish Procurement Office.

The data consists of:

1. The content of the procurement announcement, including in particular: the title of the announcement, a summary description of the announcement, a list of the parts that compose the procurement.
2. Each section of the contract shall include the title of the section, a summary description of the section, and the criteria for selecting the contractor for the section.

The data is provided in the XML document format, so it was transferred to an SQL database for the analysis. Announcements in which the contracting authority is an entity from Poland were downloaded and converted: 415 173. One announcement, as indicated above, consists of one or more parts, and each part may have zero or more quality criteria (other than price). There are 859 493 items in the downloaded database of non-price criteria, which gives an interesting value of 2.07 non-price criteria per one proceeding.

5 Research Plan

The contents of public proceedings announcements are structured in the form of an XML document. Each XML includes one announcement, which may consist of one or more parts of the announcement. Each part contains one or more criteria (the minimum is the price criterion). Preparing the data for analysis thus begins with extracting the criteria from the individual parts in each XML file.

The database contains individual criteria assigned to the proceedings, and the criterion can be:

- Price criterion (mandatory for each proceeding);
- Quality criterion (optional).

One procedure can have more than one criterion. Preparation of data for calculation includes:

- Removal of price criteria;
- Removal of duplicates.

The cleaning left 25 535 unique criterion names[1]. Such a diametric difference is due to the fact that in a significant number of tenders, price is the only criterion used and that there are criterion names that are often repeated (e.g., warranty period, delivery date).

5.1 Clustering Using Cosine Distance

Clustering was performed using the Python language and libraries available for this programming language. The sentences making up the criterion were divided into words and reduced to a basic form - lemma.

[1] Whereby "warranty term" and "warranty terms" are two different criteria for this operation.

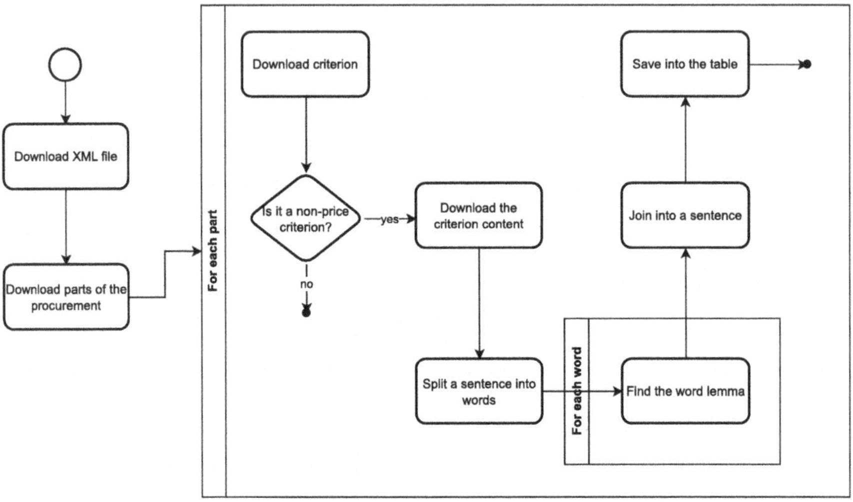

Fig. 1. Flow diagram of content preparation for processing.

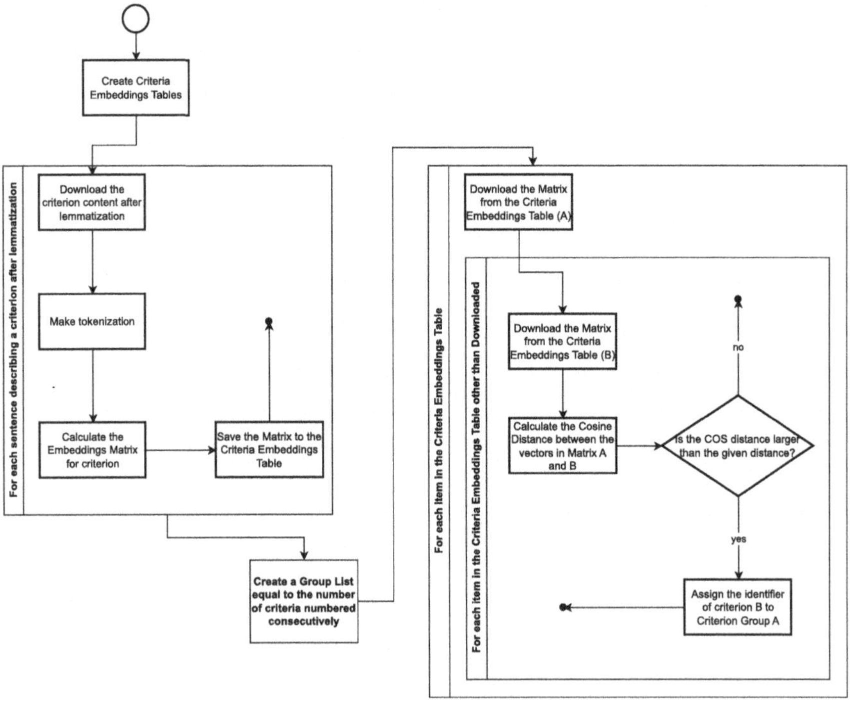

Fig. 2. Flow diagram illustrating the criteria grouping algorithm.

The Spacy medium model (pl_core_news_md) was used for lemmatization. The flow diagram of content preparation for processing used in this research is illustrated in Fig. 1.

Sentences composed of lemmas were used to create embeddings using transformers and tokenization. The BERT library with the paraphrase-multilingual-mpnet-base-v2 trained model was used to process the text. Figure 2 presents a flow diagram illustrating the criteria grouping algorithm applied in this research.

5.2 Grouping Using the Lingo Algorithm

Grouping was performed using the graphical interface provided by the authors[2]. The grouping procedure was executed using a file with post-cleaning criteria and parameters other than the default ones:

- Document coverage target: 1, to force the algorithm to try to assign all criteria to groups;
- Excluded names for groups: criterion, conditions, aspects, additional, contractors, concept, factor;
- Analysis dedicated to the Polish language;
- Maximum word document frequency: 0.8, denotes that words occurring in at least 80% of the criteria will be ignored;
- Word-level Document Frequency (DF) cut-off scaling factor: 0.05 indicates the threshold for minimum word occurrence to be included in clustering;
- Maximum tokens per document: 8, which means taking the first eight tokens (in simple terms: relevant words) of the criterion content for analysis.

As a differentiating parameter for the Lingo algorithm was chosen:

- Maximum top-level clustering passes. It is a parameter that takes into account how many times the comparison of clusters among themselves is repeated at most to determine the top-level cluster. Each successive iteration reduces the number of clusters. Setting the parameter to zero (0) results in the highest possible number of clusters.

5.3 Benchmarking Plan

Benchmarking consists of a comparative analysis of the results obtained by the baseline method with those provided by reference methods as a benchmark [3,5]. The comparative plan is to examine the effectiveness of criteria clustering with different cosine distance parameters: 0.6; 0.8, and with different settings of the "Maximum top-level clustering passes" parameter in the Lingo algorithm. The effectiveness of clustering was evaluated according to Precision, Recall, F1, and Accuracy measures with the understanding of their disadvantages [22], which, in the authors' opinion, do not negatively affect the possibility of choosing the best approach to clustering criteria.

[2] https://search.carrotsearch.com/#/workbench.

6 Results

For the purpose of comparing clustering results, a test set was created with 26 criteria forming seven groups:

1. Warranty period (5 criteria).
2. Delivery/realization times (5 criteria).
3. Environmental aspects (3 criteria).
4. Social aspects (1 criterion).
5. Time to fix the error (1 criterion).
6. Team experience (4 criteria).
7. Creativity/innovation (3 criteria).
8. Technical parameters (4 criteria).

The outcomes obtained as a result of two algorithms execution were checked in terms of the decision the algorithm made concerning the test set. For the purpose of evaluating the performance of the algorithms, an assignment was made to one of four categories:

1. True positive (TP) - two similar criteria were assigned to the same group.
2. True negative (TN) - two different criteria were assigned to different groups.
3. False positive (FP) - two different criteria were assigned to the same group.
4. False negative (FN) - two similar criteria were assigned to different groups.

For the aforementioned evaluation, coefficients were calculated:

- Precision - informs about recognition accuracy, that is, the ratio of the number of TP pairs to the sum of TP+FP pairs.
- Recall (Sensitivity) - informs how many elements from a given class were correctly recognized. It is expressed as the ratio TP/(TP+FN).
- F1 - the harmonic average for Precision and Sensitivity calculated from the formula provided in Eq. (1).

$$F1 = 2\frac{Precision \times Recall}{Precision + Recall} \tag{1}$$

Table 2 shows the results for each algorithm and the differential parameter employed.

Table 2. Comparison of the results achieved using different algorithms.

Algorithm	Cosinus distance		Lingo	
Differentiating parameter	Distance: 0.6	Distance: 0.8	Maximum top-level clustering passes: 4	Maximum top-level clustering passes: 0
Number of criteria groups identified	1414	7931	214	473
Precision score	0.74	0.73	0.6	0.74
Recall score	0.53	0.58	0.74	0.66
F1 score	0.62	0.65	0.66	0.69

7 Conclusions

The aim of the paper was to examine the feasibility of performing a grouping of criteria used in public tender proceedings in Poland. A potential application for such grouping is the creation of a dictionary of criteria (at present, contracting authorities arbitrarily formulate the content of criteria).

In the performed research, it was proven that it is possible to join criteria into groups successfully. However, further work is necessary on the precision of this grouping (perhaps using other algorithms or simply refining the parameters used). The results of the Precision, Recall, and F1 coefficients achieved allow us to conclude that 25 535 different criteria can be joined into groups ranging from 473 to 1 414 items. Further work will continue to understand how criteria shape themselves depending on the category of items or services ordered.

Acknowledgements. This research was funded in whole by the National Science Centre, Poland 2022/45/N/HS4/03050 (A.B., J.W.).

References

1. An official website of the European Union: Tenders Electronic Daily (2023). https://ted.europa.eu/TED/main/HomePage.do
2. Avasthi, S., Chauhan, R., Acharjya, D.P.: Techniques, applications, and issues in mining large-scale text databases. In: Goar, V., Kuri, M., Kumar, R., Senjyu, T. (eds.) AICTC 2019. LNCS, vol. 135, pp. 385–396. Springer, Singapore (2021). https://doi.org/10.1007/978-981-15-5421-6_39
3. Bączkiewicz, A., Kizielewicz, B., Shekhovtsov, A., Wątróbski, J., Sałabun, W.: Methodical aspects of MCDM based E-commerce recommender system. J. Theor. Appl. Electron. Commer. Res. **16**(6), 2192–2229 (2021). https://doi.org/10.3390/jtaer16060122
4. Bączkiewicz, A., Wątróbski, J.: Crispyn - a Python library for determining criteria significance with objective weighting methods. SoftwareX **19**, 101166 (2022). https://doi.org/10.1016/j.softx.2022.101166
5. Bączkiewicz, A., Wątróbski, J., Kizielewicz, B., Sałabun, W.: Towards objectification of multi-criteria assessments: a comparative study on MCDA methods. In: 2021 16th Conference on Computer Science and Intelligence Systems (FedCSIS), pp. 417–425. IEEE (2021). https://doi.org/10.15439/2021F61
6. Chen, X., Xie, H., Cheng, G., Poon, L.K., Leng, M., Wang, F.L.: Trends and features of the applications of natural language processing techniques for clinical trials text analysis. Appl. Sci. **10**(6), 2157 (2020). https://doi.org/10.3390/app10062157
7. Cinelli, M., Burgherr, P., Kadziński, M., Słowiński, R.: Proper and improper uses of MCDA methods in energy systems analysis. Decis. Support Syst. **163**, 113848 (2022). https://doi.org/10.1016/j.dss.2022.113848
8. Dotoli, M., Epicoco, N., Falagario, M.: Multi-criteria decision making techniques for the management of public procurement tenders: a case study. Appl. Soft Comput. **88**, 106064 (2020). https://doi.org/10.1016/j.asoc.2020.106064
9. El-Sayegh, S.M., Basamji, M., Haj Ahmad, A., Zarif, N.: Key contractor selection criteria for green construction projects in the UAE. Int. J. Constr. Manag. **21**(12), 1240–1250 (2021). https://doi.org/10.1080/15623599.2019.1610545

10. EUR-Lex: Directive 2014/24/EU of the European Parliament and of the Council of 26 February 2014 on public procurement and repealing Directive 2004/18/EC (Text with EEA relevance) (2022). http://data.europa.eu/eli/dir/2014/24/2022-01-01/eng
11. Giuffrida, L.M., Rovigatti, G.: Supplier selection and contract enforcement: evidence from performance bonding. J. Econ. Manag. Strategy **31**(4), 980–1019 (2022). https://doi.org/10.1111/jems.12492
12. Kenter, T., De Rijke, M.: Short text similarity with word embeddings. In: Proceedings of the 24th ACM International on Conference on Information and Knowledge Management, pp. 1411–1420 (2015). https://doi.org/10.1145/2806416.2806475
13. Khurana, D., Koli, A., Khatter, K., Singh, S.: Natural language processing: state of the art, current trends and challenges. Multimed. Tools Appl. **82**(3), 3713–3744 (2023). https://doi.org/10.1007/s11042-022-13428-4
14. Kieraś, W., Woliński, M.: Morfeusz 2-analizator i generator fleksyjny dla języka polskiego. Język Polski XCVI **I**(1), 75–83 (2017)
15. Koba, D.: Non-price criteria for evaluating tenders, vol. 2. Public Procurement Office (2023). https://www.uzp.gov.pl/baza-wiedzy/publikacje/poradnik-z-katalogiem-dobrych-praktyk-pozacenowe-kryteria-oceny-ofert
16. Lauriola, I., Lavelli, A., Aiolli, F.: An introduction to deep learning in natural language processing: models, techniques, and tools. Neurocomputing **470**, 443–456 (2022). https://doi.org/10.1016/j.neucom.2021.05.103
17. Liu, X.: Analyzing the impact of user-generated content on B2B Firms' stock performance: big data analysis with machine learning methods. Ind. Mark. Manag. **86**, 30–39 (2020). https://doi.org/10.1016/j.indmarman.2019.02.021
18. Miller, F.A., Lehoux, P.: The innovation impacts of public procurement offices: the case of healthcare procurement. Res. Policy **49**(7), 104075 (2020). https://doi.org/10.1016/j.respol.2020.104075
19. Mokarram, M., Mokarram, M.J., Gitizadeh, M., Niknam, T., Aghaei, J.: A novel optimal placing of solar farms utilizing multi-criteria decision-making (MCDA) and feature selection. J. Clean. Prod. **261**, 121098 (2020). https://doi.org/10.1016/j.jclepro.2020.121098
20. Osiński, S., Stefanowski, J., Weiss, D.: Lingo: search results clustering algorithm based on singular value decomposition. In: Klopotek, M.A., Wierzchoń, S.T., Trojanowski, K. (eds.) AINSC, vol. 25, pp. 359–368. Springer, Heidelberg (2004). https://doi.org/10.1007/978-3-540-39985-8_37
21. Parliament of the Republic of Poland: The Act of September 11, 2019. - Public Procurement Law (2019). https://isap.sejm.gov.pl/isap.nsf/DocDetails.xsp?id=WDU20190002019
22. Powers, D.M.: Evaluation: from precision, recall and F-measure to ROC, informedness, markedness and correlation. arXiv preprint arXiv:2010.16061 (2020)
23. Public Procurement Office: Public Procurement Bulletin (2023). https://ezamowienia.gov.pl/mo-client-board/bzp/list
24. Reimers, N., Gurevych, I.: Sentence-BERT: Sentence Embeddings using Siamese BERT-Networks. arXiv preprint arXiv:1908.10084 (2019)
25. Stefanowski, J., Weiss, D.: Comprehensible and accurate cluster labels in text clustering. In: Large Scale Semantic Access to Content (Text, Image, Video, and Sound), pp. 198–209. Citeseer (2007)
26. Tao, D., Yang, P., Feng, H.: Utilization of text mining as a big data analysis tool for food science and nutrition. Compr. Rev. Food Sci. Food Saf. **19**(2), 875–894 (2020). https://doi.org/10.1111/1541-4337.12540

27. Troje, D., Gluch, P.: Populating the social realm: new roles arising from social procurement. Constr. Manag. Econ. **38**(1), 55–70 (2020). https://doi.org/10.1080/01446193.2019.1597273
28. Urbieta, M., Antonelli, L., Rossi, G., do Prado Leite, J.C.S.: The impact of using a domain language for an agile requirements management. Inf. Softw. Technol. **127**, 106375 (2020). https://doi.org/10.1016/j.infsof.2020.106375
29. Wątróbski, J.: Temporal PROMETHEE II - new multi-criteria approach to sustainable management of alternative fuels consumption. J. Clean. Prod. **413**, 137445 (2023). https://doi.org/10.1016/j.jclepro.2023.137445
30. Wątróbski, J., Bączkiewicz, A., Król, R., Sałabun, W.: Green electricity generation assessment using the CODAS-COMET method. Ecol. Ind. **143**, 109391 (2022). https://doi.org/10.1016/j.ecolind.2022.109391
31. Wątróbski, J., Bączkiewicz, A., Rudawska, I.: A Strong Sustainability Paradigm based Analytical Hierarchy Process (SSP-AHP) method to evaluate sustainable healthcare systems. Ecol. Ind. **154**, 110493 (2023). https://doi.org/10.1016/j.ecolind.2023.110493
32. Wątróbski, J., Bączkiewicz, A., Sałabun, W.: New multi-criteria method for evaluation of sustainable RES management. Appl. Energy **324**, 119695 (2022). https://doi.org/10.1016/j.apenergy.2022.119695
33. Wątróbski, J., Bączkiewicz, A., Ziemba, E., Sałabun, W.: Sustainable cities and communities assessment using the DARIA-TOPSIS method. Sustain. Urban Areas **83**, 103926 (2022). https://doi.org/10.1016/j.scs.2022.103926
34. Ye, X., Lu, Y.: Automatic extraction of engineering rules from unstructured text: a natural language processing approach. J. Comput. Inf. Sci. Eng. **20**(3), 034501 (2020). https://doi.org/10.1115/1.4046333
35. Ying, H., Zhang, S., Zhao, X., Zhang, M.: How to make project plans trustworthy: an investigation based on multi-criteria contractor selection. Int. J. Log. Res. Appl. **26**(4), 460–477 (2023). https://doi.org/10.1080/13675567.2021.1961700

Author Index

M. Hernes and J. Wątróbski (Eds.): ECAI 2023, LNNS 1079, pp. 343–344, 2024.
https://doi.org/10.1007/978-3-031-66761-9

SPRINGER NATURE

GPSR Compliance

The European Union's (EU) General Product Safety Regulation (GPSR) is a set of rules that requires consumer products to be safe and our obligations to ensure this.

If you have any concerns about our products, you can contact us on ProductSafety@springernature.com

In case Publisher is established outside the EU, the EU authorized representative is:

Springer Nature Customer Service Center GmbH
Europaplatz 3
69115 Heidelberg, Germany

The manufacturer's authorised representative in the EU is Springer Nature Customer Service Centre GmbH, Europaplatz 3, 69115 Heidelberg, Germany. If you have any concerns regarding our products, please contact ProductSafety@springernature.com

Printed and bound by CPI Group (UK) Ltd, Croydon, CR0 4YY
01/10/2026
02225893-0010